INTRODUCTION TO

Classical *and* Quantum Harmonic Oscillators

INTRODUCTION TO

Classical *and* Quantum Harmonic Oscillators

S.C. BLOCH

A WILEY-INTERSCIENCE PUBLICATION
JOHN WILEY & SONS, INC.
NEW YORK • CHICHESTER • WEINHEIM • BRISBANE • SINGAPORE • TORONTO

Cover illustration depicts dispersion and absorption modes of a single-electron oscillator. Courtesy of Professor Hans G. Dehmelt, University of Washington.

This text is printed on acid-free paper.

Library of Congress Cataloging in Publication Data:

Bloch, S. C. (Sylvan Charles)
Introduction to classical and quantum harmonic oscillators / Sylvan C. Bloch.
p. cm.
"A Wiley-Interscience publication."
Includes bibliographical references and index.
ISBN 0-471-14744-3 (cloth : alk. paper)
1. Harmonic oscillators. I. Title.
QC125.2.B56 1997
530.4'16—dc21 96-44728
CIP

Printed in the United States of America

10 9 8 7 6 5 4 3 2 1

To my grandfathers
Charles David Bloch, Major, U.S. Army
Sylvan Myers, M.D.,

and my son
Lawrence Edward Bloch, J.D., M.D.

Contents

Chapter 12 Quantum Oscillators 279

Preface

If calculus is the most pervasive concept in our description of the physical world, perhaps the harmonic oscillator is a close second. Harmonic oscillators appear in mechanics, electromagnetics, electronics, optics, acoustics, and quantum mechanics, to name a few. The simple harmonic oscillator is a fundamental building block in our system of knowledge of the physical universe.

People have studied the harmonic oscillator analytically for almost 300 years, but it was recognized as a useful device for musical instruments and weapons for perhaps thousands of years. There is speculation that stringed instruments originated when soldiers and hunters discovered that some of their bows emitted pleasing sounds when plucked. Reeds, sea shells, rams' horns, and bamboo were employed as musical instruments by blowing into, or across, an opening. Stretched skins provided the first drum heads. Advances in metallurgy eventually produced bells and chimes. The historical connection between weapons and oscillators continues; high-power microwave oscillators have knocked down helicopters, the French army has experimented with acoustic anti-personnel weapons, and lasers are commonplace for aiming, guidance, and remote energy deposition.

Galileo Galilei, a talented musician, painter, craftsman, and medical student, probably performed the first quantitative measurements of a harmonic oscillator while observing the periodic motion of a great pendulum chandelier during a long sermon in the cathedral of Pisa. This sort of thing would get him in trouble later on.

Today harmonic oscillators are ubiquitous. Nobel prizes await the person who builds (or finds) the most nearly perfect oscillator, like the laser and the Mössbauer effect. Atomic oscillators enable us to make our most accurate measurements. And where would rock bands be without music synthesizers?

Simple harmonic oscillators are used to measure physical quantities over an enormous range, from the cosmic implications of the universal gravitational constant G, to the equivalence of inertial mass and gravitational mass, to the density variations deep inside planet Earth, to the sub-atomic domain of electron charge and spin.

The harmonic oscillator is easily visualized as a mechanical system. It is just a bit more abstract as an electronic device, but still easily understood. Deeper, at atomic, nuclear, and sub-nuclear levels we find phenomena that seem to behave as if they were harmonic oscillators, even though they cannot yet be visualized or because we do not really understand what is oscillating.

And yet, despite the simplicity of the oscillator, there is a richness of analytical techniques that can be used to describe it. The harmonic oscillator is an excellent system to learn these methods so you can apply them to more complex systems.

This book is an elementary introduction to harmonic oscillators with practically no steps omitted. A diskette is provided so you can experiment with various aspects of harmonic oscillators. The diskette contains worksheets that are ready to run in the Microsoft *Excel* spreadsheet. You can also import these files into *Quattro Pro* and *Lotus 1-2-3*. Some people regard spreadsheets as "power tools" that enable the user to crunch large amounts of numbers, and that is true. But spreadsheets are more than that. The newer versions contain the fast Fourier transform (FFT) and eighteen complex functions that not only free you from the tedium of calculation but they also enable you to carry out numerical procedures that, in many cases, cannot be done in closed form or with elementary functions. Science and technology require insight, imagination, reasoning, and calculation. These mutually reinforce each other, so if you can get help with calculation you have a 25 percent advantage at the start. The other three are up to you.

You can ask "what if . . .?" and see the answer. You have instant *visualization* of new results and the ability to interface with data acquisition boards to import real-world information. Your spreadsheet and your computer become an extension of your mind. I urge you to use your computer and the diskette to explore and gain an intuitive as well as a quantitative understanding. Most of the graphics in this book were done on a spreadsheet; for better graphic clarity you can see them in color on your monitor and you can see new results as you change parameters and initial conditions.

Several threads run through the fabric of this book. They include adiabatic invariance, phase space and its discrete, incompressible, and fuzzy nature, classical and quantum uncertainty principles, and the almost inevitable Aristotelian Incline from the classical domain through the semi-classical to the quantum domain.

Some years ago Professor P. A. M. Dirac and his wife Margit came to my home for dinner. Robert and Edie Dressler stopped by for a while. In the course of our conversation I remarked that I had read that Einstein said it would be enough if we could understand the electron. Dirac replied that it would be enough if students could understand the harmonic oscillator.

This book is based on the works of some of the best minds of the last three hundred years, including Galileo Galilei, Sir Isaac Newton, Louis de Lagrange, Sir William Rowan Hamilton, Leonhard Euler, the Bernoullis (Jacob, John, and Daniel), Baron Jean Baptiste Joseph Fourier, Joseph Liouville, Pierre-Simon Marquis de Laplace, Siméon Denis Poisson, C. G. J. Jacobi, Karl Friedrich Gauss, Lord Kelvin, Lord Rayleigh, Max Planck, Albert Einstein, Niels Bohr, Arnold Sommerfeld, Paul Ehrenfest, Prince Louis Victor de Broglie, David Hilbert, Erwin Schrödinger, Werner Heisenberg, Wolfgang Pauli, Max Born, Paul Adrien Maurice Dirac, and Richard Phillips Feynman. There are many others, and this work is continuing.

It may be difficult to imagine that anything new can be said now about harmonic oscillators, yet hundreds of papers are published every year on harmonic oscillators and their multitude of applications in engineering, science, and technology. We would not have the Global Positioning System without atomic clocks. Nobel prizes are still being awarded for harmonic oscillators; one of the latest involves an oscillator made with a single electron.

I have attempted to start at the beginning and follow the yellow brick road slowly and without heavy mathematical baggage. This is possible because things are easier in retrospect. Everything is obvious, once you've seen it (to paraphrase the mathematician Felix Klein). Some things that were worthy of Nobel prizes can now be done in a page or two. I have tried to stay with the fundamentals, but I have also included some interesting applications.

I would like to express my appreciation to Dr. Robert Dressler for his helpful suggestions and insights, too numerous to list. I would also like to thank Prof. Robert W. Flynn and Prof. John A. Llewellyn for their comments and assistance, and Prof. Hans G. Dehmelt for permission to reproduce his results of the single-electron oscillator. For their ever-cheerful help, I am indebted to Marilyn Cureton, Jill Lones, Art Matheny, Debbie O'Hearn, and Shauna Schullo.

Special thanks go to Jack Drucker, Robb Frederick, Perry King, executive editor George J. Telecki, and managing editor Lisa Van Horn for their assistance and their patience. I prepared this book in camera-ready form, so any mistakes are my own.

Feynman said that research is seeing what everyone else has seen, but thinking what no one else has thought. Pasteur said fortune favors the prepared mind. I hope this book will be helpful or (in accord with the oath of Hippocrates) do no harm.

Sylvan Charles Bloch

Tampa, Florida
January 1997

. . . I have had to unteach myself the difficulties, and now beg to present . . . the parts that are not hard. Master these thoroughly, and the rest will follow.

Professor Silvanus Phillips Thompson, F.R.S.

Chapter 1

Classical Perspective

Si non oscillas noli tintinnare.
Aristotle

The harmonic oscillator is a pervasive concept in science and engineering; it has been applied from sub-nuclear quarks to cosmic systems. In electronics it is used from sub-audio frequencies to the gamma-ray region. The harmonic oscillator is a simple system that can be solved exactly in both classical and quantum descriptions. That, by itself, is a major claim to fame and places the harmonic oscillator in a special category but there is much more.

An examination of current databases reveals that several hundred articles are published every year on various aspects of the harmonic oscillator. The bibliography (pages 333–358) shows some of the broad applications.

1.1 Importance of the Harmonic Oscillator

From both a practical and theoretical viewpoint the harmonic oscillator occupies a central position as the *first approximation for all vibrating systems*. As you will soon see, when any realistic system vibrates with small amplitude about its equilibrium position the system can be described as a simple harmonic oscillator. Chapter 2 shows why this is so. In learning about the harmonic oscillator you will also be learning something about methods used to analyze a variety of systems.

The analogies between classical mechanical, electrical, and acoustical systems are well known. When we solve one system we are also solving its analogs. Some quantities which play similar roles in systems are shown in Table 1-1.

Although the mechanical, acoustical, and electrical quantities are analogous, there are two major differences:

1. Displacement x is considered to be a continuous variable but electric charge q is discrete, quantized in units of the electron charge. This has a profound effect on randomness and noise in measurements and on the structure of the universe.

2. Mass m is quantized in a hiearchy of discrete masses. Electrons may be the smallest (5.11×10^5 electron-Volts), although in 1996 neutrino oscillations seemed to indicate that neutrinos may have a mass of about 0.2 eV. (One eV is the energy equivalent to a mass of 1.78×10^{-36} kg.)

Table 1-1. Analogous Quantities in Physical Systems

Mechanical	*Electrical*	*Acoustical*
F force	V emf	Pressure
x displacement	q charge	Volume displacement
v velocity	I current	Volume current
a acceleration	$\frac{dI}{dt}$ rate of change of current	Rate of change of volume current
m mass	L inductance	Inertance
κ restoring force constant	$1/C$ reciprocal of capacitance	Reciprocal of acoustic capacitance
R resistive force constant	R resistance	Acoustic resistance

Oscillators constructed with electrons trapped in quantum wells have important applications in electronics. For one thing, they make excellent little lasers. Single electrons have been confined in three-dimensional electromagnetic traps for months, and have been observed to behave as quantized harmonic oscillators; Hans Dehmelt was awarded the Nobel Prize in 1989 for this, and related, work.

We will begin our journey with mechanical oscillators because simple mechanical devices can be easily visualized. Classical mechanics has three foundations. If you start with any one you can derive the other two so it does not appear that any particular one is more fundamental than any other. For historical reasons you learn Newton's formulation of mechanics first; also with Newton's laws simple problems can be solved with easy mathematics. In Chapter 8 we will see that Lagrange's formulation is useful for complicated systems that would be very difficult to solve with Newton's laws. Hamilton's formulation (also Chapter 8) provides deeper insight into fundamental questions and a bridge to quantum theory.

Degrees of Freedom

The number of degrees of freedom of a system is the number of quantities needed to determine the velocities of all particles in the system, for any motion that does

not violate the constraints. We will see later that this number is also the number of squared terms in the expression for the total energy of the system.
For N particles and n degrees of freedom,

- Newton's laws require the solution of N second-order ordinary differential equations. Constraints are used to reduce these to n equations.
- Lagrange's equations require the solution of n second-order partial differential equations.
- Hamilton's equations require the solution of $2n$ first-order partial differential equations.

Need for a statistical description

Classical mechanics is based on the idea of a *point particle* with mass. In order to treat a group of particles, such as a baseball, planet, electrical current, or Helium gas in a balloon, it becomes necessary to introduce statistical quantities representative of the group. In this way we can reduce the problem to that of a point particle. Consider the orbit of a baseball or a planet. We couldn't possibly follow all of the particles even if we could see them. The computations would be impossible even if we knew all of the data needed to track the individual particles. However, the problem is deeper than the lack of exact data and the ability to do enormous numbers of computations. There is an inherent unpredictability, even in some classical descriptions; we will explore some of these ideas in Chapter 10.

Although we are not going to dwell on this now, before we go on let's get some concept of how much computing power we would need just to track one mole of a gas such as oxygen (a diatomic molecule). One mole contains Avogadro's number of molecules, $N_A = 6.02 \times 10^{23}$.

For each molecule we need three numbers for the initial coordinates (x_o, y_o, z_o) of the center of mass, and three numbers for the angles to specify the initial orientation of the molecule in space. We need three numbers for the initial velocity components of the center of mass, and three numbers for the initial angular velocity components. That adds up to 12 bits of data for each diatomic molecule, and we have N_A molecules.

All in all, we need $12 \times 6.02 \times 10^{23}$ bits of data. Suppose we can compute 10^{10} per second (10 gigabits/s). How long would it take just to process the initial conditions?

$$(12 \times 6.02 \times 10^{23})/10^{10} = 7.22 \times 10^{14}\ \text{s} = 2.29 \times 10^{7}\ \text{years.}$$

A monoatomic gas like helium would be easier, because we would not require data on initial angles and angular velocities. Helium would take less than 1.2×10^{7} years. At this point things look hopeless.

Part of the solution to this dilemma is to abandon the idea of following detailed motions of individual particles. As an alternative we can follow average values of the system and treat these average values as a single particle. This will provide a very fruitful approach. We introduce quantities like mass density, center of mass, moments of inertia, and r.m.s. velocity, which represent the group of particles. Then we follow the motion of the fictitious point particle with these group properties and pretty much ignore the fact that we are dealing with a large number of particles, about which we lack detailed information.

Statistical methods are more practical, time saving, and thrifty. We are going to concentrate on a single point particle until we come to Section 1.6.

1.2 Newtonian Perspective

In today's world we start with classical mechanics, try to get as good a solution as we can, and then quantize this solution. If it doesn't work, we've pushed too far. Logically we should do the quantum calculation and then go to the classical limit but most often the quantum calculation is too difficult. Intuition, skill, and luck are important. It is astonishing how logic and guessing advance our knowledge.

The Newtonian formulation is based on astronomical measurements. Very precise data had been collected by Tycho Brahe and others concerning the motion of some planets. While studying planet Mars, Johannes Kepler (a former student of Brahe) discovered that the data fit three empirical laws. For Newton planetary motion was the most pure mechanical experiment available; it didn't seem to run down and probably there was only one force that determined the motion. Newton stated three laws of motion which, together with the force law of gravitation, could derive Kepler's empirical laws, that is, a theory that fit the data. It's lucky that very accurate measurements of planet Mercury were not available.

It was a leap of faith to apply these laws that worked for planets to the motions of cannon balls and machinery, but the applications were justified by their successes. Nevertheless people searched to find the limits of Newton's laws and now we think we know.

Newton's second law is the one that gives us the equations of motion. It is a vector law, and it says that the vector sum of all of the external forces makes the momentum change in time. This implies a *principle of superposition*. Switching one force on and off doesn't affect the others. Also implicit in the laws is the concept of the *inertial coordinate system* (first law) and the universality of time as an independent variable, the same for all observers. Forces exist in equal and opposite pairs, so internal forces do not change the momentum, by Newton's third law.

In three dimensions Newton's second law can be decomposed into three ordinary, second-order differential equations in scalars by using vector components of the forces and the momentum. The solution to each equation requires two constants of integration, which correspond to the initial conditions. Newton's laws are basically laws that describe point particles; it took him 20 years to show that a planet could be treated as a point particle by his methods. That's not bad considering he had to invent calculus and keep up with his regular jobs, which included Chancellor of the Exchequer. Did we ever have a Secretary of the Treasury who could do that?

By now all of the easy problems, and a lot of the hard ones, have been solved using Newton's laws. Fortunately, the harmonic oscillator is one of these.

Inertial Coordinate System

The Newtonian formulation of classical mechanics is based on an inertial coordinate system. You usually begin by learning kinematics, which is the description of motion in this coordinate system. You learn how to describe the motion of a point in space in terms of its coordinates, velocity, and acceleration. Later you learn dynamics, which involves mass, momentum, energy, and angular momentum, and how these things affect the motion of a particle and groups of particles. Newtonian mechanics makes great use of something called a *force*, which is the time rate of change of momentum.

Time is assumed to be an independent variable, which is the same in all coordinate systems. We now have experimental evidence that this is not correct, but it's lucky that Newton didn't know this, because it would have complicated things too much. Newtonian mechanics basically involves ordinary differential equations, of order no greater than two, in an inertial system.

An inertial coordinate system is an idealization and the closest we can come to one is a coordinate system at rest relative to the fixed stars, or moving with constant velocity relative to them, as seen in Fig. 1-1.

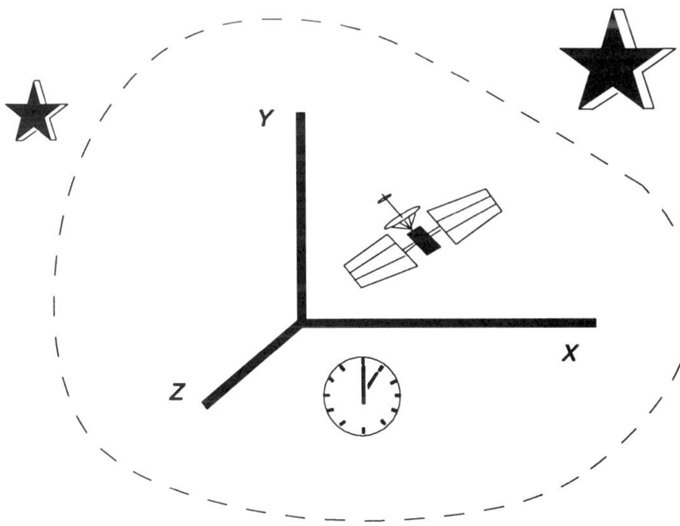

Figure 1-1. A representation of a Cartesian inertial coordinate system.

The dotted line around our coordinate system seen in Fig. 1-1 is a shield to keep out all unwanted forces due to the rest of the universe. That way we can switch on one force at a time and examine it without the complexities caused by several forces. After we know what each type of force does individually we will allow several forces to act simultaneously.

When we are not in an inertial coordinate system strange, fictitious forces seem to operate. For short distances and short times planet Earth may be used as an approximate inertial system if our measurements are not very accurate. We can still use Newton's laws in a non-inertial system if we include fictitious forces just as if they were real forces, but we have to be careful.

We don't worry about the non-inertial system while playing football or baseball on a field attached to Earth. However, imagine playing baseball on a merry-go-round with the pitcher at the center and the batter on the edge. To an observer attached to Earth the baseball appears to move in a straight line (except for a fall due to gravity) but to the pitcher and batter the baseball seems to move in a strange orbit that depends on the ball's speed, and the speed and direction of the merry-go-round rotation. We see the same kind of strange force (the Coriolis force) at work when the water spirals down the bathtub drain or when we see a satellite view of Earth with spirals of clouds around high-pressure and low-pressure regions in the atmosphere.

Until we reach Chapter 10 we will imagine that:

- We are in an inertial coordinate system, isolated from everything else in the universe, so that no forces are acting except the ones that we choose. The dotted line in Fig. 1-1 admits only forces that we choose.
- Measurements can, in principle, be made with arbitrary precision.

This is the method of elementary abstraction that strips away all of the complications of the real world and concentrates on a very few basic aspects of the thing we are trying to describe and understand. If we can describe these main features then we can later add more details that may improve our correspondence with the real world.

This inertial coordinate system is what we will use to describe the harmonic oscillator from Newton's perspective. We will see that the second thing we imagine (the possibility of arbitrary precision) is not true even from the classical viewpoint. The first thing isn't true either, but don't worry about it now. Also, to keep things simple, we are going to examine the one-dimensional harmonic oscillator.

1.3 Restoring Force

So, here we are in our one-dimensional inertial coordinate system, protected from all outside forces, even gravity, by that impenetrable dotted line. Because you already know something about Newton's laws and their applications we will not discuss motion due to a constant force.

Simple harmonic oscillators are characterized by a linear restoring force. In one dimension we can represent such a force as $F(x) = -\kappa x$, where κ is the restoring force constant and x is the displacement from equilibrium. This is the *only* kind of force that produces pure harmonic oscillations. This force is always directed toward the equilibrium position. When x is positive the force is negative; when x is negative the force is positive. This force is represented graphically in Fig. 1-2.

This type of force is not realistic for large displacements from equilibrium, because the force and the energy of the system would increase without limit. All of the forces that we know eventually die away with distance. Clearly, the linear restoring force can only hold over a restricted region, after which it must decrease.

A more realistic force is shown in Fig. 1-3, in which approximate linearity (and, therefore, approximate harmonic oscillations) exists over a restricted range of displacement, after which the force decreases in some way as x increases. Perhaps beyond some critical x_c the system breaks catastrophically. This is a common occurrence.

The linear restoring force $F(x) = -\kappa x$ bears the name *Hooke's Law*. It was first stated by Robert Hooke, a contemporary of Newton. Hooke, who attended Oxford University, was the first to use a spiral spring to regulate watches. He also constructed the first Gregorian telescope and he discovered plant cells.

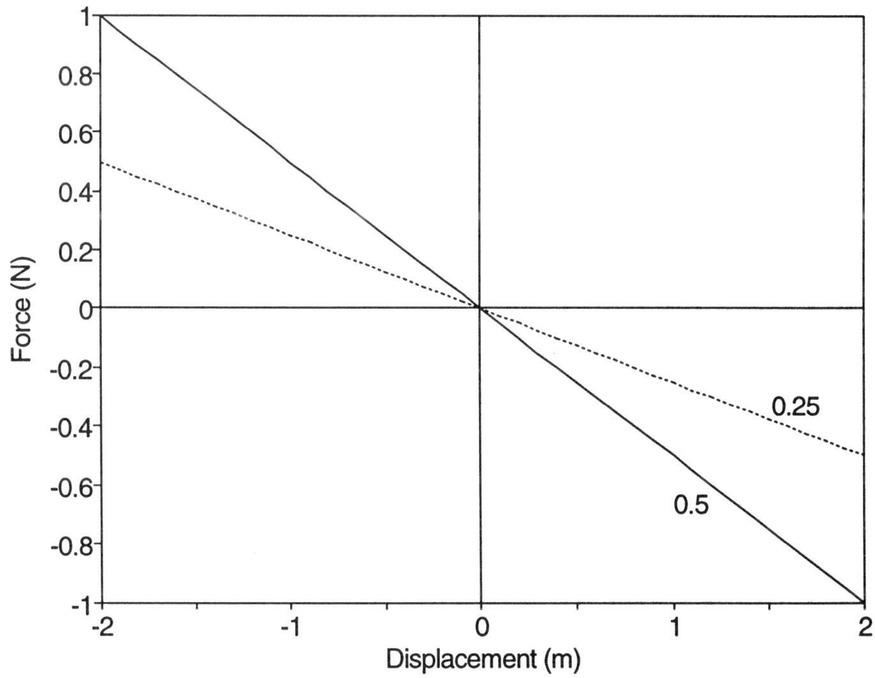

Figure 1-2. Linear restoring forces for $\kappa = 0.5$ and $\kappa = 0.25$. One force is twice as strong as the other, for the same displacement.

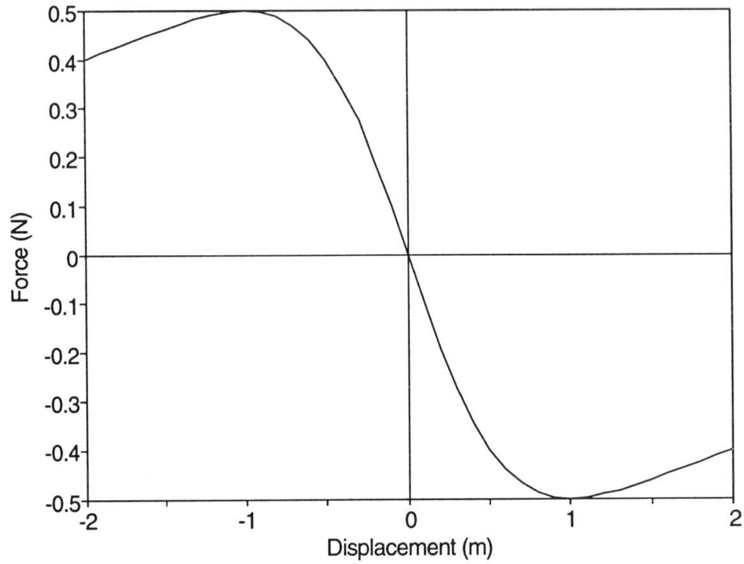

Figure 1-3. A typical realistic force that does not support pure harmonic oscillations. Harmonic oscillations are a good approximation for very small displacements from equilibrium, $x = 0$.

It is interesting to get an idea of the range of some restoring force constants, and in doing so we'll bring up a little anecdote. In 1986 Drs. Gerd Binnig and Heinrich Rohrer of IBM's Zurich laboratory shared the Nobel Prize in Physics for their invention of the scanning tunneling microscope. This worked by quantum tunneling between the surface of an electrical conductor and a fine conducting probe, and it could image individual surface atoms with undreamed-of resolution. The probability of quantum tunneling depends exponentially on distance, so it is extremely sensitive to small changes in displacement.

Later, Dr. Binnig got an idea for another kind of instrument, the atomic force microscope, while visiting at Stanford University and the IBM Almaden Research Center. This new microscope concept was completely different and it could operate even with insulating surfaces; it was based on measuring the displacement of a cantilever spring.

He discussed his idea with Dr. Christoph Gerber of IBM and Prof. Calvin Quate of Stanford, and they calculated the forces between atoms which, to their surprise, turned out to be equivalent to a spring constant of about 10 N/m. It is easy to make a leaf spring with a smaller force constant. A piece of household aluminum foil 4 mm long and 1 mm wide has a force constant of about 1 N/m. It was possible to sense atomic-sized displacements with a soft spring, imaging the suface topography at atomic resolution with a force small enough so that the atoms were not disturbed very much. They built the first prototype in a few days and the results have been magnificent. Atomic force microscopes (AFMs) are now commercially available.

It may seem remarkable that mechanical devices can measure such small displacements. Even biological mechanical devices can measure them! At the threshold of human hearing the motion of the eardrum is approximately 1/10 the diameter of the hydrogen molecule. Professor R. V. Jones has constructed capacitance micrometers with an r.m.s. sensitivity limit of 5×10^{-15} m for a recording time constant of 1 s, with a stability of approximately 10^{-12} m per day. With these he has measured atmospheric vibrations in Scotland due to thermonuclear explosions in China, and motion of Scottish buildings due to tides in the North Sea.

Later we'll discuss these numbers in more detail, but now let's examine this linear restoring force in Newton's second law.

1.4 Free Harmonic Oscillator

According to the second law the sum of all external forces $\boldsymbol{F}$ is the cause of the change in momentum $\boldsymbol{p} = m\boldsymbol{v}$. This is a vector statement: $\boldsymbol{F} = d\boldsymbol{p}/dt$. For a while we will be examining only one force in one dimension, so the second law looks like this in scalar form with the linear restoring force

$$F(x) = \frac{d}{dt}(mv) = -\kappa x \,. \tag{1-1}$$

We will assume that the mass is not a function of time so, $m\, dv/dt = -\kappa x$. From kinematics we know that

$$v = \frac{dx}{dt} \,. \tag{1-2}$$

We can restate the equation of motion as

$$m\frac{d^2x}{dt^2} = -\kappa x \,. \tag{1-3}$$

Divide both sides by m and we have a simple, but remarkable, expression,

$$\frac{d^2x}{dt^2} = -\frac{\kappa}{m}x \,. \tag{1-4}$$

This is a second order, ordinary, linear differential equation with constant coefficients, so the solution $x(t)$ has two constants of integration that are determined by the initial conditions, for example,

- One constant is determined by the value of $x(0) = x_o$, the initial displacement when $t = 0$.

- The other constant is determined by the value of $v(0) = v_o$, the initial velocity when $t = 0$.

One of the striking things about Eq. (1-4) is that it says the second derivative of x is proportional to $-x$. We know that the second derivative of a sine or cosine is the negative of itself. This suggests that we try a solution of the form

$$x(t) = A\cos(\omega t) + B\sin(\omega t). \tag{1-5}$$

When we substitute this into Newton's second law, Eq. (1-4), we obtain

$$-\omega^2 A\cos(\omega t) - \omega^2 B\sin(\omega t) = -\frac{\kappa}{m}[A\cos(\omega t) + B\sin(\omega t)]. \tag{1-6}$$

So, our assumed solution is really a solution if, and only if,

$$\omega^2 = \frac{\kappa}{m}. \tag{1-7}$$

This is the square of the angular frequency ω. Angular frequency, measured in radians/s, is convenient in calculations, but we usually measure frequency in Hz, kHz, MHz, GHz, and so on. The angular frequency is related to f (frequency in Hz) by,

$$\omega = 2\pi f, \tag{1-8}$$

because, if a sine goes through 2π radians in one second, it goes through one complete cycle, that is, it goes through all of its values one time. The period T is a related parameter given by

$$T = \frac{1}{f}. \tag{1-9}$$

The period is the time interval required for $x(t)$ to go through all of its values once. For example, if something goes through all of its values a thousand times in one second the frequency is 1 kHz and the period is 1 ms.

To determine the constants set $t = 0$ in,

$$x(t) = A\cos(\omega t) + B\sin(\omega t) \tag{1-10}$$

so,

$$x(0) = x_o = A\cos(0) + B\sin(0). \tag{1-11}$$

This indicates that $A = x_o$ but it doesn't tell us anything about B. Now let's take the first derivative of $x(t)$ to get the velocity

$$v(t) = -\omega A\sin(\omega t) + \omega B\cos(\omega t). \tag{1-12}$$

At $t = 0$, Eq. (1-12) becomes

$$v(0) = v_o = -\omega A \sin(0) + \omega B \cos(0) = \omega B . \tag{1-13}$$

Therefore, $B = v_o/\omega$, and we have the complete solution

$$x(t) = x_o \cos(\omega t) + \frac{v_o}{\omega} \sin(\omega t). \tag{1-14}$$

This is one form of solution of the classical simple harmonic oscillator. Note that we must know the initial displacement and the initial velocity.

By differentiating Eq. (1-14) we can find the velocity and acceleration:

$$v(t) = -\omega x_o \sin(\omega t) + v_o \cos(\omega t), \text{ and} \tag{1-15}$$

$$a(t) = -\omega^2 x_o \cos(\omega t) - \omega v_o \sin(\omega t) . \tag{1-16}$$

Equations (1-14), (1-15), and (1-16) provide complete information about the kinematics of the harmonic oscillator. Typical results are shown in Fig. 1-4 for specific values of x_o, v_o, and frequency.

From now on, let's distinguish the natural frequency of a free harmonic oscillator by calling it ω_o,

$$\omega_o = \left(\frac{\kappa}{m}\right)^{1/2} .$$

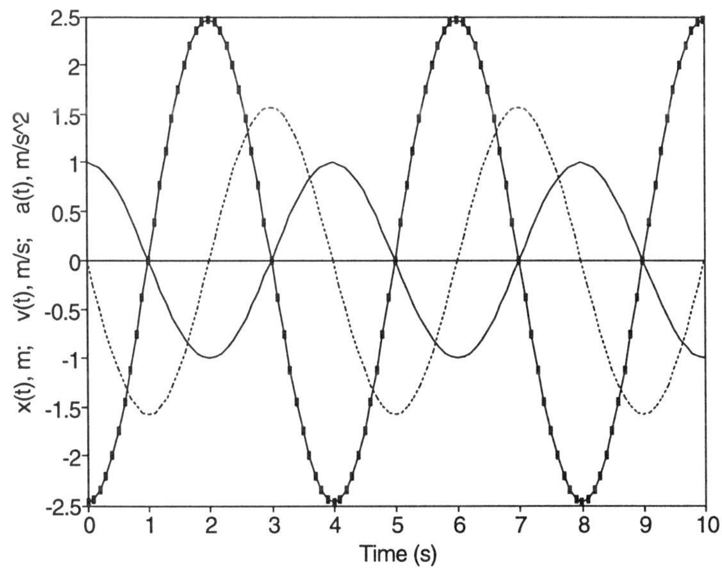

Figure 1-4. Displacement (solid), velocity (dotted), and acceleration (markers). Time interval between markers is constant. $x_o = 1$, $v_o = 0$, $f_o = 0.25$ Hz.

Magnitude and phase

An equivalent form of solution is useful if we are interested in magnitude X and phase θ. We would like an expression of the form

$$x(t) = X\cos(\omega_o t + \theta) . \tag{1-17}$$

Take the first derivative of Eq. (1-17) to find the velocity:

$$v(t) = -\omega_o X\sin(\omega_o t + \theta) . \tag{1-18}$$

Now we have two equations in X and θ. Let's use (1-17) and (1-18) at $t = 0$ to find x_o and v_o:

$$x_o = X\cos(\theta), \text{ and} \tag{1-19}$$

$$v_o = -\omega_o X\sin(\theta) . \tag{1-20}$$

Dividing (1-20) by (1-19),

$$\frac{\sin(\theta)}{\cos(\theta)} = \tan(\theta) = \frac{v_o}{-\omega_o x_o} . \tag{1-21}$$

From Eq. (1-19) and (1-20),

$$\cos(\theta) = \frac{x_o}{X} \quad \text{and} \quad \sin(\theta) = \frac{v_o}{-\omega_o X} . \tag{1-22}$$

Now square the cosine and sine and add them:

$$\cos^2(\theta) + \sin^2(\theta) = 1 = \frac{x_o^2}{X^2} + \frac{v_o^2}{\omega_o^2 X^2} . \tag{1-23}$$

Now we can obtain the magnitude from Eq. (1-23),

$$X = (x_o^2 + v_o^2/\omega_o^2)^{1/2}. \tag{1-24}$$

To summarize,

$$x(t) = X\cos(\omega_o t + \theta), \tag{1-25}$$

where X is given by Eq. (1-24) and $\theta = \arctan\left(\frac{v_o}{-\omega_o x_o}\right)$.

Phasor solution

The free harmonic oscillator is described by a linear, secondorder, ordinary differential equation with constant coefficients,

$$m\frac{d^2x}{dt^2} = -\kappa x . \tag{1-26}$$

We know that this type of equation is always satisfied by an exponential (phasor) $x(t) = X e^{-st}$, where, in general, s is a complex function. Let's rewrite Eq. (1-26) as

$$\frac{d^2x}{dt^2} = -\omega_o^2 x . \tag{1-27}$$

Now let's try our phasor solution. Every application of the time derivative simply brings down a factor of $-s$,

$$\frac{dx}{dt} = \frac{d}{dt} X e^{-st} = -sx \tag{1-28}$$

$$\frac{d^2x}{dt^2} = \frac{d^2}{dt^2} X e^{-st} = s^2 x . \tag{1-29}$$

So, Eq. (1-27) becomes

$$s^2 x = -\omega_o^2 x \tag{1-30}$$

This tells us that the exponential is a solution if, and only if,

$$s = \pm j\omega_o . \tag{1-31}$$

We already knew this, so it appears that the phasor solution is not much help at this point. Nevertheless, later we will see that phasors are quite useful.

1.5 *LC* Resonator

Let's consider a lossless inductor and capacitor in a resonant circuit as seen in Fig. 1-5. Add up the voltages around the closed loop with current i in the inductor L and charge q on the capacitor C,

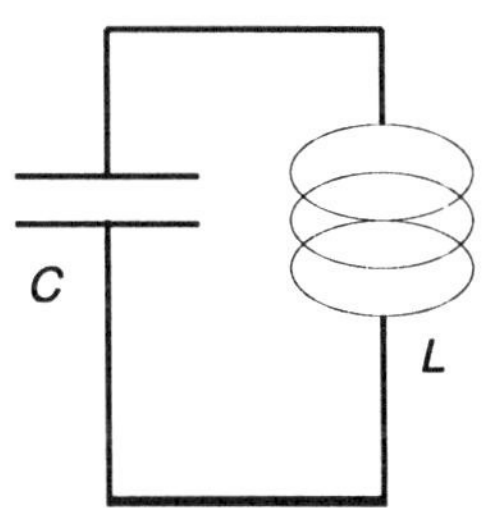

Figure 1-5. An idealized *LC* resonator with zero resistance fits our description of an undamped harmonic oscillator.

$$-L\frac{di}{dt} = \frac{q}{C}. \tag{1-32}$$

Current is defined as $i = dq/dt$, so $di/dt = d^2q/dt^2$. Now we can write Eq. (1-32) as

$$L\frac{d^2q}{dt^2} = -\frac{q}{C}. \tag{1-33}$$

Rearranging Eq. (1-33) we have an equation for a simple harmonic oscillator, similar to Eq. (1-4):

$$\frac{d^2q}{dt^2} = -\frac{1}{LC}q. \tag{1-34}$$

The natural frequency of the *LC* circuit is given by

$$\omega_o = \frac{1}{\sqrt{LC}}. \tag{1-35}$$

In Chapter 3 we will examine the effect of losses in a realistic *LC* resonator. Braginsky et al. have constructed extraordinarily low-loss superconducting microwave resonators (see Chapter 5, and Bibliography).

1.6 Journey Through the Center of the Earth

At first glance the universal law of gravitation, an inverse square force, does not appear to have much in common with our linear restoring force. The gravitational force is given by

$$\boldsymbol{F}(r) = -G\frac{Mm}{r^2}\mathbf{r}_1, \tag{1-36}$$

where the gravitational force constant G is 6.67×10^{-11} N·m^2 kg^{-2}, M and m are two point masses, r is the distance between them, and $\mathbf{r}_1$ is the unit vector along the line of centers. This force rapidly becomes weaker with distance, not stronger like $\boldsymbol{F}(r) = -\kappa\boldsymbol{r}$. Harmonic oscillators are currently used to measure the value of G.

At second glance one sees a few things in common. Both belong to the special class called *central forces*, which are associated with conservative force fields (see Chapter 2). Both are *attractive forces.* Also, for central forces the angular momentum is a constant of the motion, but that doesn't concern us now.

Equation (1-36) applies to point particles. What happens when we have a large collection of particles distributed in space, like a planet? Let's examine the gravitational force outside and inside of a spherical mass distribution of radius R and total mass M. We will use m as a small test mass to study the field at various points. It is convenient to define the average mass density ρ,

$$\rho = \frac{M}{V}, \tag{1-37}$$

where V is the volume of the sphere. For mass distributions with high symmetry like a sphere, a quick and easy way to calculate the field is by means of the divergence theorem,

$$\int \boldsymbol{F} \cdot d\boldsymbol{S} = \int \nabla \cdot \boldsymbol{F}\, dV, \tag{1-38}$$

where the surface integral on the left-hand side is the surface enclosing the volume on the right-hand side. This general theorem, for any vector, simply says that the net flow from a volume passes through the surface of the volume. To use Eq. (1-38) we need to calculate the divergence of Eq. (1-36). By analogy with electric fields, it is convenient to define a gravitational field vector $\boldsymbol{g}$ with units of N/kg or m/s^2,

$$\boldsymbol{g} = \frac{\boldsymbol{F}}{m}. \tag{1-39}$$

This is the gravitational force normalized to a unit test mass. Since the gravitational force has no θ or φ components, the divergence in spherical polar coordinates reduces to

$$\nabla \cdot \boldsymbol{g} = \frac{1}{r^2} \frac{\partial}{\partial r} (r^2 g). \tag{1-40}$$

Using Eq. (1-36), (1-37), and (1-39) we obtain

$$g = -\frac{GM}{r^2} = -\frac{G}{r^2} \frac{4}{3} \pi r^3 \rho \tag{1-41}$$

and substitute this into Eq. (1-40), with the result

$$\nabla \cdot \boldsymbol{g} = -4\pi G \rho . \tag{1-42}$$

The negative sign shows that there is always an inward flow because this force is attractive. Now we can substitute this divergence into Eq. (1-38):

$$\int \boldsymbol{g} \cdot d\boldsymbol{S} = -4\pi G \int \rho\, dV. \tag{1-43}$$

To check on what we have done, let's compute the magnitude of the gravitational field vector $\boldsymbol{g}$ at a distance $r > R$ from the center of the spherical mass. Since the field is anti-parallel to the outward normal to the surface, the left side of Eq. (1-43) is simply $g 4\pi r^2$. The anti-parallelism has been automatically accounted for, by the negative sign on the right-hand side

$$g\,4\pi r^2 = -4\pi G\int_0^R \rho dV. \tag{1-44}$$

The integral on the right-hand side has an upper limit of R, because the density ρ is zero for $r > R$. But this integral is simply the mass M, so the field vector outside the mass distribution becomes

$$g = -\frac{GM}{r^2}. \tag{1-45}$$

In view of Eq. (1-39) it appears that what we have done is consistent with the gravitational force law for a single particle of mass M based on Eq. (1-36). In other words, we have shown that the gravitational field outside of a spherical mass distribution is the same *as if* it were a point mass located at the center of the distribution. It took Newton about twenty years to prove this without using the divergence theorem. The divergence theorem is often called Gauss' theorem.

Now let's look at the field inside of the mass distribution. Imagine a frictionless tunnel drilled through the center of planet Earth. We can use Eq. (1-44) again but now $r < R$. The only thing that changes is the upper limit on the integral

$$g\,4\pi r^2 = -4\pi G\int_0^r \rho dV = -4\pi G\left(\rho\frac{4}{3}\pi r^3\right) \tag{1-46}$$

which reduces to

$$g = -\frac{4}{3}\pi G\rho r. \tag{1-47}$$

So, the force on test mass m inside the mass distribution is

$$F = mg = -\frac{4}{3}\pi G m\rho r. \tag{1-48}$$

But this is just the simple harmonic oscillator force with a restoring force constant of $4\pi Gm\rho/3$. An object with mass m dropped into the tunnel through the center of Earth (or any spherical mass distribution) would execute simple harmonic motion with a natural frequency, the square of which is given by

$$\omega_o^2 = \frac{\kappa}{m} = \frac{4}{3}\pi G\rho. \tag{1-49}$$

The mass of Earth is 5.98×10^{24} kg and the mean radius is 6.37×10^6 m. From these data Eq. (1-49) predicts that the natural frequency for the mass in the tunnel is 1.242×10^{-3} rad/s or 1.977×10^{-4} Hz. The period for the round trip is 5057.96 s or 1.405 hours, the period of the *Schuler pendulum* of inertial navigation fame.

A little thought shows that oscillatory motion would occur for any tunnel in Earth, even one that did not go directly through the center. Some people who are paid to find new ways to spend tax money have given consideration to the construction of a tunnel from England to Turkey, and from Point Paloma, Spain, to Malabata, Morocco.

1.7 Low Satellite Orbit

Consider an Earth satellite of mass m and tangential speed v in a circular orbit, and let's neglect all resistive forces like atmospheric drag. Under this assumption only gravitational force is acting on the test mass m, and it produces an acceleration toward the center of Earth. The equation of motion in an inertial system is

$$F(r) = -\frac{GMm}{r^2} = -\frac{mv^2}{r} . \tag{1-50}$$

Therefore,

$$v^2 = \frac{GM}{r} . \tag{1-51}$$

Multiplying both sides of Eq. (1-51) by r^{-2},

$$\frac{v^2}{r^2} = \frac{GM}{r^3} . \tag{1-52}$$

Since $v = \omega r$,

$$\frac{v^2}{r^2} = \omega^2 = \frac{GM}{r^3} . \tag{1-53}$$

Equation (1-53) is equivalent to Kepler's third law (the square of the period is proportional to the cube of the semi-major axis), which he obtained empirically from Tycho Brahe's careful measurements of planetary motion. Now consider a low orbit, so low that the radius of the orbit is R, the radius of Earth. For this special case Eq. (1-53) becomes

$$\omega^2 = \frac{GM}{R^3} = \frac{G\,(4/3)\pi\rho R^3}{R^3} = \frac{4}{3}\pi G\rho, \tag{1-54}$$

which is the same frequency as shown in Eq. (1-49).

- An Earth satellite with an orbit that skims Earth's surface has the same frequency and period as the harmonic oscillator constructed with a tunnel through the center of Earth.

Suppose the satellite was at the entrance to the tunnel at the time you released the mass and let it fall freely. The projection of the circular satellite orbit corresponds exactly to the position of the mass in the tunnel. When the mass in the tunnel is at Earth's center the satellite has completed 25 percent of one orbit. Both arrive simultaneously at the other end of the tunnel.

1.8 FREE-OSC, Free Oscillator Worksheet

This worksheet lets you experiment with a free harmonic oscillator. You can input the force constant, mass, and initial displacement and velocity and then view the frequency, period, displacement, velocity, momentum, and acceleration.

Enter the information as requested in Fig. 1-6. The natural frequency (rad/s) appears in cell C11, and cell F11 displays the frequency in Hz. Move the cursor around the worksheet to examine the formulas in each cell. Graphs connected to this worksheet are named *X, V, A vs. T*, *Momentum vs. T, K.E. vs. P.E., Energy vs.X,* and *Energy vs. T.* A typical result is shown in Fig. 1-4.

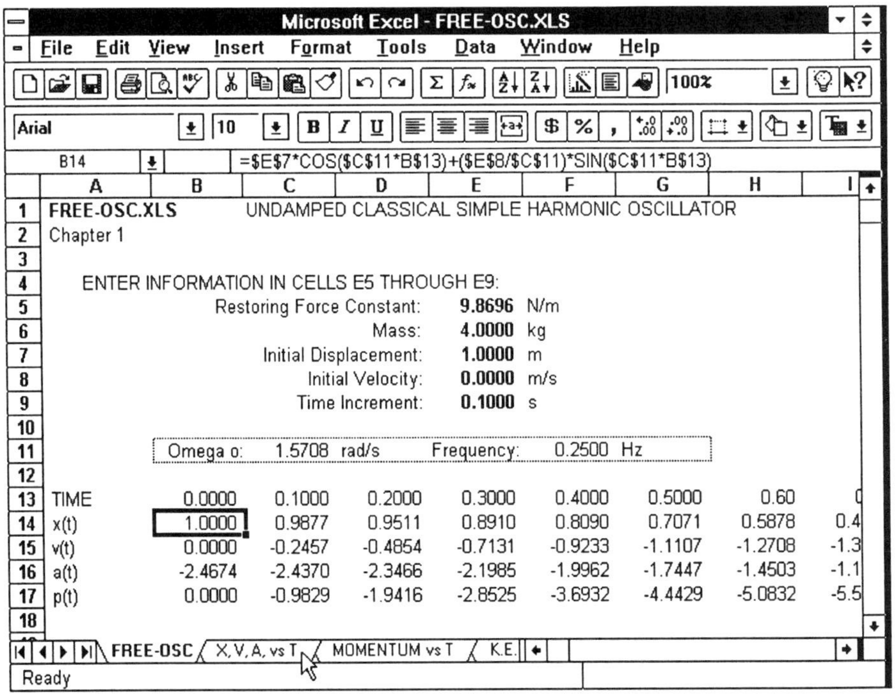

Figure 1-6. Home position for the worksheet FREE-OSC. Highlighted cell shows formula for $x(t)$. You can change constants and initial conditions by entering information in cells E5 through E9.

1.9 Phase Space, Part 1

Time appears only as a parameter in the two equations for displacement and velocity, (1-17) and (1-18). In what follows we will find it more productive to discuss displacement and momentum $p = mv$. Time can be eliminated from the parametric equations for the displacement and the momentum, to obtain $p(x)$. This is the usual convention; we could just as well find $x(p)$.

One way of eliminating time is to square and add (1-17) and (1-18), obtaining (1-57) as follows:

$$x(t) = X\cos(\omega_o t + \theta) \tag{1-17}$$

$$p(t) = mv(t) = -\omega_o mX\sin(\omega_o t + \theta) \tag{1-18}$$

$$\frac{x^2(t)}{X^2} = \cos^2(\omega_o t + \theta) \tag{1-55}$$

$$\frac{p^2(t)}{m^2\omega_o^2X^2} = \sin^2(\omega_o t + \theta) \tag{1-56}$$

$$\frac{p^2}{m^2\omega_o^2X^2} + \frac{x^2}{X^2} = 1. \tag{1-57}$$

This is in the form of the equation of an ellipse. While studying the motion of planet Mars, Johannes Kepler worked over six years before he recognized that this type of equation was an ellipse. Figure 1-7 shows the graph *Phase Space* in the worksheet PHASE-SP.

The oscillator is one-dimensional and its associated ellipse is two-dimensional. This space consisting of x and p is called *phase space* and it has twice the number of dimensions of its associated ordinary space. A two-dimensional oscillator has an associated four-dimensional phase space. For ordinary mortals, visualization of six-dimensional phase space is not easy for a three-dimensional oscillator. In phase space, as in ordinary space, there are two values of momentum for each position, and two positions for each value of momentum.

The real estate in the phase space of Fig. 1-7 has units of $\boldsymbol{p \cdot x}$ which is (kg·m/s)·m. This can be restated as (kg·m^2/s^2)·s, which is J·s. The dimension of this "area" is called *action* or *wirkung* and it has the units of Planck's constant. Often it's more convenient to measure energy in electron volts, in which case the units of this area is eV·s. We'll see more about this in Chapter 2, Section 2.10, and in Chapters 10, 11, and 12.

Consider the r.m.s. (root-mean-square) values of displacement and momentum in phase space. From Eq. (1-36),

$$x_{rms}^2 = \frac{\int_0^T X^2 \cos^2(\omega_o t + \theta)\, dt}{\int_0^T dt}, \tag{1-58}$$

where T is the period. You could actually carry out the integrations, or just use the trignonometric identity that states $\cos^2\varphi = 0.5 + 0.5\cos(2\varphi)$, so the mean value of $\cos^2\varphi = 0.5$. Either way you do it the r.m.s. value is

$$x_{rms} = \frac{X}{\sqrt{2}}. \tag{1-59}$$

Similarly, for the momentum given by Eq. (1-37),

$$p_{rms}^2 = \frac{\int_0^T (-m\omega_o X)^2 \sin^2(\omega_o t + \theta)\, dt}{\int_0^T dt} \tag{1-60}$$

so,

$$p_{rms} = m\omega_o \frac{X}{\sqrt{2}}. \tag{1-60}$$

Of course, the r.m.s. values depend on how the functions are distributed; here we are considering sines and cosines. If we were considering something else, Gaussian distributions for example, we would obtain different r.m.s. values.

Let $x_{rms} = \Delta x$ and $p_{rms} = \Delta p$. Then we can express an area of phase space as

$$\Delta p\, \Delta x = m\omega_o \frac{X}{\sqrt{2}}\left(\frac{X}{\sqrt{2}}\right) = \frac{1}{2} m\omega_o X^2. \tag{1-61}$$

The mass, frequency, and amplitude are constants. This indicates that the area $\Delta p\, \Delta x$ is a constant that depends on the parameters of the oscillator. Let's rewrite Eq. (1-61) to state it explicitly in units of Joule·seconds:

$$\Delta p\, \Delta x = \frac{0.5\, m\omega_o^2 X^2}{\omega_o} \frac{\text{J}}{\text{s}^{-1}}. \tag{1-62}$$

This may appear to be a pointless exercise, but it isn't. In Chapter 2, and in later chapters, we will see interpretations of Eq. (1-62) that have far-reaching consequences.

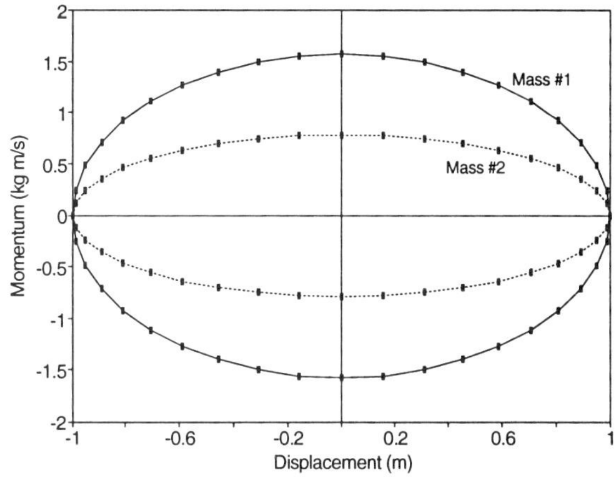

Figure 1-7. Phase space for two one-dimensional harmonic oscillators. Displacement magnitude = 1 m and frequency = 0.25 Hz. Mass #1 = 1 kg, mass #2 = 0.5 kg. Motion in phase space is clockwise. $x_O = 1$; $v_O = 0$. The time interval between markers is 0.1 s.

1.10 PHASE-SP, Phase Space Worksheet

The worksheet named PHASE-SP lets you experiment with the phase space of two one-dimensional oscillators. After you start your spreadsheet and retrieve this file you should see the upper-left-hand corner (Home position) shown in Fig. 1-8.

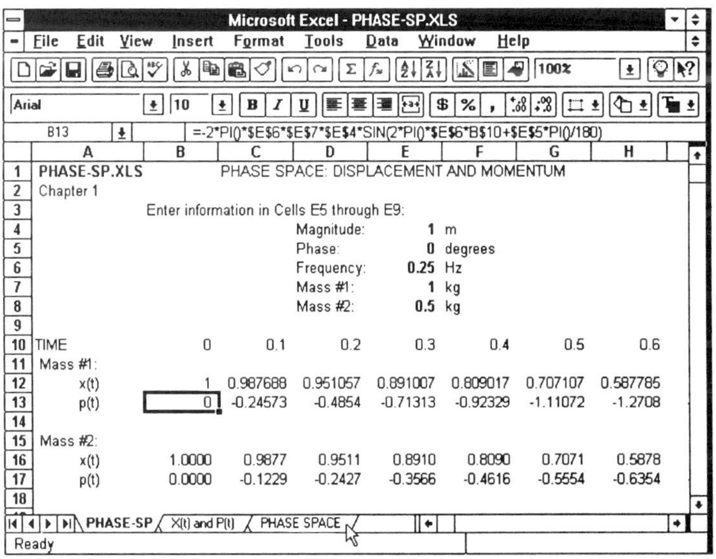

Figure 1-8. Home screen for the worksheet PHASE-SP, with two oscillators of different mass and same frequency. Two graphs are on tabs at bottom edge.

1.11 Callisto, Galileo, and French

About 400 years ago Holland was a boiling cauldron of optical invention. The Dutch spectacle-maker Zacharias Janssen is generally credited with development of the compound microscope about 1590. Hans Lippershey patented the first successful telescope in 1608 and soon almost every naval and merchant ship carried one. When the news reached Italy, Galileo designed and constructed his own telescopes and used them for astronomical observations. With his third telescope, of about 30X magnification, he discovered four moons of planet Jupiter. The moons' motions were rapid, impelling Galileo to write "... the revolutions of these planets are so speedily completed that it is usually possible to take even their hourly variations."

The moon Callisto is particularly interesting. It moves with almost constant speed in its orbit, which is almost perfectly circular. From planet Earth the plane of the orbit lies in our line of sight, so that we see essentially simple harmonic motion, the projection of circular motion on our sensors. This is analogous to the tunnel through planet Earth and the low satellite orbit.

Professor A. P. French (M.I.T.) has taken Galileo's original data (in Galileo's own handwriting) and plotted it, showing that the motion that Galileo observed was almost certainly simple harmonic motion, *the projection of uniform circular motion on a diameter of the circle in which the circular motion occurs.* You can see this and a facsimile of Galileo's original records from July through October of 1610, on pages 288 and 289 of French's book (see Bibliography).

1.12 Searching for Planets

In October, 1995, Swiss astronomers apparently discovered a planet, about the size of planet Jupiter, outside of our solar system. The planet, with a period of four days, is close to its star, 51 Pegasi, and approaches the star within a distance of about 20 percent of the mean distance from planet Mercury to our Sun. This was confirmed spectroscopically in four days of observation by a team at the Lick Observatory on Mount Hamilton, California. The spectroscopic data from October 11 to October 16, 1995, show a sinusoidal motion of 51 Pegasi along the line of sight, similar to Galileo's visual observations of the moon Callisto 385 years previously.

The discovery of planet Neptune in our solar system was one of the high points of classical physics. Based on Newton's laws, the French astronomer U. J. J. Le Verrier (or Leverrier) and J. C. Adams of Cambridge independently calculated where the unknown planet should be, based on perturbations of planet Uranus. Uranus had been visually discovered by Sir William Herschel, a professional musician and devoted astronomer. Neptune was subsequently observed on September 23, 1846.

A theory with predictive powers seems to contain elements of reality. In 1846 John Potter wrote in a letter to Joseph Henry, "I beg to inclose you two news paper notices of the new planet, lately discovered, by the French astronomer Le Verrier—discovered, do I say, when by calculations entirely his own, he predicted, this discovery, lost in space, far beyond Uranus—it is so far exceeding Common apprehension."

Most books hail the heroic work of Le Verrier and Adams as a triumph, but Ivars Peterson wrote that this triumph was an accident, because it was based on faulty data (*Newton's Clock: Chaos in the Solar System*, W. H. Freeman, New York, 1993). It was not the first, nor the last, lucky accident in the progress of human knowledge.

1.13 Franklin, Priestly, Verne, and Burroughs

In the nineteenth century, Jules Verne wrote a fantasy about a Scottish professor and his colleagues who enter an inactive volcano in Iceland and descend to Earth's core, where they discover a prehistoric environment complete with dinosaurs and cave people. In the twentieth century, Edgar Rice Burroughs (who won fame and fortune with the Tarzan novels and movies, in addition to his Martian books) wrote about a land called Pellucidar beneath Earth's surface that was remarkably like that of Jules Verne's story. A movie based on Verne's *Journey to the Center of the Earth*, directed by Henry Levin and filled with state-of-the-art special effects, was produced in 1959.

The central idea in these works of fiction is that planet Earth is a shell with a hollow core. Dinosaurs and people stand, walk, and run with their feet firmly planted on the inner surface of the shell. Oceans with sea monsters exist on the inner surface of the shell. Is this reasonable, or literary license? For one thing, the density of the spherical shell would have to be enormous to account for the known mass of planet Earth. For another thing, let's compute the gravitational field inside a spherical shell of inner radius r and outer radius R. Using the divergence theorem as before,

$$\int \boldsymbol{g}\cdot d\boldsymbol{S} = \int \nabla\cdot \boldsymbol{g}\, dV = -4\pi G \int \rho\, dV, \tag{1-63}$$

we can put on the limits for the spherical shell at the inner radius,

$$\int_0^r \boldsymbol{g}\cdot d\boldsymbol{S} = -4\pi G \int_0^r \rho\, dV. \tag{1-64}$$

So,

$$g\, 4\pi r^2 = -4\pi G \int_0^r \rho\, dV = 0, \tag{1-65}$$

which shows g is zero inside of a spherical shell because $\rho = 0$.

This is completely consistent with the unique property of an inverse square force, and is analogous with the electrostatic field, in which there is no electric field inside a closed conductor or a spherical shell of electric charge. Only the mass or charge *below* the test point contributes to the field at that point. Only the small acceleration $\omega^2 r$ due to Earth's rotation would appear as a force on mass m toward the axis of rotation. For an inner radius of 10^6 m this centripetal force is only about 0.00529 N/kg, far less than the 9.8 N/kg that we normally experience at the surface of this planet. The dinosaurs, cave people, and intrepid explorers would float around in almost a weightless condition.

This would be an interesting environment for experimenting with harmonic oscillators like a pendulum and a mass on a spring. What would be the period of a pendulum of one meter length at the inner radius in the rotating coordinate system described above?

During his researches in Philadelphia, Benjamin Franklin wrote to his friend Joseph Priestly in England that he had discovered that electric charge always appeared on the outer surface of a conductor. Priestly verified Franklin's experiments and reasoned that, like Newton's gravitation law, this implied an inverse square law for electric charges.

In the third edition of his book *The History and Present State of Electricity*, published in 1775, Priestly wrote, "May we not infer from this experiment that the attraction of electricity is subject to the same laws with that of gravitation and is therefore according to the squares of the distances; since it is easily demonstrated that were the earth in the form of a shell a body on the inside of it would not be attracted to one side of it more than another."

If Jules Verne and Edgar Rice Burroughs had received the blessings of scientific education, the world would have been deprived of several adventure novels and an entertaining movie.

What's Next?

There are profound connections between Galileo's observation of the pendulum lamp in the cathedral at Pisa and his observation of Jupiter's moon Callisto, but we will not be prepared to discuss them until Chapter 12. As a young medical student, Galileo applied his observation of the pendulum lamp to his invention of a pendulum with adjustable length to time the pulse of a patient. In this way, he could measure pulse rate in terms of the length of a pendulum. He could shorten or lengthen the pendulum to match the pulse. Changing the length of a pendulum slowly will lead us to another profound idea, adiabatic invariance. But before all that, in the next chapter we need to examine the energy of undamped oscillators.

Chapter 2

Oscillator Energy

Energy will do anything that can be done in the world.
Johann Wolfgang von Goethe

Newton's laws of motion involve vector quantities of force, displacement, velocity, acceleration, and momentum. Time is treated as a scalar. In Chapter 1 we confined our attention mostly to one-dimensional motion, which obscured the vector nature of the calculations. In this chapter we will discuss a true scalar quantity, energy.

2.1 Conservative Forces

The linear restoring force that results in harmonic oscillations is a member of the class of *conservative forces*. This fact has far-reaching implications that we will explore as we go.

There are seven ways to identify a conservative force. The seven ways may look different but they all mean the same thing.

1. The work done by the force is independent of the path.
2. The work done around any closed path is zero.
3. $(F_x dx + F_y dy + F_z dz)$ is an exact differential.
4. The force is a function only of the coordinates and $\nabla \times \boldsymbol{F} = 0$. In Cartesian coordinates this requirement can be expressed as

$$\frac{\partial F_z}{\partial y} = \frac{\partial F_y}{\partial z}, \quad \frac{\partial F_x}{\partial z} = \frac{\partial F_z}{\partial x}, \quad \frac{\partial F_y}{\partial x} = \frac{\partial F_x}{\partial y} .$$

5. A potential energy function exists and it has a definite value at every point in space.
6. The sum of the kinetic energy K and potential energy U is constant.
7. Between any two points in the path, $\Delta K = -\Delta U$.

For any scalar function of coordinates U there is a general vector identity that is easily proved in Cartesian coordinates. The identity states:

$$\nabla \times \nabla U = 0. \tag{2-1}$$

Because a conservative force has $\nabla \times \boldsymbol{F} = 0$ we could let $\boldsymbol{F} = \nabla U$. However, the gradient points in the direction of increasing potential and we observe that forces cause things to move toward lower potential so we choose a negative sign,

$$\boldsymbol{F} = -\nabla U = -\left(\mathbf{i}\frac{\partial U}{\partial x} + \mathbf{j}\frac{\partial U}{\partial y} + \mathbf{k}\frac{\partial U}{\partial z}\right). \tag{2-2}$$

This simply says that the force is the negative gradient of the potential function. Equation (2-2) provides a tremendous simplification and convenience because now we can deal with a scalar field described by $U(x,y,z)$. When we want to know the vector field associated with this scalar field, we just take the negative gradient at the point in question. To find a vector sum, we can add scalars instead of vectors.

Force Test

Let's confirm that the linear restoring force is a conservative force. To do this we will calculate $\nabla \times \boldsymbol{F}$ and check to make certain that it is always zero. Let

$$\boldsymbol{F}(x,y,z) = \mathbf{i}F_x + \mathbf{j}F_y + \mathbf{k}F_z = -\mathbf{i}\kappa_x x - \mathbf{j}\kappa_y y - \mathbf{k}\kappa_z z. \tag{2-3}$$

The different force constants make the discussion more general by allowing for a possible anisotropic characteristic. In Cartesian coordinates $\nabla \times \boldsymbol{F}$ is given by

$$\mathbf{i}\left(\frac{\partial F_z}{\partial y} - \frac{\partial F_y}{\partial z}\right) + \mathbf{j}\left(\frac{\partial F_x}{\partial z} - \frac{\partial F_z}{\partial x}\right) + \mathbf{k}\left(\frac{\partial F_y}{\partial x} - \frac{\partial F_x}{\partial y}\right). \tag{2-4}$$

Substituting the force components from Eq. (2-3) into (2-4) shows that $\nabla \times \boldsymbol{F}$ is indeed zero everywhere. The linear restoring force of the harmonic oscillator is a conservative force.

2.2 Potential Energy

Now that we are certain we are dealing with a conservative force we can look for the potential energy function from which the force can be derived. As we have seen the force is the negative gradient of a potential, $\boldsymbol{F} = -\nabla U$. In one dimension this becomes

$$F_x(x) = -\frac{\partial U}{\partial x}. \tag{2-5}$$

We can use a total derivative because we have only one independent variable,

$$F_x(x) = -\frac{dU}{dx} . \tag{2-6}$$

Now we can integrate to find

$$\int_a^b F_x(x)\,dx = -\int_a^b dU(x) . \tag{2-7}$$

The integral on the left is the definition of work W, so

$$W = -[U(b) - U(a)] = \Delta U . \tag{2-8}$$

Work can change potential energy. Positive work (work done on the system) increases potential energy and negative work (work done by the system) decreases potential energy. Potential energy is energy stored by virtue of the relative positions of the components of the system in space.

We cannot measure an absolute potential, only a potential difference. For this reason it is convenient here to define $U(0) = 0$. For electrical systems it is convenient to define planet Earth as the zero of potential, or "ground." If you can't get connected to a good Earth ground (suppose you are in an aircraft) you have to use the largest conductor you can find, like the body of the aircraft. An electrical ground must approximate an infinite source and sink for charge; the ground must be able to provide charge and absorb charge so that its potential does not change. To be effective, ground leads must be short. For high-frequency components the emf induced due to lead inductance can effectively isolate a system from ground for short times. Losing the zero of potential can have serious consequences.

Now let's find $U(x)$ associated with the linear restoring force:

$$\int F_x(x)\,dx = -\int \kappa x\,dx = -\frac{1}{2}\kappa x^2 = -U(x) . \tag{2-9}$$

So the potential energy function for the simple harmonic oscillator is

$$U(x) = \frac{1}{2}\kappa x^2 . \tag{2-10}$$

If you need to place the equilibrium position somewhere other than at the origin, say at x_o, you can use

$$F(x) = -\kappa(x - x_o). \tag{2-11}$$

Now the potential can be written as

$$U(x) = \frac{1}{2}\kappa(x - x_o)^2 . \tag{2-12}$$

Figure 2-1 shows a typical linear restoring force and its associated potential energy function. At every point the restoring force is the negative of the slope of the potential energy.

2.3 Kinetic Energy

Let's assume that we have a constant mass and all motion is in one space dimension. By Newton's second law,

$$F = \frac{dp}{dt} = m\frac{dv}{dt} . \tag{2-13}$$

We can make the substitution,

$$\frac{dv}{dt} = \frac{dv}{dx}\frac{dx}{dt} = v\frac{dv}{dx} , \tag{2-14}$$

so we can write,

$$F = mv\frac{dv}{dx} . \tag{2-15}$$

Integrate Eq. (2-15) to find

$$\int_a^b F_x(x)dx = W = \int_{v(a)}^{v(b)} mv\,dv = \frac{1}{2}mv^2(b) - \frac{1}{2}mv^2(a) = \Delta K . \tag{2-16}$$

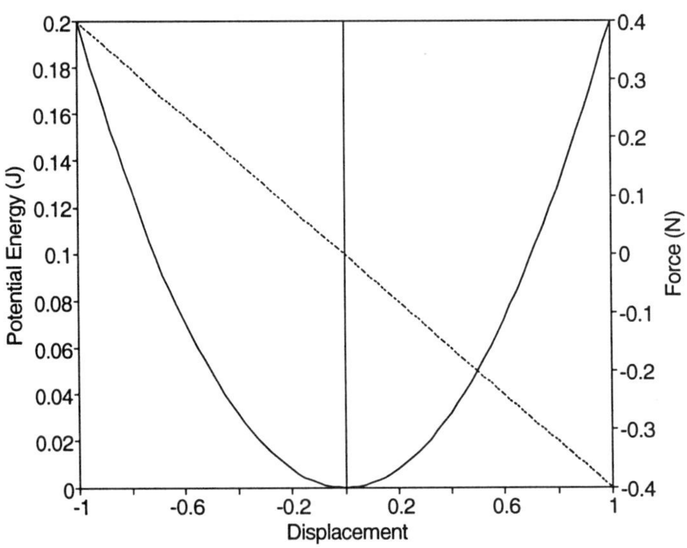

Figure 2-1. Linear restoring force (dotted line, right axis) and its associated potential energy function (solid line, left axis). Force constant $\kappa = 0.4$ N/m.

Again, the integral on the left is the definition of work W and this is completely equivalent to ΔK, the change in kinetic energy,

$$W = \Delta K. \tag{2-17}$$

Long ago kinetic energy was called *vis viva*, the force of life, because of its association with motion. It is also useful to express kinetic energy in terms of momemtum, $p = mv$,

$$\frac{1}{2}mv^2 = \frac{p^2}{2m}. \tag{2-18}$$

2.4 Conservation of Energy

In the case of a free oscillator no work is done on the system or by the system, so by Equations (2-8) and (2-17)

$$\Delta K + \Delta U = 0 \tag{2-19}$$

or,

$$\Delta(K + U) = 0. \tag{2-20}$$

So,

$$K + U = \text{a constant.} \tag{2-21}$$

The total energy, the sum of potential and kinetic, is a constant. Let's call this constant E. Using Eq. (2-18) we can write the total energy as

$$\frac{1}{2}m^{-1}p^2 + \frac{1}{2}\kappa x^2 = E. \tag{2-22}$$

Equation (2-22) is written in a peculiar form to emphasize the symmetry between momentum and displacement. Equation (2-22) is a familiar form of an equation that we have seen before. It describes an ellipse.

2.5 Phase Space, Part 2

Let's look again at Eq. (1-39) for the elliptical phase-space trajectory,

$$\frac{p^2}{(m\omega_o X)^2} + \frac{x^2}{X^2} = 1. \tag{1-39}$$

Multiply both sides by $\frac{1}{2}m\omega_o^2 X^2$,

$$\frac{p^2}{2m} + \frac{1}{2}m\omega_o^2 x^2 = \frac{1}{2}m\omega_o^2 X^2 . \tag{2-23}$$

When p is zero, at the end points where $(\pm x)^2 = X^2$, all of the energy is in potential form, so

$$\frac{1}{2}m\omega_o^2 X^2 = E . \tag{2-24}$$

Also, since $\kappa = m\omega_o^2$ it is clear that Eq. (2-22) is identical to (2-23). Note that each term in Eq. (1-39) is dimensionless, but each term in (2-23) has units of energy.

From the foregoing you can see that the linear restoring force has resulted in a system in which the total energy is constant, even though the potential energy and kinetic energy are changing in time. The total energy is conserved. The simple harmonic oscillator is a conservative system. $K + U = E = \text{constant}$. With the concept of total energy E we now have a nice way of describing the oscillator. Using Eq. (2-24) we can write

$$X^2 = \frac{2E}{m\omega_o^2} \tag{2-25}$$

so $x(t)$ can be written as,

$$x(t) = \left(\frac{2E}{m\omega_o^2}\right)^{1/2} \cos(\omega_o t + \theta) . \tag{2-26}$$

This is useful because it does not involve explicit knowledge of the initial conditions. In experiments we are often more likely to be interested in, and to measure, the energy, mass, frequency, and phase than to know the initial conditions.

Proceeding as we did in Section 1.9, we can calculate the r.m.s. values of $x(t)$ and $p(t)$ from Eq. (2-26),

$$\Delta x = x_{rms} = \frac{1}{\sqrt{2}}\left(\frac{2E}{m\omega_o^2}\right)^{1/2} \tag{2-27}$$

and

$$\Delta p = p_{rms} = mv_{rms} = m\omega_o \frac{1}{\sqrt{2}}\left(\frac{2E}{m\omega_o^2}\right)^{1/2} . \tag{2-28}$$

Therefore the phase space area $\Delta p\, \Delta x$ is given by

$$\Delta p\, \Delta x = \frac{m\omega_o}{2}\left(\frac{2E}{m\omega_o^2}\right) = \frac{E}{\omega_o} . \tag{2-29}$$

In view of Eq. (2-24) we can see that Eq. (2-29) is the same result that we obtained in Eq. (1-62), but in Chapter 1 we could not identify this in terms of the total energy. The quantity E/ω_o in Eq. (2-29) is a constant (an invariant) but in this equation it was derived for constant energy, so its full importance is not apparent. What if the energy and frequency change? We will delay investigating this question until we examine the case of constant energy in more detail.

2.6 Energy Diagrams

The energy level diagram is an important representation for many types of systems; it is not restricted to the harmonic oscillator. For the classical harmonic oscillator there is no reason why the energy levels cannot be set at any arbitrary values, so the change in energy from one level to another can take on any value.

The energy diagram for the simple harmonic oscillator is shown in Fig. 2-2. Notice how the maximum and minimum values of the displacement are controlled by the total energy. In Fig. 2-2, displacement is limited to 1 m. If you think 1 m is a rather large oscillator, think of it as 1 nm and scale the energies accordingly.

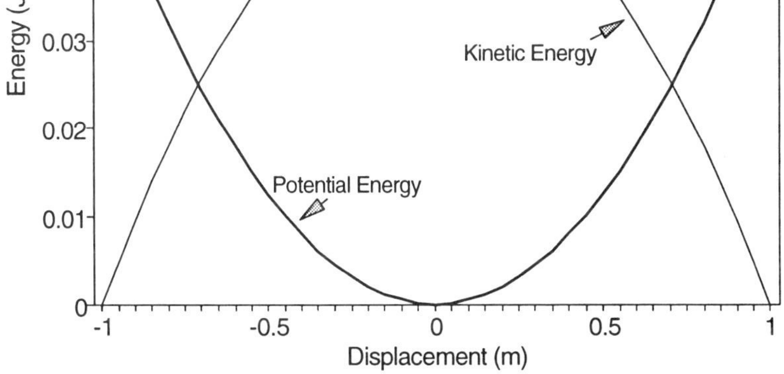

Figure 2-2. Potential, kinetic, and total energy for simple harmonic oscillator. Except for two points, potential and kinetic energy are always different. At all points their sum is constant. Time interval between markers is 0.05 s. Think of this graph as a view from the positive time axis, looking back toward the origin.

For displacement to exceed these limits kinetic energy would have to become negative, but kinetic energy is proportional to the square of the velocity. Does this mean that the velocity becomes imaginary, and what would this imply? This is one of the puzzles of quantum tunneling; how does the particle penetrate into the classically unallowed region? They are the subject of contemporary research. Quantum tunneling is the basis of our "understanding" of alpha-particle emission from nuclei, the operation of tunnel diodes, Josephson junctions, microwave absorption by ammonia, tunneling microscopes, and many other phenomena.

Does momentum become imaginary outside of the classically allowed region? What would that mean? What would the phase space look like? As we will see in Chapter 12 there is a finite probability of finding the particle in the classically unallowed region. This probability decreases exponentially with distance.

For any particular energy level the motion of the oscillator is strictly confined within the limits set by the classical amplitude. The difference between the total energy level and the potential energy curve, at any displacement, is the kinetic energy.

It is also useful to examine the time evolution of the potential, kinetic, and total energy. These are shown in Figs. 2-4 and 2-5. The kinetic energy is a function of velocity and the poential energy is a function of displacement, and velocity and displacement are funcions of time. The total energy is constant throughout time.

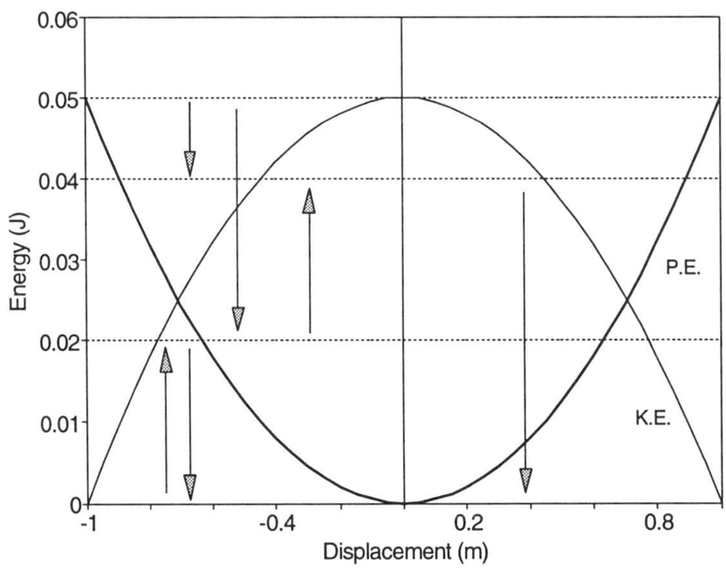

Fig. 2-3. Amplitude (maximum displacement) at each energy level is determined by energy of that level. The classical oscillator can have any arbitrary energy level and transitions between all levels are allowed. In this Figure the force constant = 0.4 N/m and the mass = 0.1 kg.

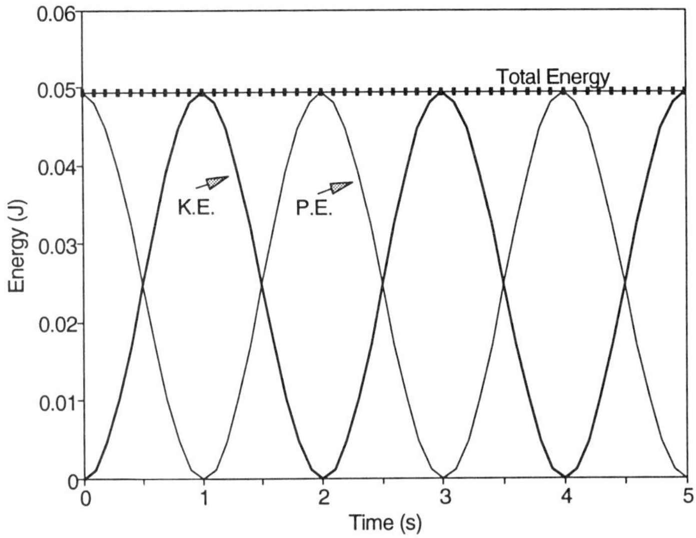

Figure 2-4. Potential, kinetic, and total energy viewed as functions of time. As the potential and kinetic energy change, their sum stays constant. Here, $m = 0.04$ kg, $\kappa = 0.01\pi^2$, $x_o = 1$m, $v_o = 0$, $\Delta t = 0.1$s between markers.

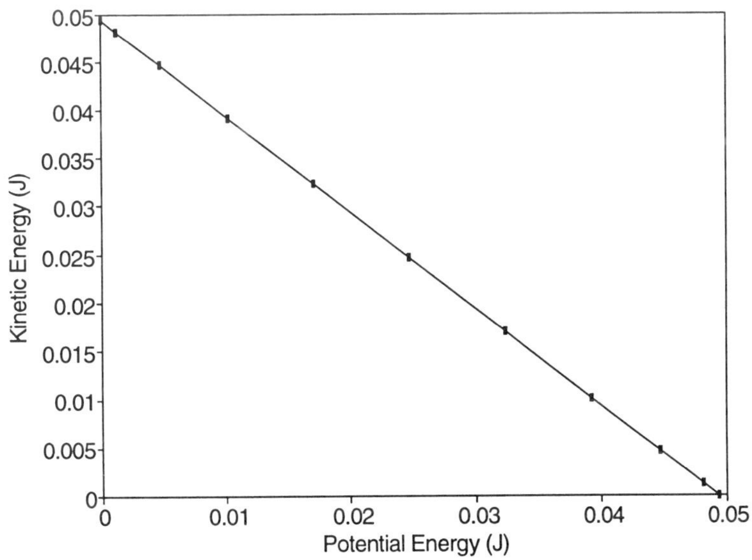

Figure 2-5. Instantaneous kinetic energy as a function of instantaneous potential energy. KE increases as PE decreases, and conversely. This graph was calculated directly from Newton's second law of motion and shows that energy is conserved instantaneously. Time interval between markers is 0.1s.

2.7 Equipartition of Energy

Equipartition of energy refers to the distribution of energy among the degrees of freedom in a system at thermodynamic equilibrium. We saw in Figs. 2-2, 2-4, and 2-5 that the instantaneous kinetic energy plus the instantaneous potential energy is equal to the total energy. Now we will see that the time-average kinetic energy plus the time-average potential energy is equal to the total energy, and each one is equal to half of the total energy. To see this for the oscillator let's calculate the time-averages directly from $x(t)$ and $v(t)$,

$$K_{av} = \left(\frac{1}{2}mv^2\right)_{av} = \frac{1}{2}m\omega_o^2X^2\left(\frac{\int_0^T \sin^2(\omega_o t+\varphi)\,dt}{\int_0^T dt}\right) = \frac{1}{2}\left(\frac{1}{2}m\omega_o^2X^2\right). \tag{2-30}$$

The time-average kinetic energy is half of the total energy. Similarly, we find the time-average potential energy is half of the total energy,

$$U_{av} = \left(\frac{1}{2}\kappa x^2\right)_{av} = \frac{1}{2}m\omega_o^2X^2\left(\frac{\int_0^T \cos^2(\omega_o t+\varphi)\,dt}{\int_0^T dt}\right) = \frac{1}{2}\left(\frac{1}{2}m\omega_o^2X^2\right). \tag{2-31}$$

Consequently, the time-average kinetic energy plus the time-average potential energy is exactly equal to the total energy,

$$K_{av} + U_{av} = E\,. \tag{2-32}$$

We will have use for this later in Chapter 10 when we calculate the r.m.s. values of velocity and displacement due to thermal motion.

Equations (2-30) and (2-31) are particularly important because they show how energy is distributed between kinetic and potential forms. In view of Eq. (2-24) we see that, in time-average, half of the total energy is in kinetic form and half of the total energy is in potential form. The total energy is equally distributed.

2.8 General Potential Energy Function

Here we are going to show why the simple harmonic oscillator can be used as the first approximation to any reasonable potential function, for small oscillations around a position of stable equilibrium. This is a particularly important point, of great practical and theoretical significance. Suppose we have a complicated potential function $U(x)$ like the one shown in Fig. 2-6.

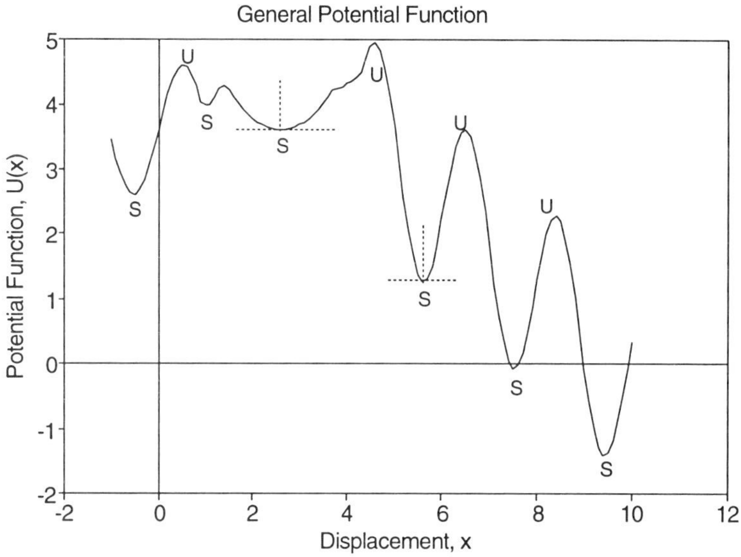

Figure 2-6. A hypothetical potential function that is associated with a complicated force. In general, simple harmonic oscillations would not occur. Minima marked S are positions of stable equilibrium. Maxima marked U are positions of unstable equilibrium. Origin of coordinate axes may be shifted to any arbitrary position, shown by dotted lines.

Positions of equilibrium are located at points where the force is zero, which occur when $dU/dx = 0$. The first derivative is zero at maxima and minima. Positions of unstable equilibrium are located at maxima of $U(x)$ because the force is always directed away from the equilibrium point. Any departure from a maximum will result in a force away from the equilibrium position.

Positions of stable equilibrium are located at minima of $U(x)$, because the force is always directed toward the equilibrium position, over a small range of x. In other words, conditionally stable equilibrium positions are identified by $dU/dx = 0$ with $d^2U(x)/dx^2$ being positive.

Let's examine any one of these minima more closely. If $U(x)$ is reasonably well behaved we can expand it in a Taylor's series about some known point x_o. Let's choose x_o to be located at one of these minima. Then Taylor's series tells us that,

$$U(x) = \frac{1}{0!}U(x_o)(x - x_o)^0 + \frac{1}{1!}\left(\frac{dU}{dx}\right)_o (x - x_o)^1 + \frac{1}{2!}\left(\frac{d^2U}{dx^2}\right)_o (x - x_o)^2 + \dots \tag{2-33}$$

The first term is a constant, and we have already seen that $U(x)$ is only determined to within an additive constant. We can set this term to zero by choosing $U(x_o) = 0$. This just shifts the x-axis up or down.

The second term in the series involves the first derivative of the potential evaluated at the position of zero force. It is zero by definition of the equilibrium position, $(dU/dx)_o = 0$. The force is zero at $x = x_o$.

The third term in the series is the *first surviving term*. Notice that the factor involving the second derivative evaluated at the equilibrium point is just a constant, which we can use as the restoring force constant κ, so the third term has the form,

$$\frac{1}{2!}\left(\frac{d^2U}{dx^2}\right)_o (x - x_o)^2 \Rightarrow \frac{1}{2}\kappa x^2 \tag{2-34}$$

This is the harmonic oscillator potential energy function if we shift the y-axis to the position x_o. This just shifts the y-axis right or left.

If we stay close to x_o the higher terms will contribute very little to the approximation for the following reasons:

- The factorial increases rapidly, decreasing the importance of higher terms.
- Higher derivatives are smaller than lower derivatives for well-behaved functions, decreasing the importance of higher terms.
- $(x - x_o)^n$ becomes very small with increasing n if we stay close to the equilibrium position. Compare $(x - x_o)^n = (0.0001)^n$ for $n = 2, 3, 4, \ldots$.

This is the justification for using the harmonic oscillator potential for things like quarks, machinery, and galaxies. The better these conditions are fulfilled, the closer the system can be approximated by the harmonic oscillator. This reasoning can easily be extended to more than one dimension and anisotropic potentials.

Departure from these conditions leads to nonlinear oscillations, a subject of extraordinary practical and theoretical interest.

2.9 Perturbed Satellite Orbit

To illustrate the preceding approximation let's consider a satellite in a circular orbit of radius r_o, measured with respect to the center of mass of a uniform spherical mass distribution, a planet. There is only one force operating, the gravitational force. Viewed by an observer in an inertial reference frame, the force of the planet on the satellite is directed toward the center of mass of the planet and it is an inverse square force. By Newton's second law, the acceleration is in the direction of the force.

An observer on the satellite, in a rotating (non-inertial) coordinate system, will also measure a fictitious force, the centrifugal force, pointing *away* from the planet,

$$F_a = \frac{mv^2}{r}, \tag{2-35}$$

where v is related to the angular velocity ω of the satellite and the distance between the centers of mass $\boldsymbol{r}$ by,

$$v = |\omega \times \boldsymbol{r}| = \omega r. \tag{2-36}$$

We can rewrite Eq. (2-35) in terms of the angular momentum $L = |\boldsymbol{r} \times m\boldsymbol{v}|$,

$$F_a = \frac{L^2}{mr^3} = -\frac{\partial U_L}{\partial r}, \tag{2-37}$$

where U_L is a potential energy function associated with angular momentum, from which the centrifugal force can be derived in the non-inertial system,

$$U_L = \frac{L^2}{2mr^2}. \tag{2-38}$$

Let's sum U_L and the gravitational potential U_g to obtain the fictitious potential U_f, shown in Fig. 2-7,

$$U_f = U_g + U_L = -\frac{GMm}{r} + \frac{L^2}{2mr^2}. \tag{2-39}$$

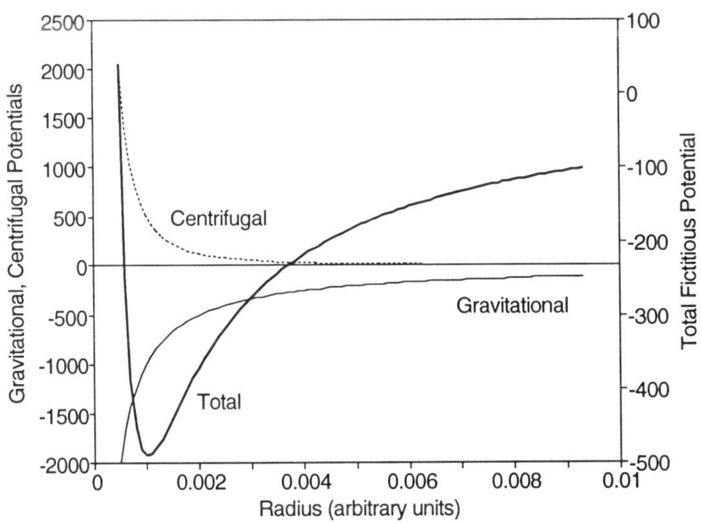

Figure 2-7. Gravitational potential, centrifugal potential, and total potential in the rotating coordinate system. The minimum value is the equilibrium position.

To an observer in the non-inertial system the satellite appears to be in dynamic equilibrium in which the gravitational force is equal and opposite to the centrifugal force,

$$\left(\frac{\partial U_f}{\partial r}\right)_o = \frac{GMm}{r_o^2} - \frac{L^2}{mr_o^3} = 0. \tag{2-40}$$

We can solve Eq. (2-40) to find the radius of the orbit, the equilibrium point in r,

$$r_o = \frac{L^2}{GMm^2}. \tag{2-41}$$

By Eq. (2-34) the second derivative evaluated at the equilibrium point is the restoring force constant κ, which is always a positive quantity for this situation,

$$\left(\frac{\partial^2 U_f}{\partial r^2}\right)_o = -\frac{2GMm}{r_o^3} + \frac{3L^2}{mr_o^4} = +\frac{(GM)^4 m^7}{L^6} = \kappa. \tag{2-42}$$

If a passing missile or meteorite imparts a small perturbation, the satellite will oscillate radially (in an elliptical orbit) with a natural frequency given by $\sqrt{\kappa/m}$ rad/s. Several energy levels are shown in Fig. 2-8.

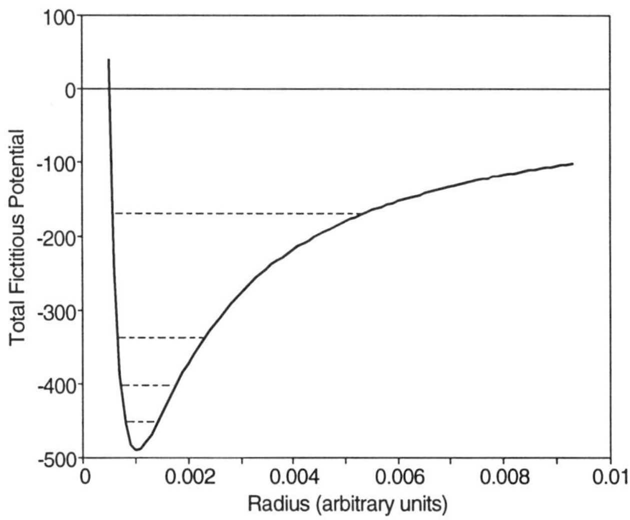

Figure 2-8. The minimum point on the fictitious potential is the position of dynamic equilibrium, a circular orbit. For small perturbations from equilibrium there are harmonic oscillations in the radial direction, observable by Doppler shift. At higher energies the potential is asymmetric about the minimum (elliptical orbits). Note how the "centrifugal barrier" rises steeply at close range.

2.10 High Power-Law Oscillators

Suppose one has a potential energy function like,

$$U(x) = Ax^n, \tag{2-43}$$

where A is a constant. Some functions like this are shown in Fig. 2-9. As n increases the potential energy function becomes more flat-bottomed and the sides become steeper near the turning points for an arbitrary energy level. In the limit of very large n the potential energy function approximates a square well or box. It's of interest to inquire how the energy levels are affected by the deformation of $U(x)$. We will see more of this in Section 11.5, but now let's consider a case where n is very large.

A Particle in a Box

Let's go to the limit as n approaches infinity. In this case the particle is practically free until it hits the "walls." Let the length of the box be L, centered on $x = 0$, and let the particle of mass m have velocity v along the x-axis. The "collisions" at the ends are perfectly elastic, consistent with a conservative force field. The total energy is approximately

$$E = \frac{1}{2}mv^2. \tag{2-44}$$

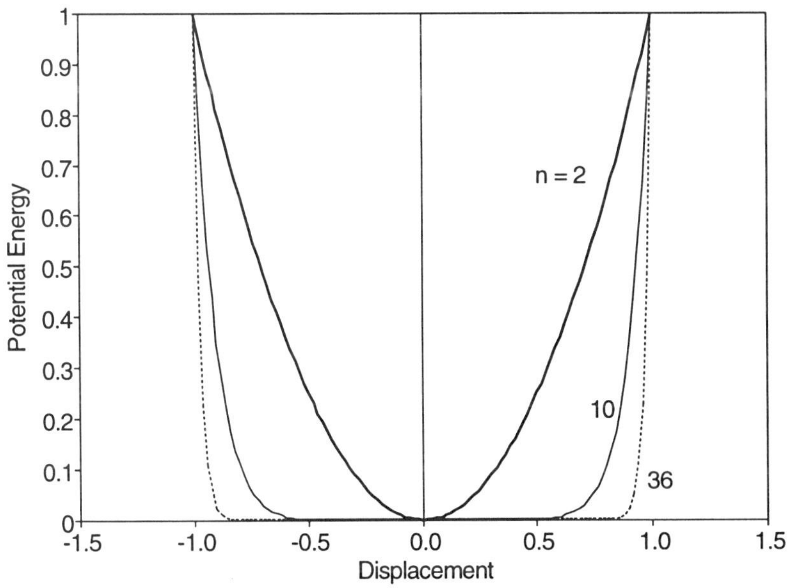

Figure 2-9. Harmonic oscillator potential energy function and some high power-law functions. As n increases $U(x)$ takes on the shape of a square well or box.

The time $T/2$ to move from $-L/2$ to $+L/2$ is

$$\frac{T}{2} = \frac{L}{v}. \tag{2-45}$$

Therefore the round-trip distance is

$$2L = vT = \frac{v}{f} = 2\pi\frac{v}{\omega}, \tag{2-46}$$

where f is the fundamental frequency in Hz and ω is in rad/s. From Eq. (2-46) the velocity is given by

$$v = \frac{\omega L}{\pi}. \tag{2-47}$$

The velocity and displacement of a particle in a box are shown in Fig. 2-10. With Eq. (2-47) the energy can be written as,

$$E = \frac{1}{2}m\omega^2\frac{L^2}{\pi^2}. \tag{2-48}$$

Compare Eq. (2-48) with a simple harmonic oscillator that has an amplitude $L/2$ and the same energy,

$$E = \frac{1}{2}m\omega_o^2\frac{L^2}{2^2}. \tag{2-49}$$

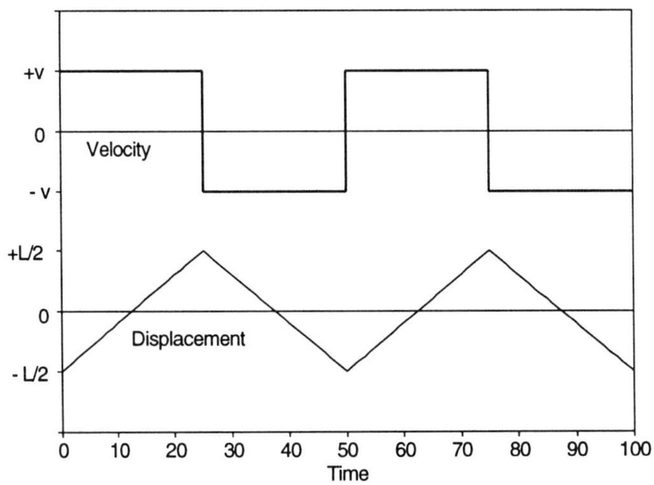

Figure 2-10. Velocity and displacement for a particle in a box, a "wall-banger." Velocity is constant until the particle hits a wall (the potential barrier).

- A simple harmonic oscillator with the same energy and the same amplitude as a particle in a box has a natural frequency that is $2/\pi$ times the fundamental frequency of the particle in the box.

We have been careful to distinguish between the natural frequency of a simple harmonic oscillator and the fundamental frequency of a particle in a box. The reason for this will become clear in Chapter 6, where we will see that a particle in a box cannot be described by a *single* frequency. The motion of a particle in a box is rich in harmonics, typical of a nonlinear or anharmonic oscillator. We will return to a particle in a box in Chapters 11 and 12 for quantum solutions, and compare these again with the harmonic oscillator. The results may be unexpected.

2.11 Adiabatic Invariance

Adiabatic invariance refers to quantities that remain constant when the energy of a system is slowly varied. If you want to visualize a specific case of the discussion that follows, think of a pendulum made with a mass on a string. The string is wound around a horizontal cylinder and the length of the string is slowly varied by unwinding, or winding up, the string on the cylinder, as in Edgar Allan Poe's short story, *The Pit and the Pendulum.* The energy and the frequency will change as the length of the string changes.

For a harmonic oscillator the total energy is,

$$E = K + U = \frac{1}{2}mv^2 + \frac{1}{2}m\omega^2x^2, \tag{2-50}$$

so the time rate of change of total energy is

$$\frac{dE}{dt} = \left[mv\frac{dv}{dt} + m\omega^2x\frac{dx}{dt}\right] + mx^2\omega\frac{d\omega}{dt}. \tag{2-51}$$

The term inside the brackets is very small. Consider the first term in the bracket. The velocity is zero when the acceleration is maximum, and the acceleration is maximum when the velocity is zero. Consider the second term in the bracket. The displacement is maximum when the velocity is zero, and the velocity is maximum when the displacement is zero. All of this follows from the phase relations among displacement, velocity, and acceleration (see Fig. 1-4).

Now consider the final term in Eq. (2-51). From equipartition of energy, Eq. (2-31), we can write,

$$U_{av} = \frac{1}{2}m\omega^2(x^2)_{av} = \frac{1}{2}E, \tag{2-52}$$

so,

$$m\omega\left(\frac{L}{2}\right)^2_{av} = 2\frac{U_{av}}{\omega} = \frac{E}{\omega}. \tag{2-53}$$

If we take the average value of Eq. (2-51) and use Eq. (2-53) we find,

$$\left(\frac{dE}{dt}\right)_{av} = m\omega\left(\frac{L}{2}\right)^2_{av}\left(\frac{d\omega}{dt}\right)_{av}. \tag{2-54}$$

So,

$$\frac{E}{\omega} = \text{constant}. \tag{2-55}$$

This is a very important result. Previously we saw that E/ω_o is a constant for an oscillator, but we held the energy constant so the frequency was also constant. Now we see that E/ω is constant even when the total energy and frequency are *not* constant. Quantum mechanics started when Max Planck discovered that, for the electromagnetic field,

$$\frac{E}{\omega} = \text{constant} = \frac{h}{2\pi}, \tag{2-56}$$

or,

$$E = hf. \tag{2-57}$$

- The adiabatic invariant E/ω implies that *energy is quantized.* This is the clue to understanding blackbody radiation, the photoelectric effect, atomic and molecular structure, and the origin and structure of the universe. We will see some results of adiabatic invariance in Chapters 11 and 12.

Note that the term *adiabatic* in mechanics usually implies a relatively slow process so that energy is almost constant over a time scale associated with the system. An adiabatic process in thermodynamics is usually rapid to minimize energy tansfer to or from a system, so energy is almost constant over a system time scale.

2.12 UNDAMPED, Free Oscillator Energy Worksheet

The worksheet associated with this section is named UNDAMPED. To access this, start your spreadsheet in the usual way and open this file. The Home position is shown in Fig. 2-11. You can experiment with undamped oscillators by entering information in the designated cells. Vary the initial conditions in any way that you wish, but be careful with the restoring force constant and the mass. If the natural frequency becomes too high the discrete data calculated in the spreadhseet will not have enough

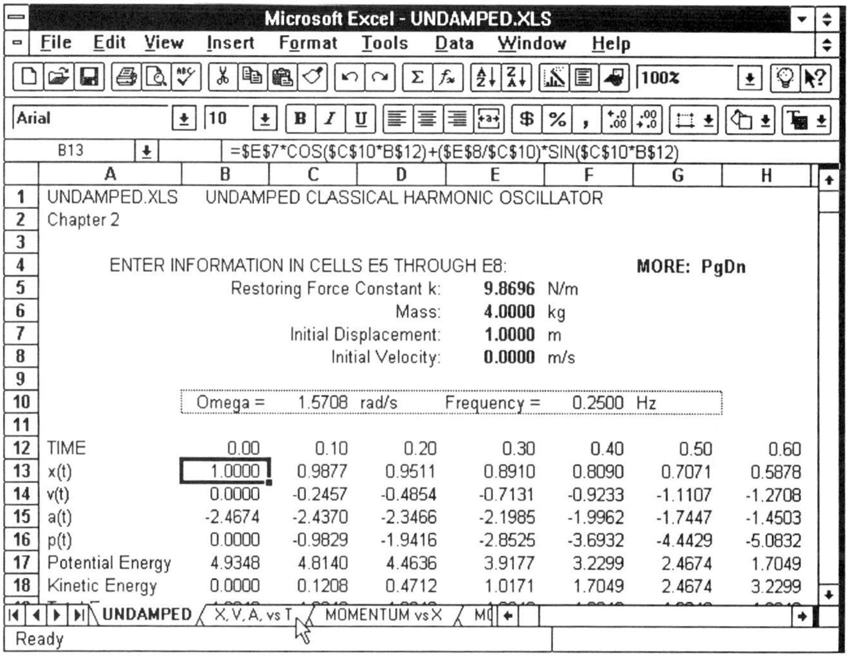

Figure 2-11. Home screen of the UNDAMPED worksheet. Press PgDn for more.

samples in one period to construct an acceptable sine wave. This is analogous to "aliasing" in digitizing analog data.All graphs connected to this worksheet can be viewed by using your mouse to click on the graph names on tabs at the bottom of the worksheet. Click on the right arrows at the lower left corner to see more graph names.

2.13 Bungee Jumpers Beware

Bungee jumping, the triumph of hope over reason, consists of attaching a static cord of fixed length D in series with an elastic cord (unstretched length L, restoring force constant κ) to the ankles of the jumper. The jumper then dives head-first from a distance H above ground, water, concrete, and so forth. One might assume that the length of the static cord D would simply add a length D to the total extension of the stretched elastic cord at the bottom. This assumption could be a fatal error.

This is a good example of conservation of energy in a harmonic oscillator. At the top all energy is in gravitational potential energy, and at the bottom all energy is in elastic potential energy (neglecting losses due to air resistance and internal friction). By conservation of energy,

$$mg(L + D + y) = \frac{1}{2}\kappa y^2 \tag{2-58}$$

where m is the mass of the jumper and y is the change in length of the elastic cord. The masses of the cords are considered negligible. We can solve for y,

$$y = \frac{mg}{\kappa} + \frac{1}{\kappa}\sqrt{m^2g^2 + 2\kappa mg(L+D)} \tag{2-59}$$

So the total distance y_{total} of the fall is,

$$y_{total} = y + L + D = L + D + \frac{mg}{\kappa} + \frac{1}{\kappa}\sqrt{m^2g^2 + 2\kappa mg(L+D)}\ . \tag{2-60}$$

The length of the static cord D adds more than D to the length of the fall. The reason for this is that the static cord allows an additional change in potential energy mgD in the system.

2.14 Summary

(a) The phase-space representation was developed to describe the relation bewteen displacement and momentum. We found that the phase-space orbit is in general an ellipse,

$$\frac{1}{2}m\omega^2x^2 + \frac{1}{2}\frac{p^2}{m} = \frac{1}{2}m\omega^2A^2 = E$$

The area of the ellipse has units of Joule·seconds or eV·seconds, which are the units of Planck's constant.

(b) We found that the phase-space orbit is an expression for the conservation of energy, with the two terms on the left-hand side being the potential energy and kinetic energy. The term on the right-hand side is the total energy E. From the solution $x(t)$ of the oscillator we then calculated potential energy, kinetic energy, and total energy and viewed them in space, time, and as functions of each other.

(c) With the concept of total energy E we can write the solution for $x(t)$ in a useful form,

$$x(t) = \left(\frac{2E}{m\omega^2}\right)^{1/2}\cos(\omega t + \theta)\ .$$

(d) When the energy of the system was slowly varied we found that the ratio,

$$\frac{E}{\omega} = \text{constant.}$$

This is an adiabatic invariant and eventually this will lead to quantization of energy. However, in Chapter 3 our next task is to apply a damping force to an oscillator.

Chapter 3

Damped Oscillators

Newton was a man who wrote a book that
neither he nor anybody else could understand.
Cambridge student

Most of what we have done in Chapters 1 and 2 involved Newton's laws with a linear restoring force $F(x) = -\kappa x = -m\omega_o^2 x$. As we have seen, this is a conservative force derivable from the potential energy function $U(x) = 0.5\kappa x^2$. For damped oscillations we need to switch on a non-conservative, velocity-dependent, resistive force that will extract energy and momentum from the oscillator. First we will investigate the consequences of this damping force, acting alone. Next, we will examine the combined effects of the restoring force and the damping force acting simultaneously. The results of this chapter, combined with the forced solutions of Chapter 4, will give the complete solutions.

3.1 Velocity-Dependent Resistive Force

Imagine that we are inside the impenetrable dotted line in Fig. 1-1 so we can switch forces on and off at will, while excluding the influences of the rest of the universe.

We can examine the forces one by one, or in any combination, because of the principle of superposition (linearity). Linearity is explicit in Newton's second law, the time rate of change of momentum is equal to the vector *sum* of the external forces. Our strategy will be to switch on additional forces, one by one. As we proceed, we can be sure that the system will respond independently to each new force.

First, we will use the simplest type of resistive force with the form $F(v) = -Rv$, where R is the mechanical resistive force constant. This force is zero unless the oscillator is in motion. This force is zero at the points of maximum displacement for an oscillator (the turning points). It is important to note that this resistive force is always *oppositely directed with respect to velocity*. This is a non-conservative force, not derivable from a potential. Let's switch off the simple harmonic oscillator force and examine only effects of the resistive force in Newton's second law,

$$\frac{d}{dt} mv = -Rv . \tag{3-1}$$

We will assume that mass m is constant. Equation (3-1) is easy because it's a first order, ordinary, linear, differential equation with constant coefficients, of the variables-separable type. Separating variables yields

$$\frac{dv}{v} = -\frac{R}{m} dt . \tag{3-2}$$

We can integrate this immediately,

$$\ln(v) = -\frac{R}{m} t + C , \tag{3-3}$$

where C is a constant of integration, which gives the value of $\ln(v)$ at time $t = 0$. Exponentiate both sides,

$$\exp[\ln(v)] = v = \exp[(-R/m)\, t + C] \tag{3-4}$$

so,

$$v(t) = v_o \exp[-(R/m)\, t], \tag{3-5}$$

where we have set $\exp[C] = v(0) = v_o$. Note that (R/m) has units of 1/time. This is the reciprocal of the *time constant* τ for non-oscillatory decay. So, whatever the initial velocity may be, it dies off exponentially and falls to a value v_o/e when $t = \tau$. What about displacement? We can integrate Eq. (3-5) to obtain $x(t)$,

$$x(t) = \int v\, dt = \int v_o \exp[-t/\tau]\, dt + C_1 . \tag{3-6}$$

Here C_1 is determined by the initial conditions. For example, suppose we start with $x(0) = x_o$ and $v(0) = v_o$ when $t = 0$. This results in,

$$x_o = -v_o\tau \exp[-0/\tau] + C_1 \tag{3-7}$$

so,

$$C_1 = v_o\tau + x_o \tag{3-8}$$

and the solution becomes

$$x(t) = v_o\tau\,[1 - \exp(-t/\tau)] + x_o . \tag{3-9}$$

The initial velocity dies off exponentially and the displacement asymptotically approaches its final value $x_o + v_o\tau$ as shown in Figs. 3-1 and 3-2. Figure 3-1 has linear scales; Fig. 3-2 has the velocity on a logarithmic scale to show the exponential form of the decay.

Time Constant and Half-Life

The time constant (time to $1/e$ of the initial value) and the half-life (time to 1/2 of the initial value) are important parameters for a decaying system. The time constant for velocity and displacement with this velocity-dependent resistive force is τ. The half-life is $\ln(2)\,\tau \approx 0.69314718\,\tau$.

Knowing $v(t)$ we can also compute momentum $p(t)$ and kinetic energy $p^2/2m$ and watch them decay. There is no potential energy function available, because the velocity-dependent resistive force is a *non-conservative* force. Kinetic energy involves p^2 so the time constant for energy decay is $\tau/2$ and the half-life is $\ln(2)\,\tau/2$.

- An exponential process always results when a system has a time rate of change that is directly proportional to the present state of the system.
- If the constant of proportionality is positive an exponential rise occurs. If the constant is negative an exponential decay occurs.
- Any point on the rise or decay can be taken as the initial value. For example, to measure the time constant for the decay of a radioactive isotope it is not necessary to go back to the beginning of the universe; you can start your clock at your convenience.

By Table 1-1, the mechanical motion here is completely analogous to the discharge of an inductance-resistance system. The displacement achieves $(1 - 1/e)$, or about 63.2 percent, of its final value in a time equal to one time constant.

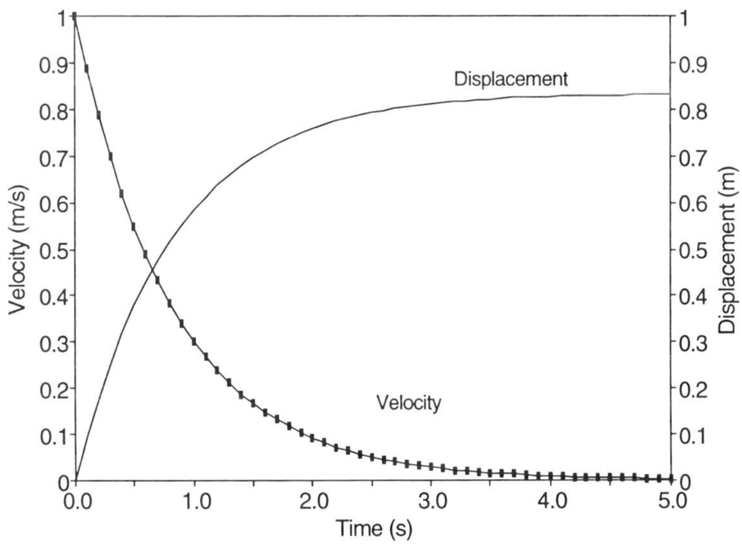

Figure 3-1. Displacement and velocity for a free mass under the influence of a velocity-dependent resistive force. The same data are shown on a semi-log plot in Fig. 3-2 to emphasize the exponential decay of $v(t)$. Time constant $\tau = 1$s.

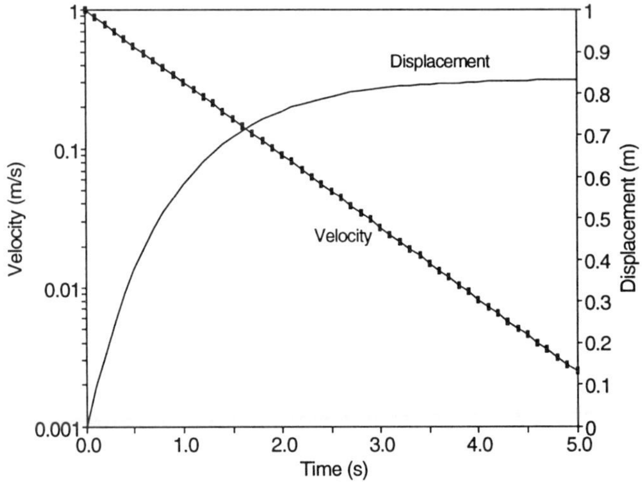

Figure 3-2. Displacement and velocity with resistive force (semi-log in velocity).

3.2 FORCE-RV, Resistive Force Worksheet

The worksheet associated with this section is named FORCE-RV. After you start your spreadsheet you can retrieve this file in the usual way. You should see the upper left-hand corner of this worksheet shown in Fig. 3-3. If you do not see this press the Home key.

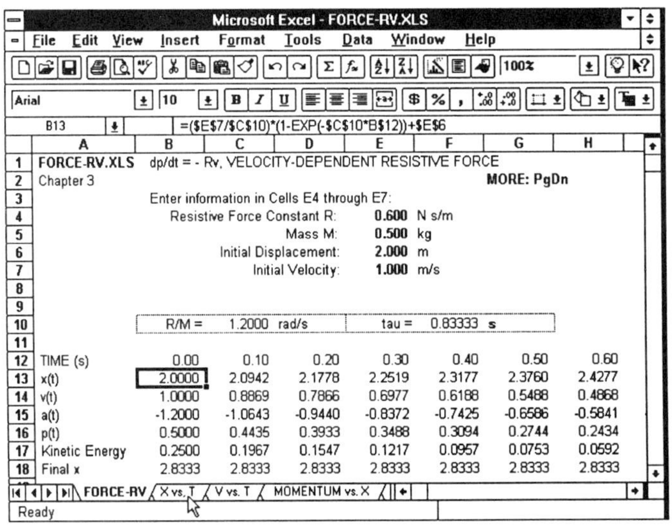

Figure 3-3. Home position for the worksheet FORCE-RV. Highlighted cell shows formula for $x(t)$. Change parameters and initial conditions by entering information in Cells E4 through E7.

The graph names connected to this worksheet are *Energy vs. T*; *Energy vs. X; Momentum vs. X; V vs. T; A vs. T*, and *X vs. T*. The last two named graphs are shown together in Fig. 3-1.

3.3 Restoring Force and Resistive Force

And now it is time to switch on two forces, each of which we have examined independently. One will be the linear restoring force $F(x) = -\kappa x$ and the other will be the resistive force $F(v) = -Rv$. When both of these are operating Newton's second law can be written as

$$m\frac{d^2x}{dt^2} = -\kappa x - Rv\,. \tag{3-10}$$

This can be rewritten in a standard form,

$$\frac{d^2x}{dt^2} + \frac{R}{m}\frac{dx}{dt} + \frac{\kappa}{m}x = 0\,. \tag{3-11}$$

or,

$$\frac{d^2x}{dt^2} + \frac{1}{\tau}\frac{dx}{dt} + \omega_o^2 x = 0\,. \tag{3-12}$$

This is a familiar linear, second-order, ordinary, differential equation with constant coefficients, so we will try our old trick of using an exponential to convert a differential equation into an algebraic equation. Let

$$x(t) = X\,e^{-st}\,,\quad \frac{dx}{dt} = -sx(t)\,,\quad \frac{d^2x}{dt^2} = s^2x(t)\,,\quad 2K = \frac{1}{\tau}\,. \tag{3-13}$$

The time-derivative operator d/dt simply brings down $-s$ each time it is applied. In general s could be complex. When we put these in Eq. (3-12) we obtain

$$s^2x(t) - 2Ks\,x(t) + \omega_o^2\,x(t) = 0. \tag{3-14}$$

This must hold for all time so we can factor out $x(t)$,

$$s^2 - 2Ks + \omega_o^2 = 0\,. \tag{3-15}$$

Lo and behold, we have a quadratic equation in s, which has the solution,

$$s = 0.5\,(-2K \pm \sqrt{4K^2 - 4\omega_o^2}\,)\,. \tag{3-16}$$

- The quantity inside the square root in Eq. (3-16) is of *critical importance* because its value determines three distinct types of solution when it is negative, zero, or positive.

Let's rewrite Eq. (3-16) slightly to obtain a little more physical insight,

$$s = -K \pm \sqrt{K^2 - \omega_o^2}\,. \tag{3-17}$$

When Eq. (3-17) is placed in the assumed solution for $x(t)$ we obtain

$$x(t) = X\,e^{-Kt} \exp[\pm \sqrt{K^2 - \omega_o^2}\; t]\,. \tag{3-18}$$

- The effective time constant ($2\tau = 1/K$) is now twice what it was when only the resistive force was operating.

Let's examine each of the three conditions individually, as K increases.

3.4 Under-Damped Oscillator

When the resistive force is small the time constant is large. The under-damped condition can be stated as

$$K = \frac{R}{2m} < \omega_o \quad \text{or} \quad \frac{1}{2\tau} < \omega_o\,.$$

In this condition the quantity inside the square root of Eq. (3-18) is negative, so the factor in the exponential is complex. Now the square root can be written as

$$j\sqrt{\omega_o^2 - K^2}\,,$$

which can be interpreted as a frequency *down-shift* due to the presence of the resistive force. Damping has apparently decreased the oscillator frequency.

- Actually, now there is *not* a single frequency. The spectrum is spread over a band of frequencies. The faster the decay in time, the broader the spectrum will be, and conversely. We will examine this classsical uncertainty principle in detail in Chapter 6.

Nevertheless, to make the equations compact it is convenient to define the frequency of the under-damped oscillator as,

$$\omega_u = \sqrt{\omega_o^2 - K^2}\,. \tag{3-19}$$

Now the solution can be written as,

$$x(t) = e^{-Kt}\left\{X_1 \exp[+j\omega_u t\,] + X_2 \exp[-j\omega_u t]\right\}. \tag{3-20}$$

Equation (3-20) can be interpreted as two exponentially damped, contra-rotating vectors, but this can be modified into a less abstract form using trigonometric identities and initial conditions x_o and v_o,

$$x(t) = e^{-Kt}\left\{x_o \cos(\omega_u t) + [(v_o + Kx_o)/\omega_u] \sin(\omega_u t)\right\}. \tag{3-21}$$

Equation (3-21) describes an exponentially damped oscillatory decay. Its spectrum has some unique properties that we will examine in Chapter 6. A little algebra can transform Eq. (3-21) into another useful form, a representation in terms of an exponentially damped, phase-shifted cosine. To see this, we can use a trigonometric identity for the sum of two angles, $\cos(a + b) = \cos(a)\cos(b) - \sin(a)\sin(b)$. Let $\cos(a) = x_o$, $\sin(a) = -(v_o + Kx_o)/\omega_u$, and $b = \omega_u t$. Then Eq. (3-21) becomes

$$x(t) = e^{-Kt} [\cos(\omega_u t + \theta)], \tag{3-22}$$

where $\theta = \arctan [-(v_o + Kx_o)/(\omega_u x_o)]$. A typical solution is shown in Fig. 3-4. The phase space representation is particularly informative, and is shown in Fig. 3-5. You can think of Fig. 3-5 as looking back along the time axis at the projection of the motion on the p-x plane. Figure 3-6 shows the phase relations among displacement, velocity, and acceleration.

You can experiment with different initial conditions by entering information in the worksheet named UNDRDAMP, whose Home screen is shown in Fig. 3-7.

Figure 3-4. Time evolution of $x(t)$ for $x_o = 1$, $v_o = 0$, $K = 0.25$. The exponential

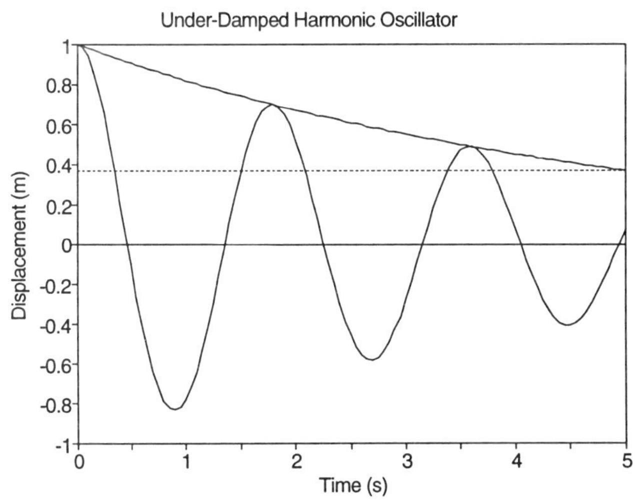

decay envelope touches the curve slightly off the maximum points. The horizontal dotted line is the $1/e$ value. In the UNDRDAMP worksheet this graph is named *X vs. T*. All graphs in this section have the parameters shown in Fig. 3-7.

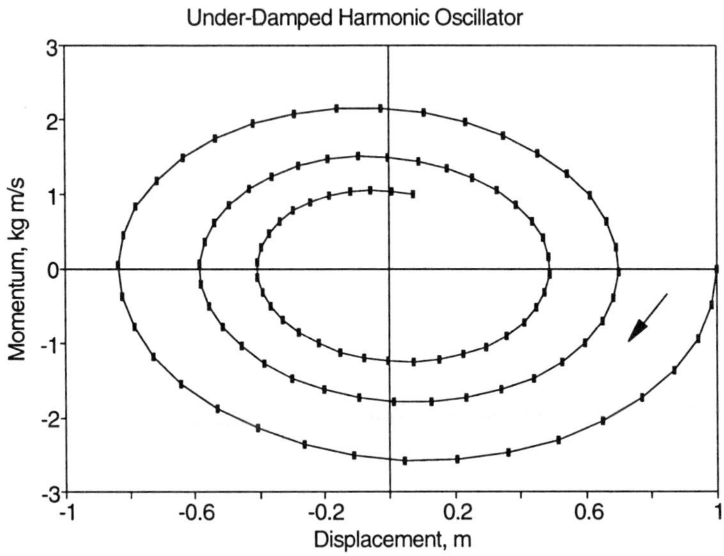

Figure 3-5. Phase space view of the under-damped oscillator shown in Fig. 3-4. The time interval between markers is 0.1 s. Graph name: *Momentum vs. X*.

Phase relations among displacement, velocity, and acceleration are shown in Fig. 3-6.

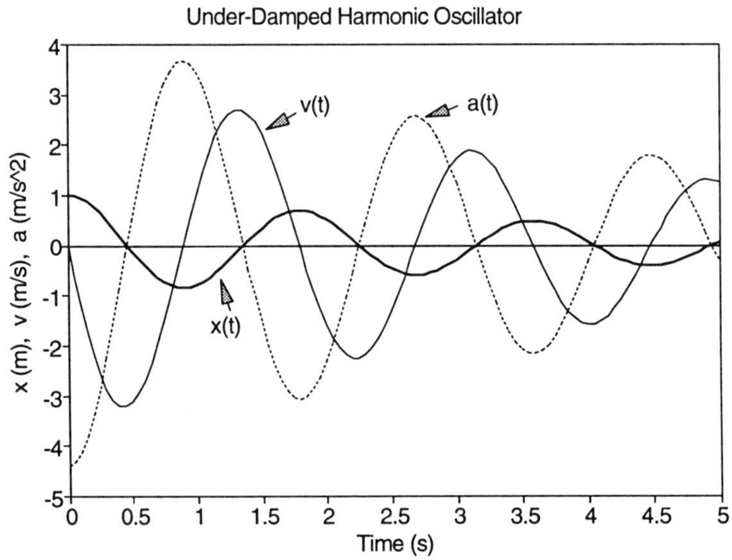

Figure 3-6. Under-damped oscillator displacement, velocity, and acceleration for same conditions shown in Fig. 3-4.

UNDRDAMP, Under-damped Worksheet

The worksheet associated with this section is named UNDRDAMP. After starting your spreadsheet retrieve this file and you should see the upper left-hand corner (Home position) shown in Fig. 3-7.

The graph names connected to this worksheet are *X, V, A vs T; X vs T; Power Loss; Momentum vs X; Momentum vs T; KE & PE vs X; Energy vs X; Energy vs T;* and *E & PE vs X*.

Energy and Power Loss

The velocity-dependent resistive force causes ordered mechanical or electrical energy to be transformed into thermal energy. The velocity is zero at the end points and maximum near (but not exactly at) the equilibrium position, so energy loss is zero at the ends and maximum near the center. The instantaneous power loss $P(t)$ is given by,

$$P(t) = \mathbf{F} \cdot \mathbf{v} \quad \text{Watts.} \tag{3-23}$$

Consequently, the power loss is not a simple exponential decay. The total energy at every point in time is given by the sum of the instantaneous kinetic and potential energy, and the time evolution of the total energy is not a simple exponential path. The power loss and energy are shown in Figs. 3-8, 3-9, and 3-10.

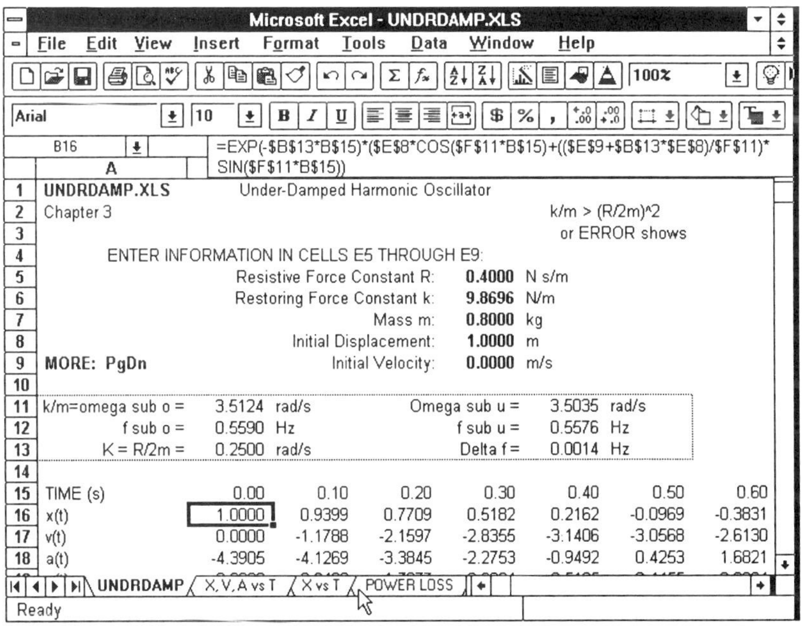

Figure 3-7. Home screen of the worksheet UNDRDAMP. Change parameters and initial conditions by entering information in Cells E5 through E9.

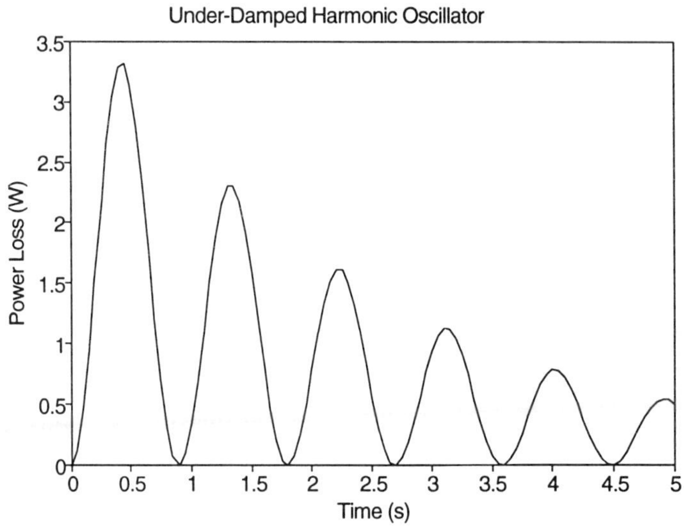

Figure 3-8. Power loss for the oscillator shown in Figs. 3-4 and 3-5. The power loss is zero when the oscillator is instantaneously stopped at the turning points, and it is a maximum when the velocity is maximum close to the equilibrium position. Graph name: *Power Loss.*

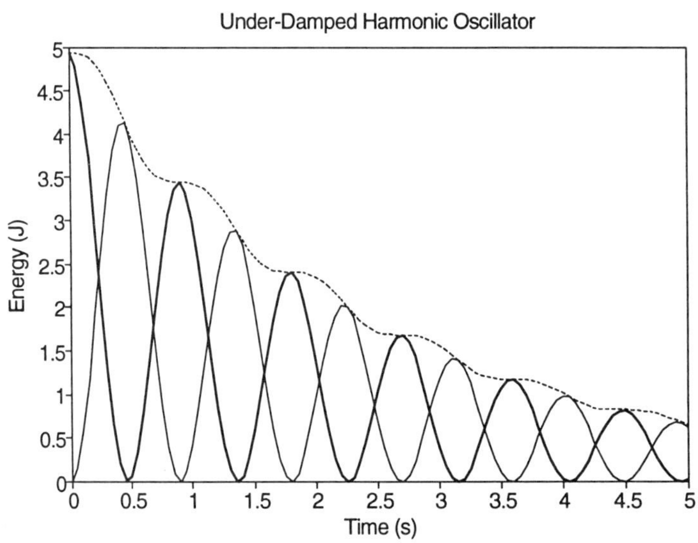

Figure 3-9. The total energy decreases due to the velocity-dependent resistive force. When the oscillator is instantaneously stopped at the turning points the total energy is in potential form and the energy is not changing. *Thick line:* Potential energy. *Thin line:* Kinetic energy. *Dotted line:* Total energy.

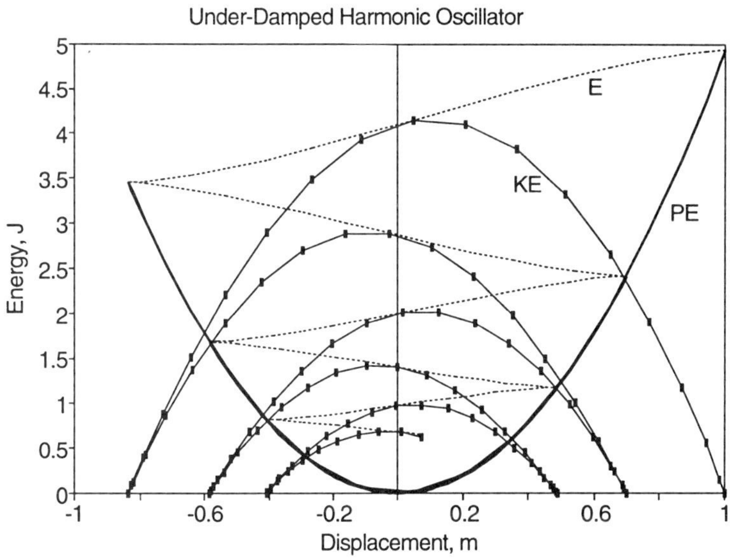

Figure 3-10. Energy diagram for the under-damped harmonic oscillator. Time increment between markers is 0.05s. Note turning points where total energy is in potential form, i.e., kinetic energy is zero. Total energy decreases continuously, except at the turning points where velocity is zero. *Dotted line:* Total energy. *Line with markers:* Kinetic energy. *Thick line:* Potential energy.

3.5 Critically Damped Oscillator

When the time constant is adjusted exactly to $K = \omega_o$ in Eq. (3-18) have,

$$K = \frac{R}{2m} = \omega_o \quad \text{or} \quad \frac{1}{2\tau} = \omega_o .$$

We will call this the critically-damped condition. Now the square root in Eq. (3-18) is zero, and our previous solution, Eq. (3-21), becomes simpler. As $\omega_u \to 0$, $\cos(\omega_u t) \to 1$ and $\sin(\omega_u t)/\omega_u \to t$, Eq. (3-21) becomes

$$x(t) = e^{-Kt}\,[x_o + (v_o + Kx_o)t]. \tag{3-24}$$

The system does not oscillate and it approaches equilibrium in the shortest time. As with the over-damped condition (Section 3.6) there is no oscillation as the system returns to its equilibrium position. This is why critical damping is desirable in analytical balances, moving-pointer meters, aircraft, vehicles, and architectural structures. For $x_o = 0$, Eq. (3-24) is the envelope of the waveform used in the Loran-C radio navigation system.

- At equilibrium a finite temperature implies a residual r.m.s. velocity and displacement around equilibrium for a classical oscillator (Chapter 10). In Chapter 13 we will see that a quantum oscillator has a non-removable zero-point energy.

The time evolution of displacement for the critically damped case is shown in Fig. 3-11 and the phase space evolution is shown in Fig. 3-12 for the same initial conditions and constants. It is difficult to set exact critical damping, $K = \omega_o$.

CRITDAMP, Critically Damped Worksheet

The worksheet associated with this section is named CRITDAMP. Start your spreadsheet, open this file, and you should see the Home position shown in Fig. 3-13. The graph names connected to this worksheet are:
X vs. T
V & A vs. T
Power Loss vs. X
Power Loss vs. T
Momentum vs. X
Energy vs. X
Energy vs. T
Enter information as requested in Cells E4 through E7. Calculated values based on this information are displayed in the box, in Cells E9 through E12. Experiment with other values of initial velocity and displacement.

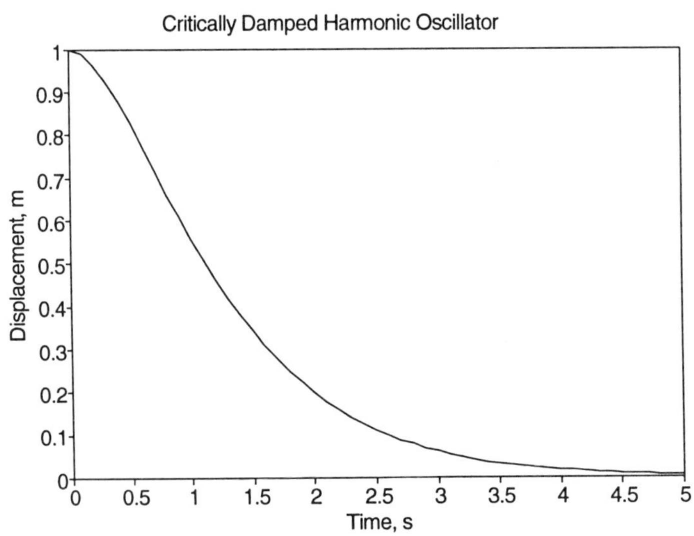

Figure 3-11. Approach to equilibrium for the critically damped oscillator for $x_o = 1$, $v_o = 0$, $K = 1.5$ in Eq. (3-24). Graph name: *X vs. T*.

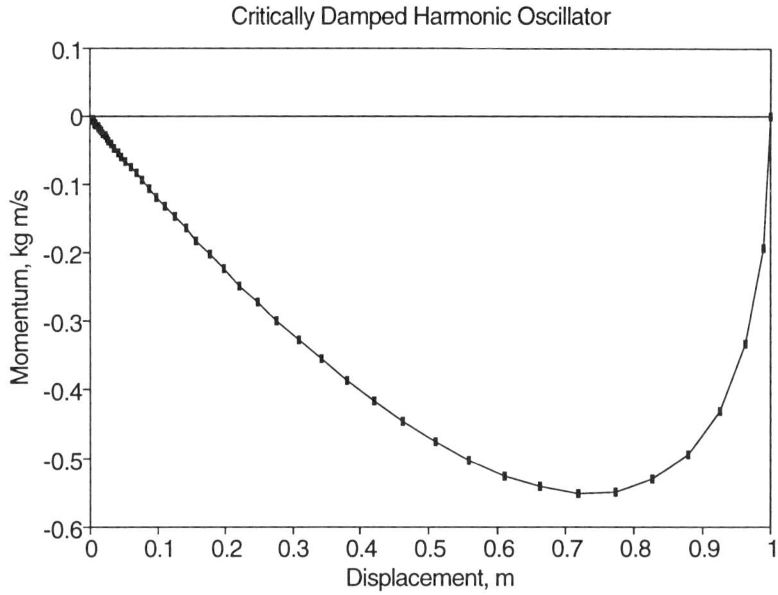

Figure 3-12. Evolution in phase space for the same conditions shown in Fig. 3-11. Time interval between markers is 0.1 s.

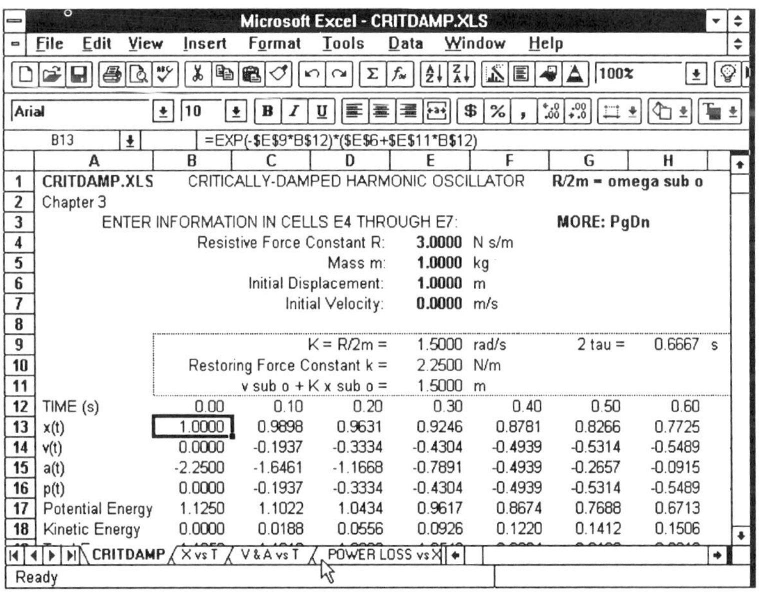

Figure 3-13. Home position for CRITDAMP. Highlighted cell shows formula for $x(t)$. Change parameters and initial conditions by entering information in Cells E4 through E7.

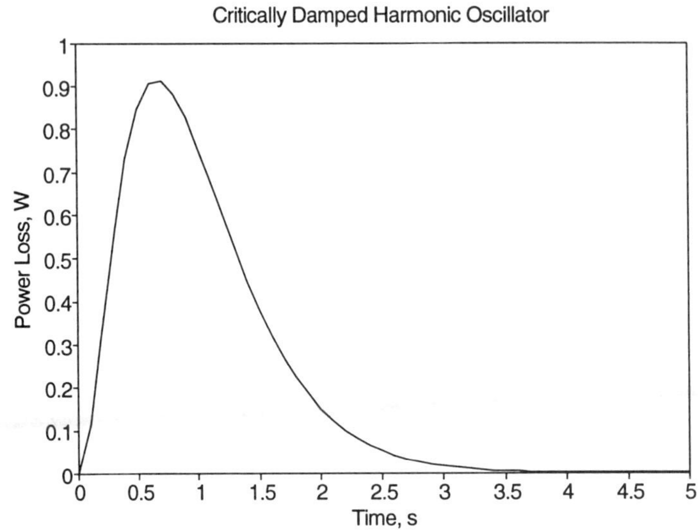

Figure 3-14. Power loss for the critically-damped case, for $x_o = 1$ and $v_o = 0$, and $K = 1.5$, the same conditions shown in Figs. 3-11, 3-12, and 3-13.

3.6 Over-Damped Oscillator

For the condition,

$$K = \frac{R}{2m} > \omega_o \quad \text{or} \quad \frac{1}{2\tau} > \omega_o$$

there is no oscillation and the system approaches equilibrium more slowly than in the critically damped case. As $\omega_u \to j\omega_m$ Eq. (3-21) becomes

$$x(t) = e^{-Kt}\{x_o \cosh(\omega_m t) + [(v_o + Kx_o)/\omega_m] \sinh(\omega_m t)\}, \qquad (3\text{-}25)$$

where $\omega_m = \sqrt{K^2 - \omega_o^2}$. The hyperbolic cosine and sine involve real exponentials,

$$\cosh(\theta) = \frac{e^{+\theta} + e^{-\theta}}{2} \qquad (3\text{-}26)$$

and,

$$\sinh(\theta) = \frac{e^{+\theta} - e^{-\theta}}{2}. \qquad (3\text{-}27)$$

From any initial conditions of displacement and velocity the oscillator sags sluggishly toward its final equilibrium position. Energy is lost until the oscillator comes to rest. For temperature above absolute zero, there is some residual r.m.s. velocity and displacement about the equilibrium point (see Chapter 10). The approach to equilibrium and the phase space evolution are shown in Figures 3-15, 3-17, and 3-18 for the same initial conditions ($x_o = 1$, $v_o = 0$) that were used in the critically damped case, Figs. 3-11 and 3-12.

The velocity and acceleration of the over-damped oscillator are given by,

$$v(t) = \frac{dx}{dt} = -Kx(t) + e^{-Kt}[\omega_m x_o \sinh(\omega_m t) + (v_o + Kx_o)\cosh(\omega_m t)]$$

and

$$a(t) = \frac{dv}{dt} = -Kv(t) - Ke^{-Kt}[\omega_m x_o \sinh(\omega_m t) + (v_o + Kx_o)\cosh(\omega_m t)] + \omega_m^2 x(t).$$

The subscript *m* is used for "molasses" because the quintessential over-damped oscillator is a pendulum or mass-and-spring moving in molasses.

OVERDAMP, Over-Damped Worksheet

The worksheet for over-damping is named OVERDAMP and can be retrieved in the usual way. The Home position of this worksheet is shown in Fig. 3-15. The graph names connected to this worksheet are: *X vs. T; V & A vs. T; Power Loss vs. X; Power Loss vs. T; Momentum vs. X; Energy vs. X*; and *Energy vs. T.*

Power loss as a function of time is shown in Fig. 3-18. Compare Figs. 3-8 and 3-14 for the under-damped and critically damped cases.

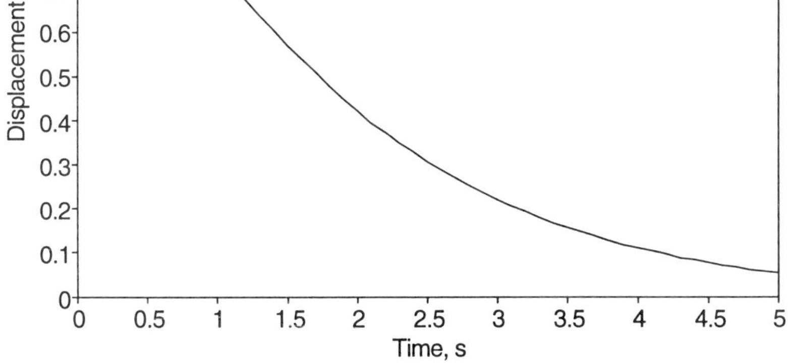

Figure 3-15. Approach to equilibrium for parameters shown in Fig. 3-16.

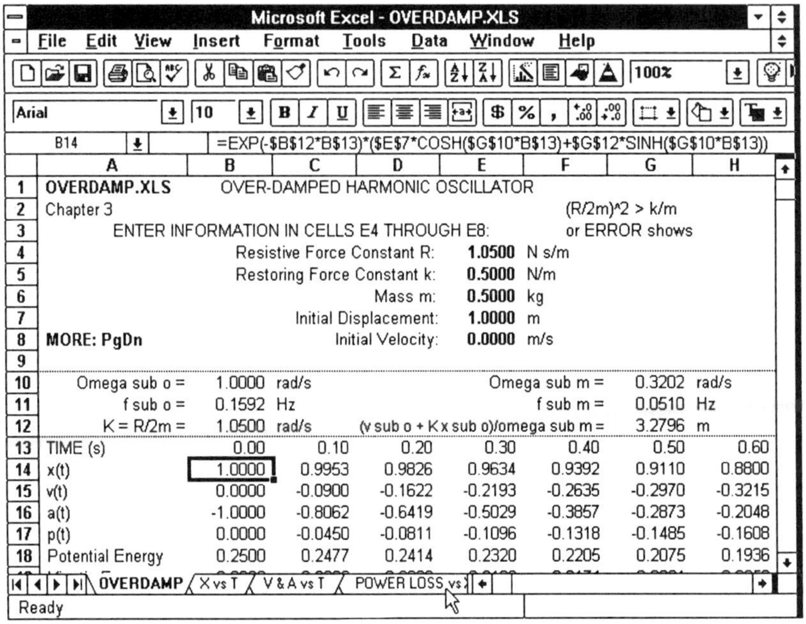

Figure 3-16. Home position for OVERDAMP. You can change parameters and initial conditions by entering information in Cells E4 through E8.

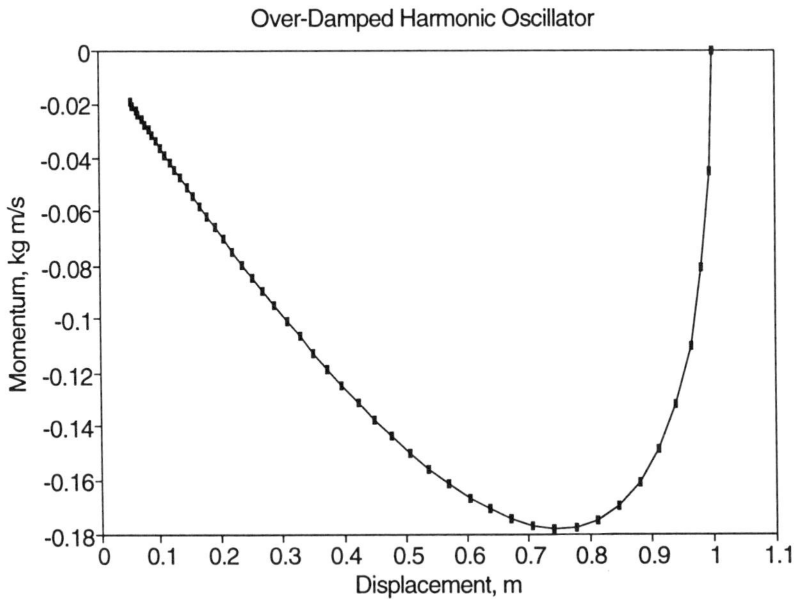

Figure 3-17. Phase space evolution of the over-damped oscillator for conditions shown in Fig. 3-16. Time interval between markers is 0.1s.

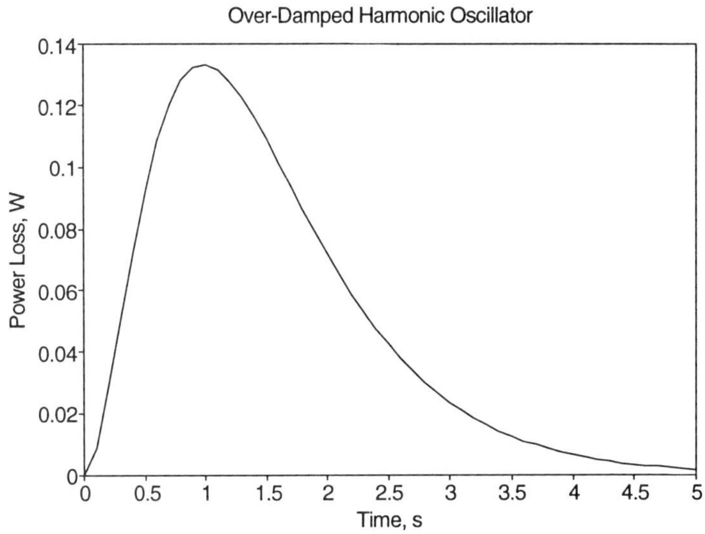

Figure 3-18. Power loss for the over-damped oscillator, for the conditions shown in Fig. 3-16 in Cells E4 through E8.

3.7 Summary of Free, Damped Oscillators

We have treated the under-damped, critically damped, and over-damped oscillators as three separate cases in order to display their unique properties. The solutions of each type are mathematically distinct, but there is a smooth transition from one case to another. For example, in an *LCR* circuit it is easy to vary the resistance to make the transition continuously from one case to another. The critically damped case separates the under-damped and over-damped cases, but achieving exact critical damping is difficult. Close to the transition region it is often not clear what type of oscillator you are dealing with because of inaccurate knowledge of the values of L, C, and R, or m, κ, and R.

Let's look again at Eq. (3-11),

$$\frac{d^2x}{dt^2} + \frac{R}{m}\frac{dx}{dt} + \frac{\kappa}{m}x = 0. \tag{3-11}$$

As before, let $K = R/2m$ and assume a solution $x(t) = X\exp[-st]$, where s may be complex. We then have the standard form,

$$\frac{d^2x}{dt^2} + 2K\frac{dx}{dt} + \omega_o^2 x = 0 \tag{3-28}$$

and its associated indicial equation,

$$s^2 - 2Ks + \omega_o^2 = 0. \quad (3\text{-}29)$$

The character of the roots of this quadratic equation determine the three cases:

- complex conjugate roots ⇔ under-damped;
- real, equal roots ⇔ critically damped; and
- real, unequal roots ⇔ over-damped.

Table 3-1 lists the solutions for the three cases, for initial conditions x_o and v_o.

Table 3-1 Summary of Free, Damped Oscillators

Case	*Roots*	$x(t)$
Under-damped	Complex $\omega_o > K$	$e^{-Kt}\{x_o \cos(\omega_u t) + [(v_o + Kx_o)/\omega_u]\sin(\omega_u t)\}$
Critically damped	Real, equal $\omega_o = K$	$e^{-Kt}[x_o + (v_o + Kx_o)t]$
Over-damped	Real, unequal $\omega_o < K$	$e^{-Kt}\{x_o \cosh(\omega_m t) + [(v_o + Kx_o)/\omega_m]\sinh(\omega_m t)\}$

Note: $K = R/2m$, $\omega_o = \sqrt{\kappa/m}$, $\omega_u = \sqrt{\omega_o^2 - K^2}$, $\omega_m = \sqrt{K^2 - \omega_o^2}$

Transition from Under-Damped Oscillation

Under-damped oscillators change continuously into critical damping as $\omega_u \to 0$. (Note that the second term in the under-damped oscillator involves the limit of $\sin(\omega_u t)/\omega_u$, which is t.) Under-damped oscillators change continuously through critical damping into over-damping as $\omega_u \to j\omega_m$, in accordance with the identities,

$$\cos(j\omega_m t) = \cosh(\omega_m t) \quad \text{and} \quad \sin(j\omega_m t) = j\sinh(\omega_m t). \quad (3\text{-}30)$$

Transition from Over-Damped Oscillation

Over-damped oscillators change continuously into critical damping as $\omega_m \to 0$. Over-dampimg changes continuously through critical damping into under-damping as $\omega_m \to j\omega_u$, in accordance with the identities,

$$\cosh(j\omega_u t) = \cos(\omega_u t) \quad \text{and} \quad \sinh(j\omega_u t) = j\sin(\omega_u t). \quad (3\text{-}31)$$

Figure 3-19 shows ω_i as a function of K, where $i = u$ for the under-damped oscillator and $i = m$ for the over-damped oscillator.

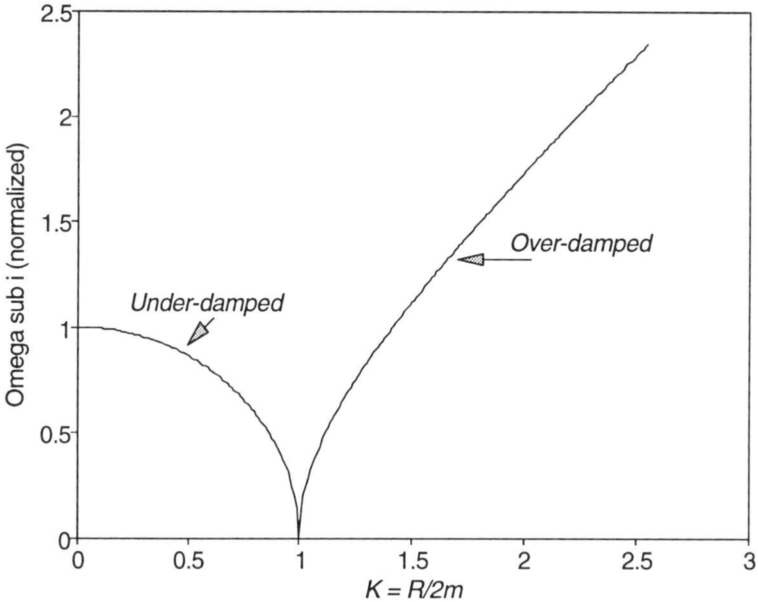

Figure 3-19. Sensitivity of ω_i to K. Resonant frequency is normalized to $\omega_o = 1$. Note how ω_i changes as K goes from $K < \omega_o$ (under-damped) through $K = \omega_o = 1$ (critically damped) to $K > \omega_o$ (over-damped). See Table 3-1.

Notice the extraordinary sensitivity of ω_i to small changes in K near critical damping. This is graphic evidence about the difficulty of setting a system exactly to the critically damped condition, $\omega_o^2 = K^2$, or in fundamental terms, $\kappa/m = R/2m$. This requires an exact knowledge of κ, R, and m. Because we can never know these quantities exactly (to an infinite number of decimal places), it is unlikely that an oscillator can ever be set exactly to critical damping, but we can come close enough for almost all practical purposes.

Charles Augustin Coulomb (French carpenter, stonecutter, army officer, prison inmate, underwater construction engineer, and King's director of hospital reform) used a damped torsion oscillator in 1785 to measure the electrostatic force constant $1/(4\pi\varepsilon_o) \approx 9 \times 10^9$ N·m^2/C^2, and to establish the inverse square nature of the force.

Henry Cavendish (English physicist and chemist) was born in France and attended Cambridge University. In 1798 he used a damped torsion oscillator to determine the average density of planet Earth. We will see the details of this measurement in Section 5.15, and how it is now used to measure one of the most important constants in the universe, the gravitational force constant G.

The gravitational attraction between small masses is very weak so his oscillator required a very small restoring torque constant, too small to be measured directly. However, Cavendish was able to measure the torque constant κ indirectly in terms of the moment of inertia I and frequency of the oscillator, $\omega^2 = \kappa/I$. Cavendish also discovered hydrogen, identified it as an element, and showed that water is a compound of hydrogen and oxygen.

What's Next?

In the real world, oscillators are subjected to a variety of driving forces, some of which are periodic and deterministic, and some of which are transient or random. Often the total force is a combination of all types. Our task now is to investigate the effects of these forces. Chapter 4 will be devoted to deterministic forces; we will defer discussing random forces until Chapter 10.

Figure 3-20 shows an oscillator under the influence of random impulses produced by the torques of the rats' feet as the rats run on the wheel.

Figure 3-20. Self-graphing oscillator. Careful analysis reveals that the rats must be running backwards. Also, the word *cheeze* on the blackboard has an unconventional spelling. (Courtesy of PASCO Scientific, Roseville, CA.)

Chapter 4

Forced Oscillators

If I have ever made any valuable discoveries, it has been owing more to patient attention, than to any other talent.
Sir Isaac Newton

Now it is time to add another force to our oscillator, an external driving force. The two forces that we have applied up to now, the linear restoring force and the velocity-dependent resistive force, are unchanging in form, although their relative strengths may be different in different oscillators. In contrast, the external driving forces to which oscillators are subjected are functions of time. In general they may be very complicated functions and of infinite variety, perhaps even random noise.

In this chapter we will, except for two examples, restrict our attention to forces of a single, but arbitrary, frequency. This is much more general than it appears at first. Later we will justify this statement. It will become clear that if you know the oscillator's response as a function of frequency then you can know the response to almost every force you are likely to encounter.

It is difficult to overstate the importance of the forced oscillator; the results of this chapter have wide applications. We will show some examples of the classical dielectric constant and index of refraction.

As in Chapter 3 we will begin by switching on one force at a time, starting with a mass that is free but subject to an external driving force. After this we will add the velocity-dependent resistive force to the driving force, and then we will add the linear restoring force to make a complete forced, damped oscillator.

We will calculate the time evolution by direct use of Newton's laws of motion and we will obtain the frequency domain representation using phasors. The frequency domain description and its applications will be explored in more detail in Chapter 5, and in Chapter 6 using Fourier methods.

4.1 Forced Free Mass

To examine the motion of a free mass let's apply a driving force of the form,

$$F(t) = F_o e^{j\omega t}, \tag{4-1}$$

where F_o is a constant magnitude and ω is the frequency (rad/s) of the force. When we plug this into Newton's second law for a free particle we obtain

$$m\frac{dv}{dt} = F_o e^{j\omega t}. \tag{4-2}$$

Integrating Eq. (4-2) yields

$$\int m\, dv = \int F_o e^{j\omega t}\, dt + mv_o. \tag{4-3}$$

At present we are more interested in velocity than momentum, so divide by m,

$$v(t) = \frac{F_o}{j\omega m}e^{j\omega t} + v(0) = -j\frac{F_o}{\omega m}e^{j\omega t} + v_o. \tag{4-4}$$

This result is shown in Fig. 4-1.

- Note that the amplitude of the velocity is a function of frequency; it is inversely proportional to the frequency of the driving force.
- The velocity and momentum lag 90° behind the force.

Now let's integrate once more to find the displacement,

$$\int v\, dt = \int \frac{dx}{dt}dt = \int dx = -j\frac{F_o}{\omega m}\int e^{j\omega t}\, dt + \int v_o\, dt + x_o. \tag{4-5}$$

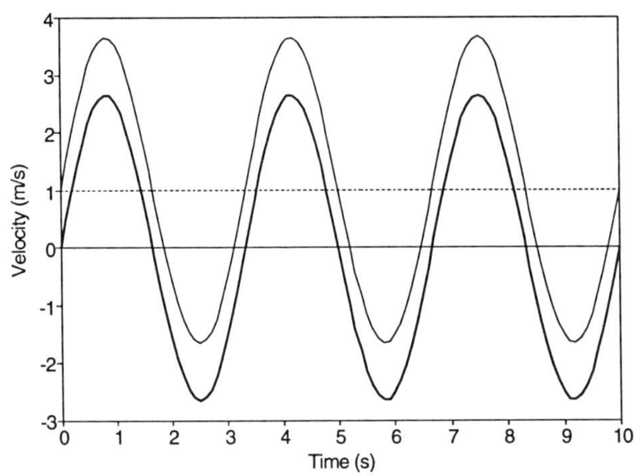

Figure 4-1. Velocity of a forced free mass. (See Fig. 4-5 for parameters.) *Dotted line:* Force-free velocity. *Thick line:* Forced velocity. *Thin line:* Total velocity. In this figure, $v_o = 1$ m/s.

The time evolution of the displacement for the forced, free mass is completely determined by the initial conditions, the mass, and the magnitude and frequency of the force,

$$x(t) = -\frac{F_o}{m\omega^2} e^{j\omega t} + v_o t + x_o . \tag{4-6}$$

This result is shown in Fig. 4-2.

- The amplitude of the forced displacement is inversely proportional to the square of the driving frequency.
- The displacement is 180° out of phase with the driving force.

This typical inertial or self-inductive characteristic is shown in Fig. 4-3. Another view of forced motion is shown in Fig. 4-4. Note the relative phases of displacement, velocity, and force.

As we have seen, the velocity lags the driving force by $\pi/2$, so the average work done by the driving force is zero. To confirm this, calculate the average value of $\boldsymbol{F}\cdot\boldsymbol{v}$ over a complete cycle.

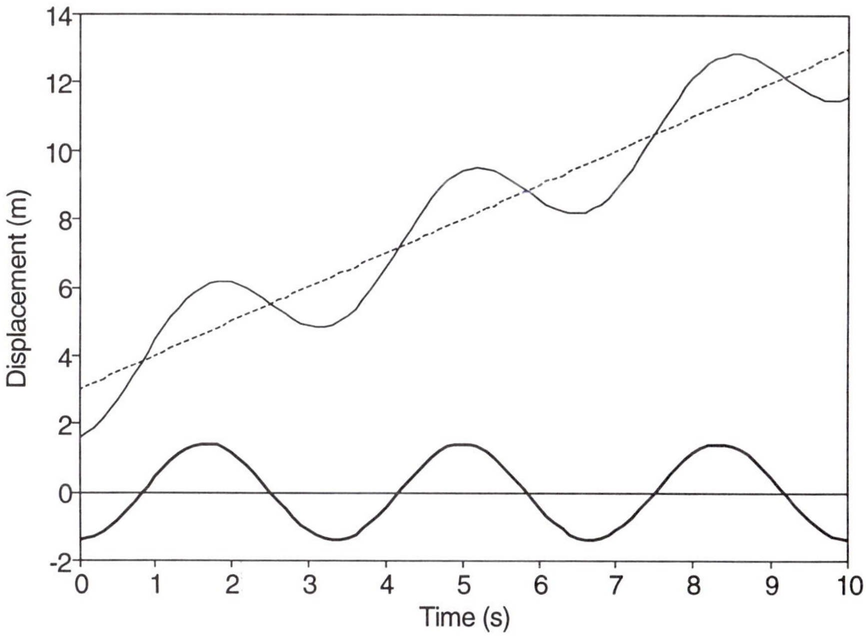

Figure 4-2. Displacement for a free mass with same conditions as in Fig. 4-1. Here, $x_o = 3$ m, $v_o = 1$ m/s. *Dotted line:* Force-free displacement. *Thick line:* Forced $x(t)$. *Thin line:* Total $x(t)$.

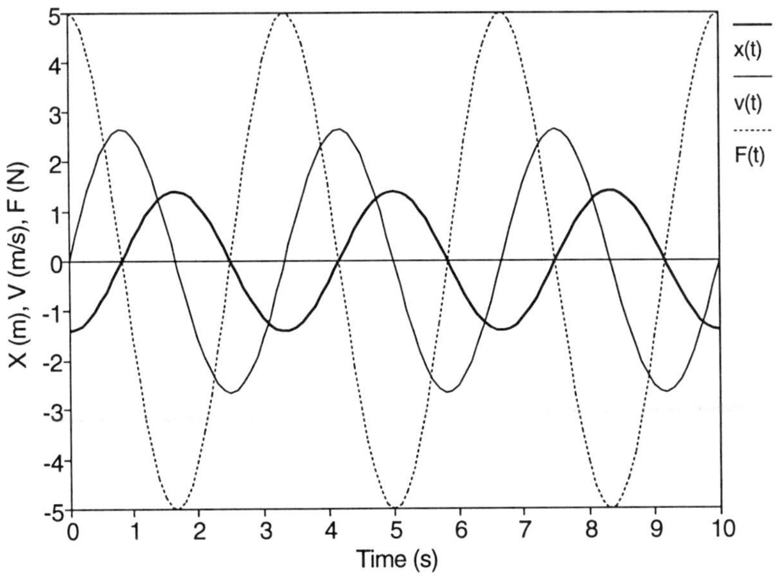

Figure 4-3. Displacement, velocity, and driving force for a free mass for same conditions as in Figs. 4-1 and 4-2.

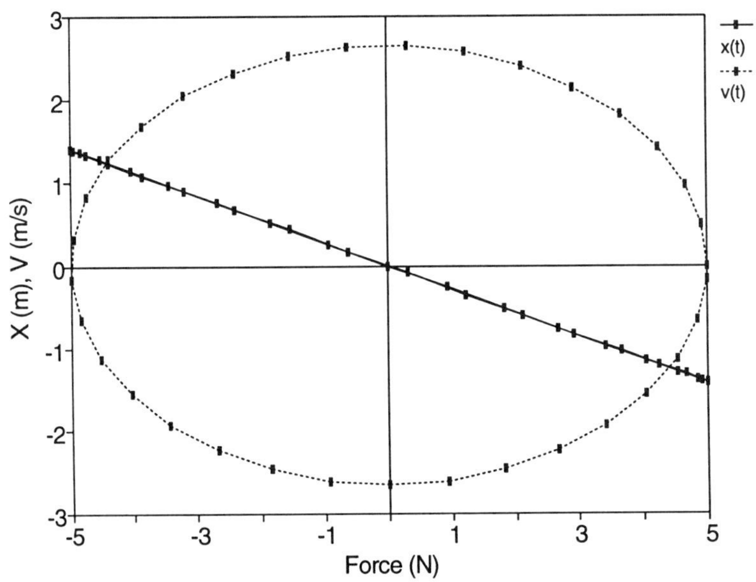

Figure 4-4. Displacement and velocity as functions of the driving force for same conditions as in previous figures. Time marks line up vertically.

FREEMASS, Worksheet for a Forced Free Mass

This worksheet lets you experiment with a forced free mass described in Eq. (4-4) and (4-6). The Home screen is shown in Fig. 4-5.

Permittivity and Index of Refraction, Damped Free Charges

The velocity, Eq. (4-4), can provide us with the current density $\boldsymbol{J}$ (amperes/m^2), which is useful in magnetic and electromagnetic calculations,

$$J = nqv, \tag{4-7}$$

where n is the volume density of charges (number/m^3) and q is the charge of mass m. If random motion is present, v in Eq. (5-7) refers to the drift velocity.

The displacement, Eq. (4-6), provides us with the dipole moment $\boldsymbol{d}$, which is useful in electromagnetic radiation calculations, but we will use it in this example to calculate the permittivity of a collection of free charges, and the relative dielectric constant and index of refraction. It will also bring forth an important statement about a natural frequency of a type not previously discussed here, the plasma frequency. The unusual feature of the plasma frequency is that it depends only on the *density* of charges, n. It does not depend on the ratio of some potential energy constant to some kinetic energy constant, nor does it depend on boundary conditions.

The electric dipole moment (the first moment of the charge distribution) is

$$\boldsymbol{d} = q\boldsymbol{r}. \tag{4-8}$$

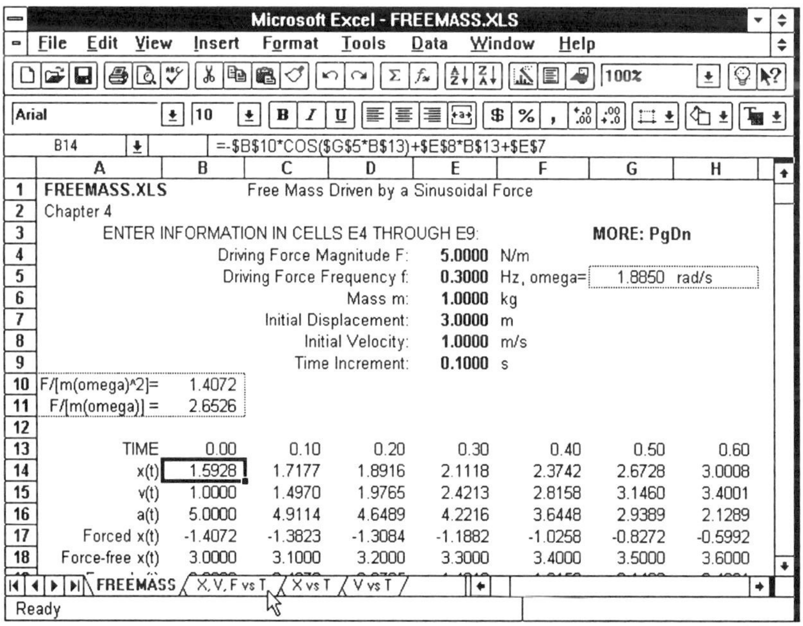

Figure 4-5. Home screen of the FREEMASS worksheet.

For simplicity let's continue our discussion in one dimension, so the electric dipole moment can be expressed as

$$d = qx. \quad (4\text{-}9)$$

Let's be specific and assume that the charge in motion is an electron, with $-e = -1.602 \times 10^{-19}$ Coulomb and mass $m = 9.109 \times 10^{-31}$ kg. Then the force due to the electric field $E_o \exp[j\omega t]$ is

$$F_o \exp[j\omega t] = -e\, E_o \exp[j\omega t]. \quad (4\text{-}10)$$

We seldom know the initial conditions in Eq. (4-6) and usually these are not important because it is the ordered electron motion due to the field that is of interest. So, let's examine the time-dependent part of the dipole moment due to the field,

$$d(t) = -e\, x(t) = --e\frac{-e\, E_o}{m\omega^2} e^{j\omega t} = -\frac{e^2\, E_o}{m\omega^2} e^{j\omega t}. \quad (4\text{-}11)$$

- Note that the field-dependent part of the dipole moment does not depend on the sign of the charge because it depends on the *square* of the charge. The negative sign in Eq. (4-11) indicates that, with no damping, the induced dipole moment is 180° out of phase with the electric field for any type of charge.

If there are n electrons per cubic meter then the dipole moment per unit volume is

$$P = nd, \quad (4\text{-}12)$$

where P is the *electric polarization* with units of Coulomb/m^2. In electromagnetic theory there is an extremely important vector $\mathbf{D}$ called the electric displacement; it relates the electric field and the polarization,

$$\mathbf{D} = \varepsilon_o \mathbf{E} + \mathbf{P} \quad (4\text{-}13)$$

where $\varepsilon_o = 8.85 \times 10^{-12}$ farad/m is the vacuum permittivity. Equation (4-13) is a vector equation and there is no reason, in general, for $\mathbf{E}$ and $\mathbf{P}$ to be in the same direction. However, we will only consider the simple case in which they are in the same direction so we can treat it in one dimension,

$$D = \varepsilon_o E + P. \quad (4\text{-}14)$$

The electric displacement can also be expressed in terms of ε, the permittivity due to the presence of matter,

$$D = \varepsilon E. \quad (4\text{-}15)$$

Equating (4-14) and (4-15) we have,

$$\varepsilon E = \varepsilon_o E + P. \quad (4\text{-}16)$$

Now let's use Equations (4-11) and (4-12) in (4-16),

$$\varepsilon E = \varepsilon_o E - \frac{n e^2}{m\omega^2} E. \tag{4-17}$$

The relative permittivity or dielectric constant is defined as $\varepsilon_{rel} = \varepsilon/\varepsilon_o$, so let's divide (4-17) by ε_o and eliminate E at the same time, since this must hold for all fields,

$$\varepsilon_{rel} = \frac{\varepsilon}{\varepsilon_o} = 1 - \frac{n e^2}{\varepsilon_o m \omega^2}. \tag{4-18}$$

- The dielectric constant of the vacuum is 1. The presence of free charges of any sign always *lowers* the dielectric constant.
- At the special frequency for which the second term in Eq. (4-18) is 1 the dielectric constant goes to zero. For specified values of charge and mass this frequency depends only on the number density.

Examine the second term in Eq. (4-18). It is dimensionless, as the dielectric constant is dimensionless. The factor $ne^2/\varepsilon_o m$ has units of frequency squared. This special frequency is called the "plasma frequency" ω_p. For electrons this frequency is,

$$\omega_p = \left(\frac{n e^2}{\varepsilon_o m}\right)^{1/2} \text{ rad/s} \quad \text{or} \quad f_p \approx 8.979\sqrt{n} \text{ Hz}, \tag{4-19}$$

where n has units of electrons/m^3. As a rule of thumb, 9 GHz corresponds to 10^{12} electrons/cm^3 or 10^{18} electrons/m^3. When the electric field has the frequency ω_p the dielectric constant is zero. For greater electron density the dielectric constant is negative and waves cannot propagate. This effect is observable in reflection of radio waves from the ionosphere, communications blackout during space vehicle re-entry, and reflection of electromagnetic waves from meteor trails.

The index of refraction is commonly defined as

$$\mathrm{n} = \frac{c}{v_p}, \tag{4-20}$$

where c is the speed of light in vacuum and v_p is the phase speed of light in the presence of matter. (Don't confuse the index of refraction n with the number density n.) The speed of light in vacuum is given by

$$c = \frac{1}{\sqrt{\varepsilon_o \mu_o}} \tag{4-21}$$

where μ_o is the vacuum magnetic permeability, $4\pi \times 10^{-7}$ volt·second/amp·meter, or weber/amp·m. The speed of light in non-magnetic matter is given by,

$$v_p = \frac{1}{\sqrt{\varepsilon\,\mu_o}}. \tag{4-22}$$

The index of refraction is given by the ratio of (4-21) to (4-22),

$$\mathrm{n} = \frac{c}{v_p} = \left(\frac{\varepsilon}{\varepsilon_o}\right)^{1/2} \tag{4-23}$$

$$\mathrm{n} = \frac{c}{v_p} = \left(\frac{\varepsilon}{\varepsilon_o}\right)^{1/2} = \left(1 - \frac{ne^2}{\varepsilon_o m\omega^2}\right)^{1/2} = \left(1 - \frac{\omega_p^2}{\omega^2}\right)^{1/2}. \tag{4-24}$$

- The speed of light in the presence of free electric charges exceeds the speed of light in vacuum. At the plasma frequency v_p becomes infinite. This is not a problem, however, because v_p corresponds to the *phase velocity* ω/k, which does not convey energy or information. (Here k is the wave number $2\pi/\lambda$.) A closer examination of the group velocity $\partial\omega/\partial k$ shows that the group velocity goes to zero at the plasma frequency.

4.2 Forced Damped Mass

Time Domain

Next, let's turn on our velocity-dependent resistive force and let it act simultaneously with the driving force. For variety we are going to use a completely real force and calculate the displacement directly. Newton's second law can now be written as

$$m\frac{d^2x}{dt^2} = F_o\cos(\omega t) - R\frac{dx}{dt}. \tag{4-25}$$

Now things are getting more interesting. Rewrite Eq. (4-25) so that all of the terms involving displacement are on the left side,

$$m\frac{d^2x}{dt^2} + R\frac{dx}{dt} = F_o\cos(\omega t). \tag{4-26}$$

With $F_o = 0$ the reduced equation is equivalent to one we have already solved in Chapter 3, although we did it there for velocity instead of displacement,

$$m\frac{d^2x}{dt^2} + R\frac{dx}{dt} = 0. \tag{4-27}$$

This is the variables-separable type that we have previously encountered in Eq. (3-1). In Chapter 3 we found that the solution is given by a real exponential,

$$x(t) = v_o\tau\,[1 - \exp(-t/\tau)] + x_o, \tag{3-5}$$

where $R/m = 1/\tau$. To obtain the complete solution we also need the particular solution for $F_o \neq 0$. Because of linearity this solution must have the same frequency as the driving force so let's try,

$$x(t) = A\cos(\omega t) + B\sin(\omega t). \tag{4-28}$$

Substitute this in (4-25) and after a little algebra we find,

$$x(t) = -\frac{(F_o/m)\,\tau^2}{1 + \omega^2\tau^2}\cos(\omega t) - \frac{(F_o/m)\,(\tau/\omega)}{1 + \omega^2\tau^2}\sin(\omega t). \tag{4-29}$$

The complete solution is the sum of the particular solution Eq. (4-29) and the solution of the reduced equation Eq. (3-5). The transient solution of Eq. (3-5) provides information on how the system starts and stops when the driving force is applied and removed but it is not as interesting to us at this point as the steady-state solution, Eq. (4-29). The way we have written it, the cosine term is in phase with the driving force and the sine term is orthogonal (in time) to the driving force. Note very carefully the interplay of the parameters in the amplitudes of the in-phase and quadrature terms.

Unlike the forced free mass, the presence of the damping term in the forced damped mass permits a permanent flow of energy and momentum from the force field (the source) to the system. The damping term implies that this energy and momentum then flow from the system to a sink. As an example, consider free electrons colliding with atoms and molecules in the presence of an electromagnetic field. The atoms and molecules provide the sink. Collisions among electrons do not change momentum because they are identical, indistinguishable particles.

Frequency Domain (Phasor Solution)

Let's try a complex exponential as a solution,

$$x(t) = X\,e^{j\omega t},\quad \frac{dx}{dt} = j\omega x(t),\quad \frac{d^2x}{dt^2} = -\omega^2 x(t),\quad F(t) = F_o\,e^{j\omega t}. \tag{4-30}$$

Substituting these in Eq. (4-26) we obtain

$$-m\omega^2 X e^{j\omega t} + j\omega R X e^{j\omega t} = F_o\,e^{j\omega t}, \tag{4-31}$$

and this must be true for all time, so,

$$-\omega^2 X + j\frac{\omega}{\tau}X = \frac{F_o}{m}. \tag{4-32}$$

We can solve this for $X(\omega)$ and separate the real and imaginary parts by multiplying the numerator and denominator by the complex conjugate,

$$X(\omega) = \frac{(F_o/m)}{-\omega^2 + j\,\omega/\tau} = (F_o/m)\left[\frac{-\omega^2 - j\,\omega/\tau}{\omega^4 + (\omega/\tau)^2}\right], \tag{4-33}$$

which can be simplified to

$$X(\omega) = (F_o/m)\left[\frac{-\tau^2 - j\,\tau/\omega}{1 + (\omega\tau)^2}\right]. \tag{4-34}$$

Equation (4-34) is in Cartesian or rectangular coordinates in terms of the real and imaginary parts. It is often useful to express complex quantities in polar form in terms of magnitude and phase. The phase angle is given by

$$\theta(\omega) = \arctan\left(\frac{\text{Im}}{\text{Re}}\right) = \arctan\,(1/\omega\tau). \tag{4-35}$$

The magnitude can be determined by the square root of the sum of the squares of the real and imaginary parts, or as follows:

$$X^*(\omega)X(\omega) = |X(\omega)|^2 = \left[\frac{F_o/m}{-\omega^2 + j\omega/\tau}\right]\left[\frac{F_o/m}{-\omega^2 - j\omega/\tau}\right], \tag{4-36}$$

so,

$$|X(\omega)|^2 = \frac{(F_o/m)^2}{\omega^4 + (\omega/\tau)^2}. \tag{4-37}$$

Finally, the magnitude is given by

$$|X(\omega)| = \frac{(F_o/m)\,(\tau/\omega)}{[1 + (\omega\tau)^2)]^{1/2}}. \tag{4-38}$$

Graphs of Eq. (4-35) and (4-38) are shown in Fig. 4-6.

Dielectric Constant and Index of Refraction, Damped Free Charges

It is easy to find the dielectric constant and index of refraction for colliding free charges if we proceed as we did in Section 4.1. Now energy and momentum transfers due to collisions produce a complex dielectric constant and a complex index of refraction. Let's summarize in Table 4-1 some of the important relations among electromagnetic field quantities and the displacement and velocity of charged particles. Radiation is associated with these time-changing vectors.

Table 4-1 Connections to the Electromagnetic Field

Electric dipole moment, $\boldsymbol{d}$	$\boldsymbol{d} = q\boldsymbol{x}$
Electric polarization, $\boldsymbol{P}$	$\boldsymbol{P} = n\boldsymbol{d} = nq\boldsymbol{x}$
Current density, $\boldsymbol{J}$	$\boldsymbol{J} = nq\boldsymbol{v}$
Conductivity, σ	$\boldsymbol{J} = nq\boldsymbol{v} = \sigma\boldsymbol{E} = (\sigma_r + j\,\sigma_i)\boldsymbol{E}$
Electric displacement, $\boldsymbol{D}$	$\boldsymbol{D} = \varepsilon\boldsymbol{E} = \varepsilon_o\boldsymbol{E} + \boldsymbol{P}$
Electric susceptibility, χ_e	$\boldsymbol{P} = \varepsilon_o\chi_e\,\boldsymbol{E}$
Relative permittivity or dielectric constant, ε_{rel}	$\varepsilon_{rel} = \dfrac{\varepsilon}{\varepsilon_o} = 1 + \chi_e$

Proceeding as we did in Eq. (4-11), we can find the dipole moment using Eq. (4-34),

$$d(\omega) = -eX(\omega) = -\frac{eE_o}{m}\left[\frac{-\tau^2 - j\,\tau/\omega}{1 + (\omega\tau)^2}\right]. \tag{4-39}$$

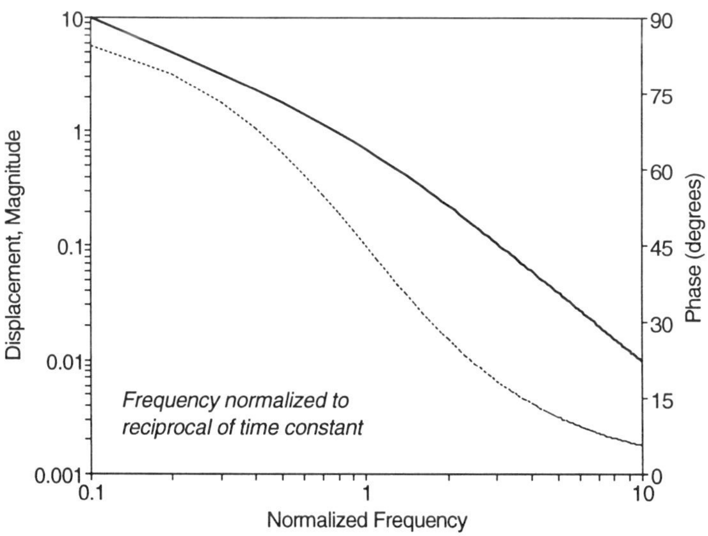

Figure 4-6. Magnitude and phase of forced, damped mass. Phase is relative to driving force. *Thick line:* Magnitude. *Dotted line:* Phase.

As before, this leads to the permittivity, but now it is complex,

$$\varepsilon = \varepsilon_o + \frac{n\,e^2}{m}\left[\frac{-\tau^2 - j\,(\tau/\omega)}{1 + (\omega\tau)^2}\right]. \tag{4-40}$$

This can be written in a simpler form by factoring out ω^{-2} from the numerator,

$$\varepsilon = \varepsilon_o + \frac{n\,e^2}{m\omega^2}\left[\frac{-(\omega\tau)^2 - j\,(\omega\tau)}{1 + (\omega\tau)^2}\right]. \tag{4-41}$$

Let $W = \omega\tau$ and Eq. (4-41) becomes

$$\varepsilon = \varepsilon_o - \frac{n\,e^2}{m\omega^2}\left[\frac{W^2 + j\,W}{1 + W^2}\right]. \tag{4-42}$$

Note that the real part of Eq. (4-42) is an even function and the imaginary part is an odd function. Complex functions with an even real part and an odd imaginary part are known as Hermitian functions; Hermitian operators play an important role in quantum mechanics. The real and imaginary parts of Eq. (4-42) are connected by the Hilbert transform (see Chapter 6). We have not used any sophisticated tricks in obtaining this equation but the results are far-reaching; this is an expression of causality. The permittivity is a *causal response* function. More about this later.

4.3 General Solution for Forced Damped Mass

Until now we have been concerned with a driving force of a single frequency characterized by $F(t) = F_o \cos(\omega t)$. The special form of Eq. (4-25) makes it possible to obtain a general solution for any type of force for which we can compute a certain integral, as we will now see. Let's start with,

$$m\frac{dv}{dt} + Rv = F(t). \tag{4-43}$$

Divide by m to obtain a standard form,

$$\frac{dv}{dt} + \frac{R}{m}v = \frac{F(t)}{m}. \tag{4-44}$$

As usual let,

$$\frac{R}{m} = \frac{1}{\tau}. \tag{4-45}$$

Equation (4-44) is in a form that is amenable to solution with an integrating factor $e^{t/\tau}$ as follows:

$$e^{t/\tau}\frac{dv}{dt} + e^{t/\tau}\frac{1}{\tau}v = e^{t/\tau}\frac{F(t)}{m}. \tag{4-46}$$

The integrating factor allows us to rewrite this as

$$\frac{d}{dt}(e^{t/\tau}v) = e^{t/\tau}\frac{F(t)}{m}, \tag{4-47}$$

so we can immediately integrate and find,

$$\int d\,(e^{t/\tau}v) = \int e^{t/\tau}\frac{F(t)}{m}\,dt + C, \tag{4-48}$$

where C is a constant of integration,

$$e^{t/\tau}v = \int e^{t/\tau}\frac{F(t)}{m}\,dt + C. \tag{4-49}$$

Now we have the formal solution,

$$v(t) = e^{-t/\tau}\int e^{t/\tau}\frac{F(t)}{m}\,dt + Ce^{-t/\tau}. \tag{4-50}$$

We could integrate $v(t)$ to find $x(t)$. Of course, to be useful one must be able to carry out the integrations. The real world is filled with complicated, often random, forces that are not integrable in terms of known functions. There are some standard test functions in common use. Among these test functions, the impulse, unit step, and sine are easy to work with and give considerable insight into linear physical systems.

Impulse or Delta Function Response

The impulse response of a system is an extremely important function that can *completely characterize the system*. The reason for this will become clear in Chapters 6 and 7, but for now let's see how to calculate it for the forced damped mass. Let,

$$F(t) = F_o\,\delta(t), \tag{4-51}$$

where

$$\int \delta(t)\,dt = 1, \tag{4-52}$$

if the integral includes the point $t = 0$, and where

$$\int \delta(t)\,dt = 0, \tag{4-53}$$

otherwise. Let's calculate the momentum of the damped mass as a result of an impulsive force,

$$p(t) = e^{-t/\tau} \int_{-\infty}^{t} e^{t/\tau} F_o \, \delta(t) \, dt. \tag{4-54}$$

The integral has a non-zero value only at $t = 0$, so

$$p(t) = e^{-t/\tau} F_o \, [e^{t/\tau} \tau]_0 = F_o \, \tau e^{-t/\tau}. \tag{4-55}$$

Note that $F\tau$ has units of momentum. The integral $\int F(t)\, dt$ is the *impulse* magnitude of the classical impulse-momentum theorem, $\int \boldsymbol{F}(t)\, dt = \Delta \boldsymbol{p}$ (in vector form).

- The time-domain response (momentum) of a velocity-dependent damped mass to an impulse force is an exponentially decaying momentum.

The impulse response of a system is often called the Green's function. George Green, a self-taught English mathematician who died in 1841, is buried in Westminster Abbey with Newton, Maxwell, and Kelvin. In Chapters 6 and 7 we will see that the impulse response gives the complete frequency response information about a linear physical system, by means of the Fourier transform, and the inverse Fourier transform of the frequency response gives the impulse response in the time domain.

Random Impulse Response

Excitation of a system by a sequence of random impulses is a problem of considerable practical and theoretical interest (see Fig. 3-20). The solution to this problem is known as Campbell's theorem; because it involves statistical concepts we are going to defer discussion until Chapter 10.

Step Function Response

The unit step function response of a system is another important function that can *completely characterize the system*. Let's use a driving force of the form,

$$F(t) = F_o \, u(t), \tag{4-56}$$

where

$$u(t) = 0 \text{ for } t < 0 \tag{4-57}$$

$$u(t) = 1 \text{ for } t \geq 0. \tag{4-58}$$

Now we can calculate the response to the step function force,

$$p(t) = e^{-t/\tau} \int_{-\infty}^{t} e^{t/\tau} F_o \, u(t) \, dt \tag{4-59}$$

or,

$$p(t) = e^{-t/\tau}\,[F_o\,\tau\,e^{t/\tau}]_0^t = F_o\,\tau\,e^{-t/\tau}\,(e^{t/\tau} - 1), \tag{4-60}$$

which reduces to,

$$p(t) = F_o\,\tau\,(1 - e^{-t/\tau}). \tag{4-61}$$

- The time-domain response (momentum) of a damped mass to a step function stimulus is similar to the current rise in a series *LR* circuit.

The unit step function is often called the *Heaviside function*, after the eccentric English genius Oliver Heaviside.

Sinusoidal Response

We have already used a sinusoidal driving force in Eq. (4-25) to compute the displacement in Eq. (4-29). Now we will use our general solution, Eq. (4-50), to compute the momentum and velocity. Let the driving force be

$$F(t) = F_o\cos(\omega t), \tag{4-62}$$

so the momentum can be immediately written as,

$$p(t) = e^{-t/\tau}\int_{-\infty}^{t} e^{t/\tau}\,F_o\cos(\omega t)\,dt. \tag{4-63}$$

The integral is in the standard form,

$$\int e^{ax}\cos(bx)\,dx = e^{ax}\,\frac{[a\cos(bx) + b\sin(bx)]}{a^2 + b^2}, \tag{4-64}$$

so the forced part of the momentum is

$$p(t) = \frac{F_o\,[\tau^{-1}\cos(\omega t) + \omega\sin(\omega t)]}{\tau^{-2} + \omega^2}. \tag{4-65}$$

In terms of $v(t)$ this can be written a,

$$v(t) = \frac{(F_o/m)\,[\tau^{-1}\cos(\omega t) + \omega\sin(\omega t)]}{\tau^{-2} + \omega^2}. \tag{4-66}$$

From this we can find the current density, electric polarization, dielectric constant, and index of refraction as before, but now these quantities have both in-phase and quadrature components. The effect of damping has naturally introduced absorption or transfer of energy from the field.

4.4 Power Transfer

Let's calculate the instantaneous power transferred from an electric field to a charge, for $\boldsymbol{F} = -\, e\boldsymbol{E}_o \cos(\omega t)$

$$\boldsymbol{F}\cdot\boldsymbol{v} = \frac{e^2}{m} E_o^2 \cos(\omega t) \frac{[\tau^{-1} \cos(\omega t) + \omega \sin(\omega t)]}{\tau^{-2} + \omega^2} \tag{4-67}$$

or

$$\boldsymbol{F}\cdot\boldsymbol{v} = \frac{e^2 E_o^2}{m} \frac{[\tau^{-1} \cos^2(\omega t) + \omega \cos(\omega t) \sin(\omega t)]}{\tau^{-2} + \omega^2}\,. \tag{4-68}$$

The time-average power is obtained by integrating over an integer number of periods with the result that only the in-phase power survives. The average value of the cosine-squared is 0.5, so,

$$(\boldsymbol{F}\cdot\boldsymbol{v})_{av} = \frac{e^2 E_o^2}{m} \frac{1}{2\tau(\tau^{-2} + \omega^2)} = \frac{e^2 E_o^2}{2m} \frac{\tau}{1 + \omega^2\tau^2}\,. \tag{4-69}$$

- Notice that the time-average power falls to half maximum or –3 dB for $\omega\tau = 1$. At very high frequencies little power is absorbed from the field.
- The forced, damped charge acts like a high-pass filter. For a filter, the special frequency for which $\omega = \tau^{-1}$ is called the *corner frequency*.
- The time-average power absorbed from the field source is exactly equal to the time-average power extracted by the damping force $(-\, R\boldsymbol{v}\cdot\boldsymbol{v})_{av}$ and delivered to the energy sink.

DAMPMASS, Worksheet for Forced, Damped Mass

The Home screen for this worksheet is shown in Fig. 4-7. You can experiment with different values of force magnitude and frequency, mass and damping constant. Graphs connected to this worksheet are named *x, v, F vs. t*; *x vs. t*; *v vs. t*; and *x Mag & Ph*.

4.5 Forced Damped Oscillator

Now we are ready to turn on all three forces,

$$m \frac{d^2x}{dt^2} + R \frac{dx}{dt} + \kappa x = F_o\, e^{j\omega t} \tag{4-70}$$

and divide by m,

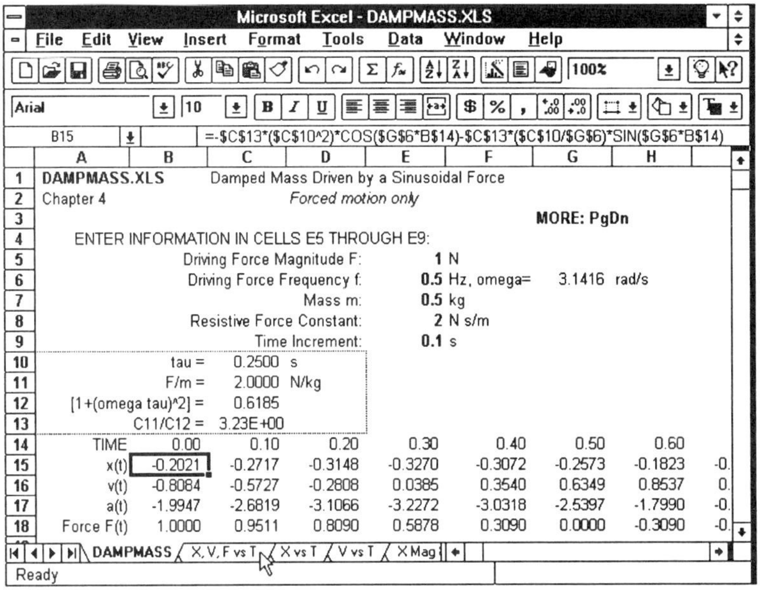

Figure 4-7. Worksheet for experimenting with a forced, damped mass.

$$\frac{d^2x}{dt^2} + \frac{R}{m}\frac{dx}{dt} + \frac{\kappa}{m}x = \frac{F_o}{m}e^{j\omega t}. \tag{4-71}$$

As before, let's rewrite this using,

$$\frac{R}{m} = \frac{1}{\tau} = 2K \quad \text{and} \quad \frac{\kappa}{m} = \omega_o^2$$

$$\frac{d^2x}{dt^2} + 2K\frac{dx}{dt} + \omega_o^2 x = \frac{F_o}{m}e^{j\omega t}. \tag{4-72}$$

We already have the transient solution (for the homogeneous equation) of Eq. (4-72), given by Equations (3-20), (3-21), or (3-22). Now all we need is the forced part (the particular solution) and we can add the two solutions to obtain the complete solution. Because this is a linear, ordinary differential equation with constant coefficients we know that the forced solution will have the same frequency as the forcing function but we need to find the magnitude and the phase.

Displacement

As before, we are going to use a phasor solution. In Chapter 6 we will see why this is called "the poor man's Fourier transform." Let the displacement, velocity, and acceleration be represented by the phasors,

$$x = X e^{j\omega t}, \quad \frac{dx}{dt} = j\omega x, \quad \frac{d^2x}{dt^2} = -\omega^2 x, \tag{4-73}$$

and substitute this into Eq. (4-72) obtaining

$$-\omega^2 x + 2j\omega K x + \omega_o^2 x = \frac{F_o}{m} e^{j\omega t}. \tag{4-74}$$

This must hold for all time so the complex exponential cancels, yielding

$$X(-\omega^2 + 2j\omega K + \omega_o^2) = F_o/m.$$

Solving for the complex magnitude $X(\omega)$,

$$X(\omega) = \frac{F_o/m}{\omega_o^2 - \omega^2 + 2j\omega K} \tag{4-75}$$

and separating into real and imaginary parts,

$$X(\omega) = \frac{F_o}{m}\left[\frac{\omega_o^2 - \omega^2}{(\omega_o^2 - \omega^2)^2 + (2K\omega)^2} - j\frac{2K\omega}{(\omega_o^2 - \omega^2)^2 + (2K\omega)^2}\right] \tag{4-76}$$

we find the phase angle,

$$\theta(\omega) = \arctan\left(\frac{\text{Im}}{\text{Re}}\right) = \arctan\left(\frac{-2K\omega}{\omega_o^2 - \omega^2}\right). \tag{4-77}$$

Note that for $\omega_o > \omega$ the real part is positive and becomes negative for $\omega_o < \omega$. The imaginary part is always negative. All that is left to do for the displacement is to find the magnitude. First, multiply Eq. (4-75) by its complex conjugate,

$$|X(\omega)|^2 = \left[\frac{F_o/m}{(\omega_o^2 - \omega^2) + j2K\omega}\right]\left[\frac{F_o/m}{(\omega_o^2 - \omega^2) - j2K\omega}\right], \tag{4-78}$$

and take the square root,

$$|X(\omega)| = \frac{F_o/m}{\sqrt{(\omega_o^2 - \omega^2)^2 + (2K\omega)^2}}. \tag{4-79}$$

Later we will need the time-average value of $[X(\omega)e^{j\omega t}]^2$ when we calculate the energy resonance. Care must be exercised when multiplying complex quantities when they are used for convenience to represent real quantities. Don't fall into the trap of thinking that the time-average is zero, because a little more thought shows that the time-average value of any phasor is $0.5\,|X|^2 = 0.5\,X^*X$, where the asterisk indicates the complex conjugate.

Displacement Resonance

The frequency for which maximum displacement occurs is the *resonant frequency for displacement*, ω_x. To find the resonant frequency take the first derivative of Eq. (4-79), set it to zero, and solve for ω,

$$\omega_x = \sqrt{\omega_o^2 - 2K^2}\,. \tag{4-80}$$

Several things should be noted:

- Damping always shifts the resonant frequency *down* from the natural frequency of the undamped oscillator.
- Damping decreases the maximum displacement at resonance.
- Phase lag is almost zero at low frequencies and asymptotically approaches –180° relative to the force at high frequencies.

The phase and magnitude of Eq. (4-77) and (4-79) are plotted in Figs. 4-8 and 4-9. Systems with small damping can fail due to large resonant displacements. Recall opera singers breaking glass objects and the collapse of the Tacoma Narrows bridge.

Thousands of years ago Roman soldiers discovered empirically that they had to break step when marching over bridges to avoid inducing resonant effects. One can only speculate about how many bridges were destroyed and how many soldiers fell into rivers before they figured this out.

Large amounts of energy can be stored in systems with small damping. Such systems respond to a narrow band of frequencies and they can pick these out of a plethora of other frequencies. While this may be desirable in a radio, TV, microwave cavity, or laser it should be avoided in architecture and civil engineering.

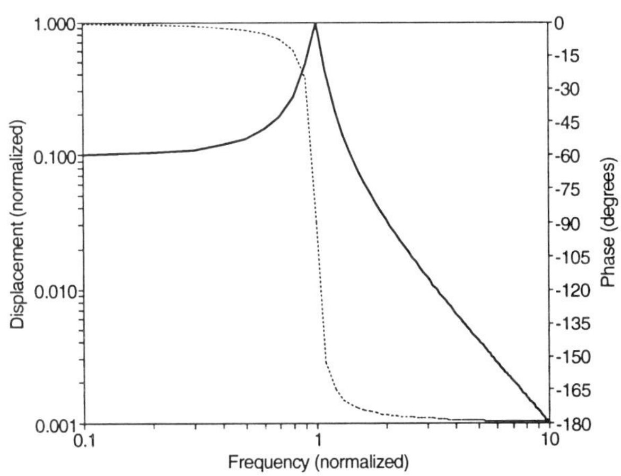

Figure 4-8. Magnitude and phase of displacement for a forced, damped oscillator. $K = 0.05$. *Thick line:* Magnitude. *Dotted line:* Phase, measured relative to driving force. Compare Fig. 4-10 for velocity resonance.

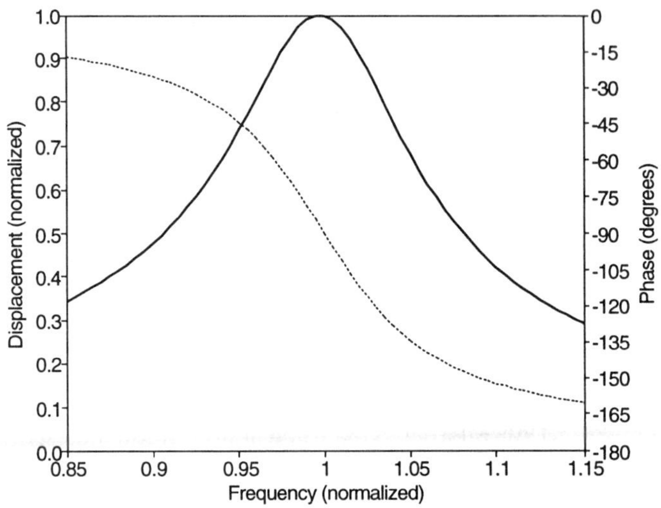

Figure 4-9. Magnitude and phase of displacement, $K = 0.05$, close to resonance. *Thick line:* Magnitude. *Dotted line:* Phase relative to driving force.

Velocity

The velocity is of interest in electrical and electromagnetic applications because it determines the current; the current is the source of the magnetic field and the time-varying current connects to radiation. Proceeding as before let,

$$v = V e^{j\omega t}, \quad \frac{dv}{dt} = j\omega v, \quad x = \int V e^{j\omega t}\, dt = \frac{v}{j\omega} \tag{4-81}$$

and substitute these into Eq. (4-72), obtaining the complex magnitude,

$$V(\omega) = \frac{\omega F_o/m}{2K\omega + j(\omega^2 - \omega_o^2)}. \tag{4-82}$$

As usual, separate the real and imaginary parts,

$$V(\omega) = \omega \frac{F_o}{m}\left[\frac{2K\omega - j(\omega^2 - \omega_o^2)}{(\omega^2 - \omega_o^2)^2 + (2K\omega)^2}\right] \tag{4-83}$$

to find the phase angle,

$$\varphi(\omega) = \arctan\left[\frac{-(\omega^2 - \omega_o^2)}{2K\omega}\right] \tag{4-84}$$

and the magnitude,

$$|V(\omega)| = \frac{\omega F_o/m}{\sqrt{(\omega^2 - \omega_o^2)^2 + (2K\omega)^2}} \quad . \tag{4-85}$$

Velocity Resonance

From Eq. (4-85) the *velocity resonance* occurs at the frequency of the undamped oscillator,

$$\omega_v = \omega_o \, , \tag{4-86}$$

so there is no down-shifting for velocity resonance. The distinction between displacement resonance and velocity resonance is sometimes overlooked.

- Velocity resonance occurs at a frequency higher than the displacement resonance.
- Maximum time-average power is absorbed from the driving force at ω_v.
- The velocity is in phase with the driving force at velocity resonance, by Eq. (4-84).
- Maximum velocity (at resonance) is $F_o/2mK$ by Eq. (4-85).

The velocity resonance is shown over a wide range in Fig. 4-10, and close to resonance in Fig. 4-11.

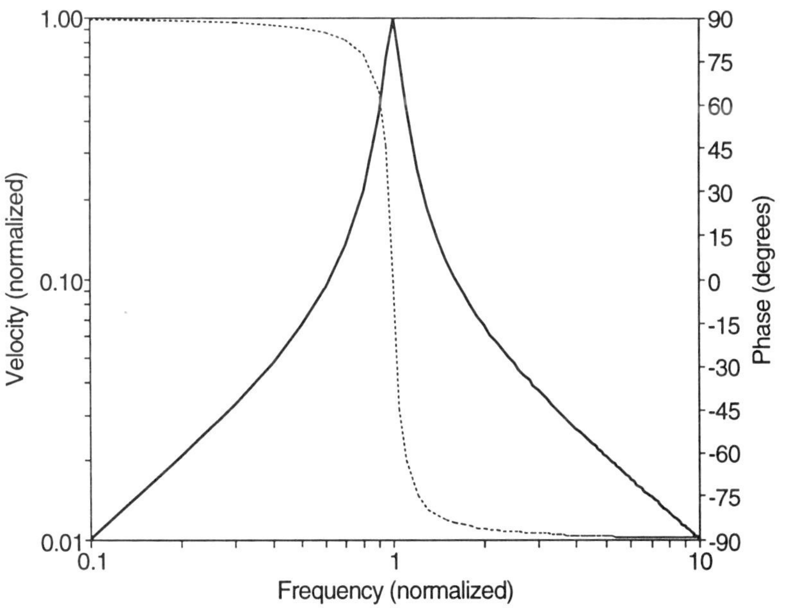

Figure 4-10. Velocity resonance for $K = 0.05$. Note symmetry of magnitude, in constrast to asymmetry of displacement magnitude. Compare Fig. 4-8.

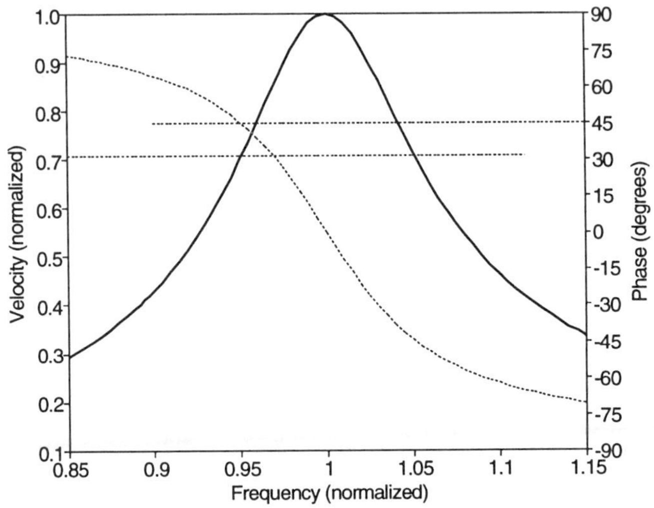

Figure 4-11. Velocity resonance for $K = 0.05$, close to resonance. "Half-power point" $v = 0.707 v_{max}$ and 45° phase shift occur at $\omega = \omega_o \pm K$.

Power Resonance

In the dynamic steady state, power flows from the driving source through the oscillator and into the sink via the damping force. Energy does not continue to accumulate in the oscillator after the steady state is achieved.

It is easy to show that the time-average power absorbed from the force field is exactly equal to the time-average power extracted by the damping force,

$$P(\omega) = (\boldsymbol{F} \cdot \boldsymbol{v})_{av} = (-R\boldsymbol{v} \cdot \boldsymbol{v})_{av} . \tag{4-87}$$

Using Eq. (4-85), this can be written as

$$P(\omega) = \frac{(F_o^2/m)K\omega^2}{(\omega_o^2 - \omega^2)^2 + (2K\omega)^2} . \tag{4-88}$$

Power resonance occurs at ω_o. A typical power resonance is shown in Fig. 4-12, for the same damping as in previous figures.

Energy Resonance

Because the kinetic energy resonance is related to velocity, and the potential energy resonance is related to displacement, they achieve maxima at different frequencies. The total time-averaged energy is the sum of the potential energy and kinetic energies of the oscillator,

$$E(\omega) = \frac{1}{2} m\omega_o^2 |X|^2_{av} + \frac{1}{2} m|V|^2_{av} . \tag{4-89}$$

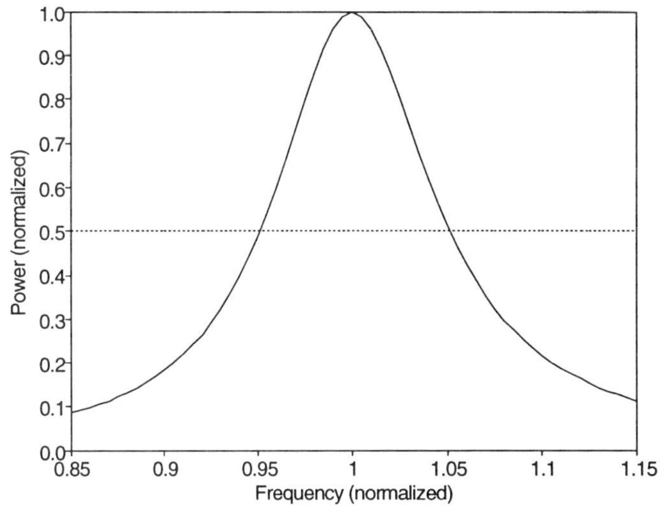

Figure 4-12. Power resonance for $K = 0.05$, close to resonance. Note symmetry about the resonant frequency, because power is based on velocity resonance.

$$E(\omega) = \frac{0.25\,(F_o^2/m)(\omega^2 + \omega_o^2)}{(\omega_o^2 - \omega^2)^2 + (2K\omega)^2} \tag{4-90}$$

Figure 4-13 shows the normalized energy resonance for $K = 0.05$.

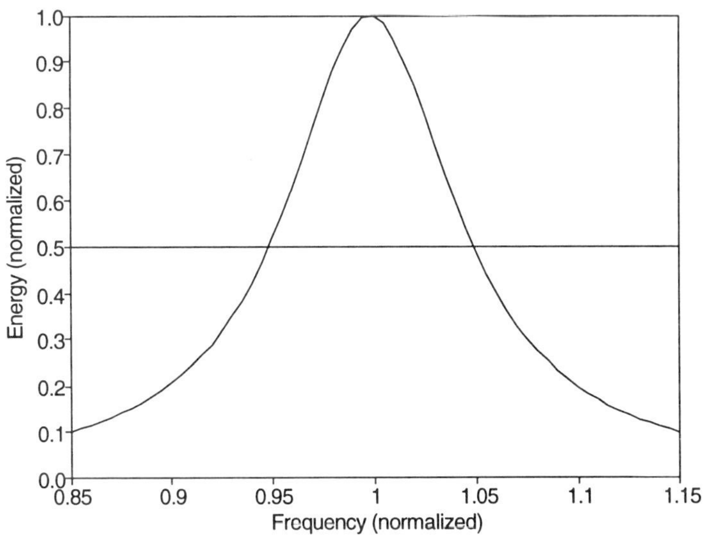

Figure 4-13. Energy resonance for $K = 0.05$. Bandwidth is $2K$.

4.6 Permittivity for Bound Charges

Proceeding as before with free charges and free, damped charges, we can calculate the induced electric dipole moment $d(\omega)$ for a harmonically-bound, damped charge $(-e)$ with mass m. Using Eq. (4-75) we have

$$d(\omega) = -e\,X(\omega) = \frac{e^2 E_o}{m}\frac{1}{(\omega_o^2-\omega^2)+2j\omega K}. \tag{4-91}$$

As before, this leads us to the complex permittivity,

$$\varepsilon = \varepsilon_o + \frac{ne^2}{m}\frac{1}{(\omega_o^2-\omega^2)+2j\omega K}, \tag{4-92}$$

and the relative permittivity or dielectric constant,

$$\varepsilon_{rel} = \frac{\varepsilon}{\varepsilon_o} = 1 + \frac{ne^2}{m\varepsilon_o}\frac{1}{(\omega_o^2-\omega^2)+2j\omega K}. \tag{4-93}$$

- Previously we observed that free (or free, damped) charges of any sign always *decrease* the permittivity and index of refraction. Now we see that harmonically bound charges of any sign always *increase* the permittivity and index of refraction. The phase velocity of light is greater in vacuum and ionized gases than it is in solids, liquids, and neutral gases.
- Near resonance, increased damping *decreases* power absorption from the field. By Eq. (4-88), $P(\omega_o) \propto 1/K$. Away from resonance, increased damping *increases* power absorption (see Fig. 4-14).

Lorentz Line Shape

Close to resonance and for small damping, $\omega_o >> K$, the energy resonance takes on a particularly simple form,

$$L(\omega) = \frac{K^2}{(\omega_o-\omega)^2+K^2}. \tag{4-94}$$

This is called the *Lorentz line shape*, and we will see in Chapter 6 that it is exactly the Fourier transform of a complex oscillatory exponential decay,

$$f(t) = \exp[-Kt]\exp[j\omega_o t] = \exp[-(K-j\omega_o)t]. \tag{4-95}$$

We can regard $K - j\omega_o$ as a complex frequency. The Lorentz line shape, normalized to a maximum of 1, is shown in Fig. 4-15. Notice how the bandwidth (full width at half-maximum) increases as K increases; bandwidth is $2K$.

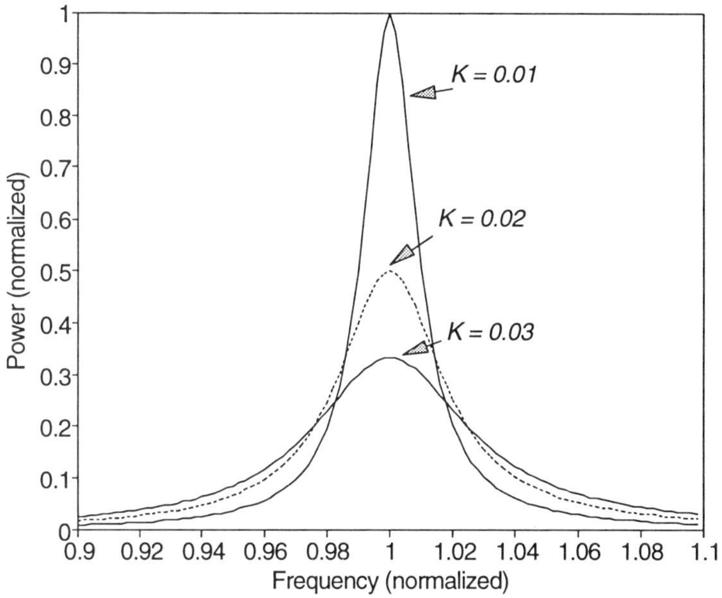

Figure 4-14. Power absorbed from the energy source and delivered to the energy sink in the steady state, for three values of K.

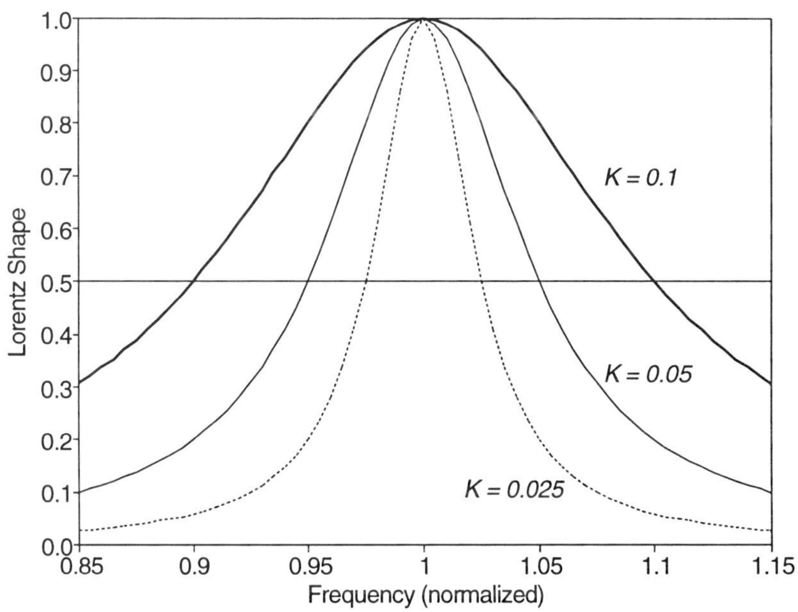

Figure 4-15. Lorentz line shape normalized to unity. This is also the Breit-Wigner line shape in nuclear physics. Bandwidth is $2K$.

4.7 FORCEOSC, Forced Oscillator Worksheet

This worksheet lets you experiment with forced oscillators in the frequency domain. Figure 4-16 shows the Home screen.

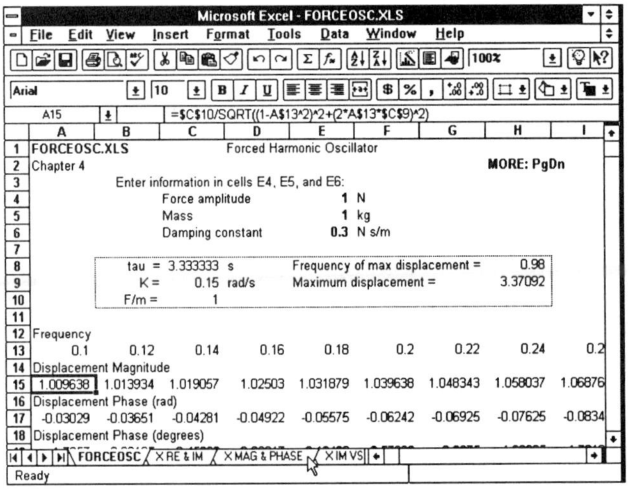

Figure 4-16. Home screen of the FORCEOSC worksheet.

4.8 Summary

Tables 4-2 and 4-3 provide an overview of the classical forced oscillator. Formal, complete, time-domain solutions are given in Table 4-2. (Also, see Table 3-1.)

Table 4-2. Forced, Damped Oscillators in the Time Domain

Case	*Conditions*	$x(t)$
Under-Damped	$\omega_o > K$ $\omega_u = \sqrt{\omega_o^2 - K^2}$	$x(t) = e^{-Kt}\{x_o \cos(\omega_u t) + [(v_o + Kx_o)/\omega_u] \sin(\omega_u t)\}$ $+ [e^{-Kt} \sin(\omega_u t)/m\omega_u] \int e^{+Kt} F(t) \cos(\omega_u t)\, dt$ $- [e^{-Kt} \cos(\omega_u t)/m\omega_u] \int e^{+Kt} F(t) \sin(\omega_u t)\, dt$
Critically Damped	$\omega_o = K$	$x(t) = e^{-Kt}[x_o + (v_o + Kx_o)t]$ $+ m^{-1} t\, e^{-Kt} \int e^{+Kt} F(t)\, dt - m^{-1} e^{-Kt} \int t\, e^{+Kt} F(t)\, dt$
Over-Damped	$\omega_o < K$ $\omega_m = \sqrt{K^2 - \omega_o^2}$	$x(t) = e^{-Kt}\{x_o \cosh(\omega_m t) + [(v_o + Kx_o)/\omega_m] \sinh(\omega_m t)\}$ $+ \dfrac{\exp[(-K + \omega_m)t]}{2m\omega_m} \int \exp[-(-K + \omega_m)t]\, F(t)\, dt$ $- \dfrac{\exp[(-K - \omega_m)t]}{2m\omega_m} \int \exp[-(-K - \omega_m)t]\, F(t)\, dt$

Table 4-3. Summary of Forced Oscillator Resonances ($K = 1/2\tau = R/2m$)		
Forced Under-Damped	*Line Shape*	*Phase Relative to Driving Force*
Displacement Resonance $\omega_x = \sqrt{\omega_o^2 - 2K^2}$	$\lvert X(\omega)\rvert = \dfrac{F_o/m}{\sqrt{(\omega_o^2 - \omega^2)^2 + (2K\omega)^2}}$	$\theta(\omega) = \arctan\left(\dfrac{-2K\omega}{\omega_o^2 - \omega^2}\right)$
Velocity Resonance $\omega_V = \omega_o = \sqrt{\kappa/m}$	$\lvert V(\omega)\rvert = \dfrac{\omega\, F_o/m}{\sqrt{(\omega_o^2 - \omega^2)^2 + (2K\omega)^2}}$	$\varphi(\omega) = \arctan\left(\dfrac{-(\omega^2 - \omega_o^2)}{2K\omega}\right)$
Power Resonance $\omega_{pwr} = \omega_o = \sqrt{\kappa/m}$	$P(\omega) = \dfrac{(F_o^2/m)K\omega^2}{(\omega_o^2 - \omega^2)^2 + (2K\omega)^2}$	(Power absorbed from external source and transferred to environment)
Energy Resonance $\omega_E = \omega_o = \sqrt{\kappa/m}$	$E(\omega) = \dfrac{0.25\,(F_o^2/m)(\omega^2 + \omega_o^2)}{(\omega_o^2 - \omega^2)^2 + (2K\omega)^2}$	(Total energy in oscillator)
Lorentz Shape $\omega_L = \omega_o = \sqrt{\kappa/m}$	$L(\omega) = \dfrac{K^2}{(\omega_o - \omega)^2 + K^2}$	(Normalized line shape, small damping, close to resonance)

What's Next?

We are not finished with forced oscillators. In Chapter 5 we are going to examine resonances of under-damped oscillators in more detail and introduce the important concepts of Q, impedance, and admittance. There is also another type of energy loss for charged particles, radiation damping. Radiation damping does not have a simple velocity-dependent force law like $-Rv$, which we have used until now.

We will find that Q relates oscillator properties in the frequency domain to properties in the time domain and leads naturally into the Fourier transform in Chapter 6. The Fourier transform gives a different and general perspective, powerful mathematical methods, and clues to the fundamentals of uncertainty principles.

We could have used the Laplace transform for the solutions of the differential equations in the present chapter, but transforms are more abstract and it seems natural to defer this general method to Chapter 6. Chapter 6 will give us the tools that we need to see how harmonic oscillators fit into the general theory of linear systems, leading naturally into Chapter 7.

In Chapter 7, we will use Table 4-3 to provide oscillators' responses to Dirac delta-function driving forces, $F(t) = F_o\delta(t)$. For example, consider the critically-damped oscillator in Table 4-3, and assume that $x_o = 0$ and $v_o = 0$ at $t = 0$. The second integral is zero because $F(t) = F_o\delta(t) = 0$ unless $t = 0$, but t is a factor in the integral so the integral is zero for all time. On the other hand, the first integral is 1 N·s so the term is

$$x(t) = m^{-1} t\, e^{-Kt} \int_0^t e^{+Kt}\, F_o\delta(t)\, dt = (F_o/m) t e^{-Kt}. \tag{4-96}$$

Equation (4-96) is the displacement impulse response of a critically damped oscillator. It is not merely a coincidence that this impulse response is the waveform of the Loran-C navigation system.

When we use these Green's functions (responses to δ-functions) with the Fourier methods of Chapter 6, we will have complete descriptions in the time and frequency domains, and deeper insights into the workings of oscillators. But first, let's look at Chapter 5 for the practical, useful, and simple ideas that we group together and call Q.

Chapter 5

Q, R, X, Y, and *Z*

... when you cannot measure it, when you cannot express it in numbers, your knowledge is of a meagre and unsatisfactory kind...

Lord Kelvin

We have seen that damping makes an oscillator respond to a band of frequencies instead of a single frequency. In this chapter we will discuss Q, a quantitative measure of the spread in frequency response. In the sense that the quality of an oscillator is measured by how well it generates, or responds to, a narrow range of frequencies, then Q is a measure of this quality. From this it follows that Q is also a measure of the rate of change of phase in the frequency domain, and the oscillatory amplitude decay in the time domain. Also, we will discuss some related, important, and practical concepts of reactance (X), admittance (Y), and impedance (Z) for a forced, damped mass and for the forced, damped oscillator.

In a still deeper sense, Q is related to a fundamental uncertainty relationship of the universe which we will treat classically here and later in Chapter 6. The associated quantum uncertainty relation will be discussed in Chapter 12. Searching for the ultimate Q has brought forth many technological and scientific advances, some of which resulted in Nobel prizes.

5.1 What Is Q?

The letter Q has become a universal symbol for a figure of merit relating energy stored and energy lost in a physical system. Estill Green has traced the historical development of the concept [*American Scientist* **43**, 584–594 (1955)]: "If scientists were as fair as Humpty Dumpty in paying words extra for overtime work, the ubiquitous Q would come near the top of the payroll. It all started with K. S. Johnson [at Western Electric Company's Engineering Department, which became Bell Telephone Laboratories in 1925, then AT&T Bell Laboratories, and now Lucent Technologies]. Many scientists have contributed new words to our vocabulary. To only one, however, has come the distinction of elevating a letter of the alphabet into a word of everyday use in many and diverse fields. Little did Johnson dream, when he first used the symbol Q to represent the ratio of reactance to effective resistance in a coil [inductor] or a condenser [capacitor], that within a span of some 30 years this same symbol would be commonly used to describe an attribute of such dissimilar things as a resonant circuit, a spectral line, a mechanical vibrator, and a bouncing ball. The story of this expanding usage of the 17th letter of the alphabet makes an interesting study for the scientific etymologist." The story of Q has continued and grown in the years since Green's article was published.

Q describes some important attributes of an oscillatory physical system in the time domain and the frequency domain, without much detailed knowledge of the system, and without much mathematics. Q is not difficult to measure, and it lets you compare different systems in terms of a single, dimensionless number.

Fundamentally, Q is now defined as

$$Q = 2\pi \times \frac{\text{average energy stored per period}}{\text{energy loss per period}}. \tag{5-1}$$

This is only useful for an under-damped oscillator, and it is most useful when the damping is small. Let's see how this applies specifically to a mechanical oscillator and an *LCR* resonator. For simplicity, let's look at an oscillator from one maximum displacement, $x(t_1) = A_1$, to the next maximum displacement $x(t_2) = A_2$. The kinetic energy is zero at these turning points, so the total energy is in potential form. Under these conditions, and for small damping, the average total energy stored per period is

$$E_{av} \approx \frac{0.5\,\kappa A_1^2 + 0.5\,\kappa A_2^2}{2} \approx 0.5\,\kappa A_1^2. \tag{5-2}$$

Similarly, the energy loss (ΔE) per period (T) s,

$$\Delta E = 0.5\,\kappa A_1^2 - 0.5\,\kappa A_2^2 = 0.5\,\kappa\,(A_1^2 - A_1^2 e^{-KT}). \tag{5-3}$$

Because the damping is small, the exponential can be expanded to yield,

$$\Delta E = 0.5\,\kappa A_1^2\,(1 - e^{-KT}) \approx 0.5\,\kappa A_1^2\,(1 - 1 + KT) \approx 0.5\,\kappa K T A_1^2 \tag{5-4}$$

Using Eqs. (5-4) and (5-2) in Eq. (5-1), we can express Q as

$$Q = 2\pi \times \frac{E_{av}}{\Delta E} = 2\pi \times \frac{1}{2KT}. \tag{5-5}$$

Other expressions for Q are useful. Since $K = R/2m$, $\omega_o^2 = \kappa/m$, $2K = \Delta\omega$, and $T = 1/f = 2\pi/\omega$, we can write

$$Q = \left(\frac{m\kappa}{R^2}\right)^{1/2} = \frac{\omega_o}{2K} = \frac{\omega_o}{\Delta\omega}. \tag{5-6}$$

For the LCR resonator, $m \to L$ and $R \to R$, so the corresponding expression for Q becomes

$$Q = \frac{\omega_o L}{R_{eff}} = \frac{\text{inductive reactance}}{\text{effective resistance}} = \frac{\omega_o}{\Delta\omega} = \frac{f_o}{\Delta f} = \frac{\sqrt{L/C}}{R_{eff}} \; . \tag{5-7}$$

Losses in the capacitor are often (but not always) negligible compared with losses in the inductor. The resistance R in Eq. (5-7) is the *effective resistance,* which may be greater than the dc resistance. As frequency increases, the skin depth becomes smaller, so the effective resistance generally increases. Losses due to induced emf's in nearby conductors, induced polarization currents in lossy dielectrics, and radiation loss (radiation "resistance") also contribute to the effective resistance. Equation (5-7) is one of the most familiar expressions for Q. Note importance of the L/C ratio.

Measurement of Q by $f_o/\Delta f$ is quick and easy, but you must be careful to distinguish between the unloaded Q and the loaded Q. Remember, you always measure loaded Q, so take care that your measuring instruments and other connecting circuits have *much* greater impedance than the resonator.

5.2 Frequency Domain

Consider the power resonance from Eq. (4-88) and Table 4-2,

$$P(\omega) = \frac{(F_o^2/m)K\omega^2}{(\omega_o^2 - \omega^2)^2 + (2K\omega)^2} \; . \tag{5-8}$$

At resonance, $\omega = \omega_o$, the power attains its maximum value $P(\omega_o) = F_o/(4mK)$. For small damping ($\omega_o >> K$) the power falls to half of its maximum value when $(\omega - \omega_o) = \pm K$. The full frequency interval between the upper half-maximum and the lower half-maximum is $2K = \Delta\omega$.

The two half-maximum power points correspond to –3 dB points, because $\text{dB} = 10 \log (P/P_r)$, where the log is base 10 and P_r is the reference power, which is the maximum value (0 dB) in this case. In other words, the full frequency width at half-maximum power (FWHM) is the bandwidth, and this corresponds to the frequency interval between the upper and lower –3 dB points. This is a standard expression of Q and bandwidth for mechanical, electrical, and acoustical systems,

$$Q = \frac{f_o}{\Delta f} = \frac{\omega_o}{\Delta\omega} \; . \tag{5-9}$$

Because power is proportional to the square of the amplitude, a –3 dB point corresponds to maximum amplitude/$\sqrt{2} \approx 0.707 \times$ maximum amplitude.

In Chapter 4 we found that the half-maximum power points correspond to ±45° phase shifts (see Fig. 4-11), so the full frequency interval between the ±45° points can serve as a measure of bandwidth and Q. Figures 5-1 and 5-2 show measurement of Q graphically. *Caution:* Sometimes you may find the term "half-width" used with a resonator. This confusing terminology usually (but not always) refers to the full width at half-maximum power (FWHM). Be careful.

In Eq. (3-19) and Tables 3-1 and 4-3 we used the symbol ω_u to characterize the under-damped oscillator. Now we can write ω_u in terms of Q as

$$\omega_u = \sqrt{\omega_o^2 - K^2} = \omega_o\sqrt{1 - (K^2/\omega_o^2)} = \omega_o\left(1 - \frac{1}{4Q^2}\right)^{1/2}. \tag{5-10}$$

In Figs. 5-1 and 5-2 the upper half-power and –3 dB point is 1.01 and the lower half-power and –3 dB point is 0.99, so $\Delta\omega = 0.02$. This corresponds to a Q of 50. Notice that the ±45° points also occur at the half-power points, and the phase is approximately linear close to ω_o.

Rate of Change of Phase

Consider the phase of the displacement given by Eq. (4-77) and Table 4-2, Using Eq. (5-9) with $\Delta\omega = 2K = \omega_o/Q$, we can express Eq. (4-77) as

$$\theta(\omega) = \arctan\left(\frac{-2K\omega}{\omega_o^2 - \omega^2}\right) = \arctan\left(\frac{-\omega_o\omega}{Q(\omega_o^2 - \omega^2)}\right). \tag{5-11}$$

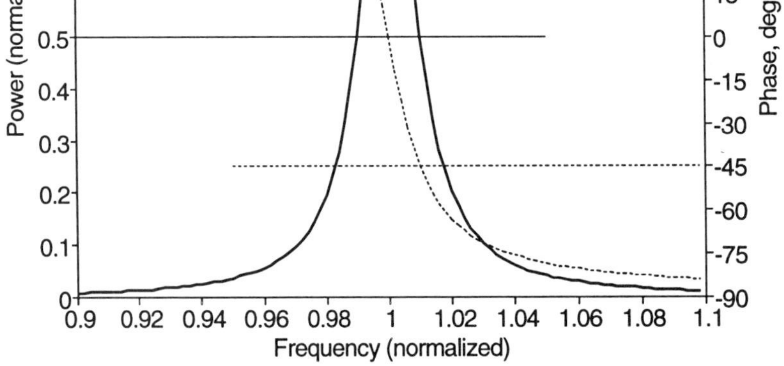

Figure 5-1. Measurement of Q in terms of half-power points and ±45° points.

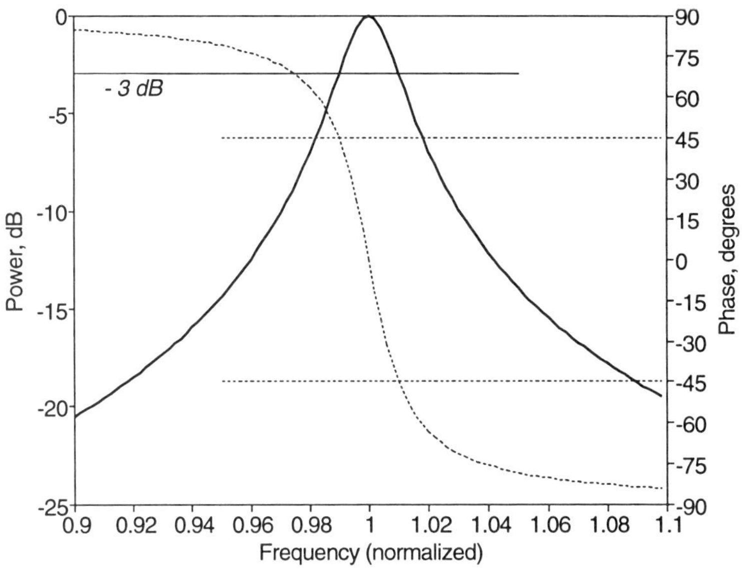

Figure 5-2. Q measurement by –3 dB points and ±45° points.

To find $d\theta/d\omega$ we need to take the derivative of the arctangent. Let

$$y = \frac{-\omega_o \omega}{Q} (\omega_o^2 - \omega^2)^{-1}. \tag{5-12}$$

So,

$$\frac{dy}{d\omega} = -\frac{\omega_o}{Q} (\omega_o^2 - \omega^2)^{-1} + \frac{\omega_o \omega}{Q} (\omega_o^2 - \omega^2)^{-2} (-2\omega). \tag{5-13}$$

Since,

$$d \arctan(y) = \frac{dy}{1 + y^2}, \tag{5-14}$$

then we can write

$$\frac{d\theta}{d\omega} = \frac{d}{d\omega} \arctan(y) = \frac{dy/d\omega}{1 + y^2}. \tag{5-15}$$

Substituting our values of y and $dy/d\omega$ from Eqns. (5-12) and (5-13), we obtain

$$\frac{d\theta}{d\omega} = \frac{-\omega_o (\omega_o^2 - \omega^2) - 2\omega_o \omega^2}{Q(\omega_o^2 - \omega^2) + \omega_o^2 \omega^2 / Q}. \tag{5-16}$$

Near resonance, $\omega^2 \approx \omega_o^2$, and we have

$$\frac{d\theta}{d\omega} \approx \frac{-2Q}{\omega_o},$$

so we finally obtain the desired expression,

$$Q = \frac{\omega_o}{2}\left|\frac{d\theta}{d\omega}\right|_o. \tag{5-17}$$

In Eq. (5-17) the derivative is evaluated at resonance. In actual measurements you must use a small frequency interval near resonance, $\Delta\theta/\Delta\omega$. As the frequency interval increases, the measured value of Q appears to decrease because of the curvature of the arctangent.

Figure 5-2 shows details of Q measurement by $\Delta\theta/\Delta\omega$. A similar measurement of Q can be made using the phase of velocity or current [see Eq. (4-84) and Table 4-2]. The velocity or current phase is given by

$$\varphi(\omega) = \arctan\left(\frac{-(\omega^2-\omega_o^2)}{2K\omega}\right), \tag{5-18}$$

and the corresponding Q is obtained as,

$$Q = \frac{\omega_o}{2}\left|\frac{d\varphi}{d\omega}\right|_o. \tag{5-17}$$

Figures 5-3 and 5-4 show power resonances and phases for several values of K.

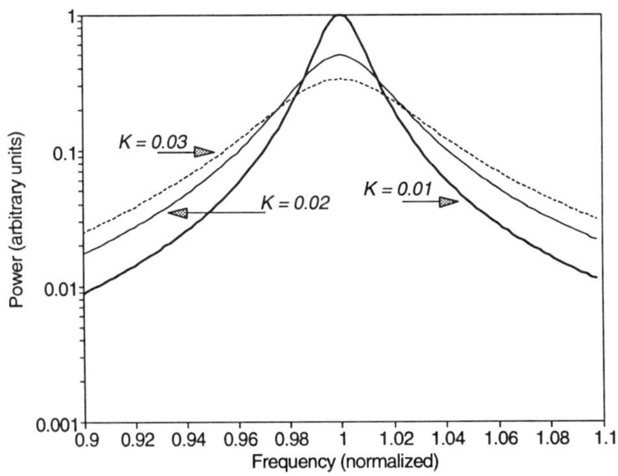

Figure 5-3. Power resonance on a logarithmic scale for $Q = 50$ *(thick line)*, $Q = 25$ *(thin line)*, and $Q = 16.67$ *(dotted line)*.

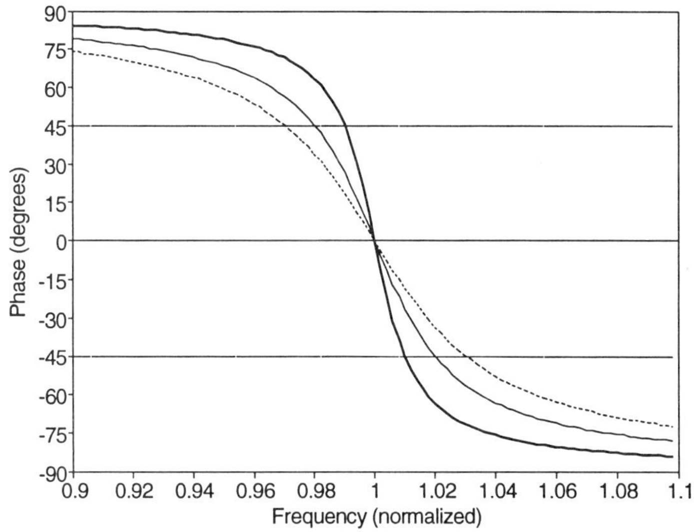

Figure 5-4. Phase of velocity or current for the same values of Q shown in Fig. 5-3. Notice the phase linearity close to resonance, and the increasing nonlinearity as you move away from resonance.

Loss Tangent and Dissipation Factor

The *loss tangent* and *dissipation factor* are often used to describe bulk dielectric and magnetic materials. The loss tangent is the reciprocal of Q,

$$\tan(\delta) = \frac{1}{Q}. \tag{5-19}$$

(Don't confuse the angle δ used here with the Dirac delta function!) In the case of a dielectric with a complex permittivity ($\varepsilon = \varepsilon_r + j\,\varepsilon_i$) the loss tangent is given by

$$\tan(\delta) = \frac{\varepsilon_i}{\varepsilon_r}. \tag{5-20}$$

The dissipation factor D is the angle $\delta = \arctan(1/Q)$. At microwave frequencies, dielectrics are often measured with a high-Q cylindrical resonant cavity, in the TM_{010}, for example. A thin cylinder of the material under test is placed in the cavity, and the dielectric properties of the material are calculated from the resonance frequency shift and the change in Q from the values of the empty cavity.

Some loss tangents of dielectric and magnetic materials are shown in Table 5-1. Compare the loss tangents of the high-quality materials in Table 5-1 with resin-molded chip tantalum electrolytic capacitors [$\tan(\delta)$ of 0.04 to 0.08], and aluminum electrolytic capacitors [$\tan(\delta)$ of 0.08 to 0.17].

Table 5-1. Dielectric Constants and Loss Tangents of Some Materials

Material	$\varepsilon_r/\varepsilon_o$	*Loss Tangent*
Microwave dielectric, type ATD-20	100.0	< 0.0003
Resonator dielectric, type ATD-40R	38.0	0.0002
Record head slider dielectric, type ATD-48S. NiZn ferrite	48.0	0.0002
Polytetrafluoroethylene (PTFE) (Teflon)	2.1 60 Hz – 10 GHz	< 0.0005
Fused silica	3.85	0.0001
Mg-Mn microwave ferrite, type 12-1750. Line width at –3 dB: 17.9 kA/m	12.3 (9 – 10 GHz)	< 0.0005

5.3 Time Domain

Using Eq. (5-10) we can write the solution for $x(t)$ in terms of Q. For simplicity, and without loss of generality, let's take the initial conditions as $x_o = 0$ and $v(0) = v_o$. Equation (3-21) becomes,

$$x(t) = \exp[-(\omega_o/2Q)\,t]\left\{(v_o/\omega_o)\,[1 - 1/(4Q^2)]^{-1/2}\right\}\sin[\omega_o\sqrt{1 - 1/(4Q^2)}\;t]\,. \quad (5\text{-}20)$$

which looks a bit cumbersome, but which enables us to measure Q quite simply in the time domain. Consider the maximum displacements on two succesive swings, Let the first displacement be A_n and let the second displacement be A_{n+1}. We can express the amplitude decrement (Δ) per period (T) as the ratio,

$$\Delta = \frac{A_{n+1}}{A_n} = \frac{\exp[-(\omega_o/2Q)(t+T)]}{\exp[-(\omega_o/2Q)t]}\,, \quad (5\text{-}22)$$

so,

$$\Delta = \exp[-(\omega_o/2Q)T]\,. \quad (5\text{-}23)$$

Using the period T from Eq. (5-20), we can express Δ as

$$\Delta = \exp\left[\frac{-\pi}{Q\sqrt{1-1/(4Q^2)}}\right]. \tag{5-24}$$

For $Q^2 > 10$, which certainly describes the range of interest here, Eq. (5-24) becomes

$$\Delta \approx \exp[-\pi/Q]. \tag{5-25}$$

Now we can measure Q in the time domain as

$$Q = \left|\frac{\pi}{\ln(\Delta)}\right|. \tag{5-26}$$

Figure 5-5 shows measurements of Q by the logarithmic decrement.

High-Q systems have their good points and their bad points. Some of their good points can also be bad points in some situations. Some good points are high frequency stability and energy storage capacity, but these were bad in the case of the collapse of the suspension bridge across Tacoma Narrows in the state of Washington, November 7, 1940, just four months after the bridge was opened. The bridge was set into oscillation on a clear day by ordinary air currents, and eventually the high-Q oscillations caused catastrophic failure. In 1989, an earthquake (7.1 on the Richter scale) near Loma Prieta, caused resonance oscillations (≈9 rad/s) in the Nimitz Freeway at Oakland, California, 100 km away. A 1.4-km segment of the Freeway collapsed, killing 67 people. The mechanical impedance (see Section 5.11) of the infrastructure of this section of the Freeway was unusually low. This is thought to have contributed to the large velocity-resonance oscillations.

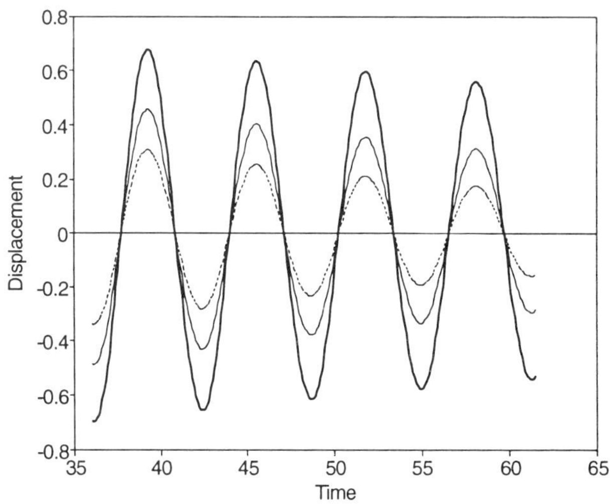

Figure 5-5. Q measurement by logarithmic decrement. Decrement (Δ) is the ratio of two adjacent maxima, A_n/A_{n+1}. *Thick line:* $Q = 50$. *Thin line:* $Q = 25$. *Dotted line:* $Q = 16.67$. Same Q values are in Figs. 5-3 and 5-4.

5.4 *Q* of Various Oscillators

The range of known Q values for oscillators is about 10^{26}. Some selected systems are shown in Table 5-2.

Table 5-2. Typical *Q*

Oscillator	*Range of Q*
LCR resonator	$1 - 10^3$
Mechanical pendulum	$1 - 10^5$
Microwave resonant cavity	$10^4 - 10^5$
Quartz crystal (selected)	$10^6 - 10^7$
Rotation of planet Earth	10^{13}
Laser, maser	10^{14}
Superconducting microwave cavity	10^{15}
Mössbauer effect, $^{26}Fe^{57}$, $^{77}Ir^{191}$	$10^{11} - 10^{26}$ (?)

It may seem odd to include planet Earth in Table 5-2, but this is reasonable if you consider the energy loss per rotation of the planet. Energy loss processes include friction due to Earth's moon and Sun gravitationally induced periodic tidal motion in water and land masses. Random bombardment by meteorites changes Earth's mass. These inelastic collisions produce changes in Earth's kinetic energy and angular momentum. This is somewhat similar to Fig. 3-20, where the rats' feet hit the wheel randomly. However, in Fig. 3-20 most of the collisions result in a transfer of energy from the rats to the wheel, and presumably only a few out-of-phase rats would receive energy and angular momentum from the wheel.

Planets, moons, and stars can be described by many other values of Q depending on the particular mode of oscillation. For example, planet Earth rings like a bell in transverse and longitudinal modes when it is excited by earthquakes, volcanic eruptions, nuclear weapon tests, and meteor impacts. Planet Earth had observable vibrations for about 30 days following an earthquake in 1960 that was located in Chile. This earthquake had a magnitude of 8.3 on the Richter scale.

R. H. Miller (University of Chicago) and B. F. Smith (NASA Ames Research Center) reported in 1996 that their computer models of about 4×10^5 masses interacting through gravitational force indicate that galaxies (typically more than 10^9 stars) may vibrate in characteristic modes. These oscillations are predicted to have periods of millions to hundreds of millions of years and could continue practically undamped for thousands of millions of years.

Standing acoustic wave oscillations are known to exist on the surface of our Sun; more than than 10^7different modes have been identified. The Sun's surface appears to oscillate with a period of about five minutes.

High-Q superconducting microwave cavities are made by plating superconducting material on nearly perfect single crystals of sapphire. The Q is affected by minute crystal imperfections such as dislocations and impurities. Even different isotopes of aluminum in the sapphire can be considered as impurities. Using mostly one isotope has been found to increase Q significantly because nuclei then have mostly the same mass.

Dielectric resonator oscillators (DROs) are common in microwave devices. Filling a resonant cavity with a low-loss dielectric can increase Q (and therefore frequency stability) significantly because losses in metallic conduction paths are reduced as the size of the cavity is reduced.

5.6 Resonant Detection, Mössbauer Effect

The Mössbauer effect, the oscillator with the largest known Q, was used by R. V. Pound and G. A. Rebka, Jr. to measure the effect of gravity on photons ["Apparent weight of photons," *Phys. Rev. Letters* **4**, 337–341 (1960)]. The Harvard group used a tower in the Jefferson Physical Laboratory, where they carried out a series of experiments with their source at the top of the tower and a series with their source at the bottom. In this way, photons would fall toward Earth, like Galileo dropping balls from the leaning Tower of Pisa, and photons would be projected up, like throwing a ball to someone in a tree.

Since photons move with the speed of light, an energy change would be observed in terms of a frequency up-shift when falling and a down-shift when rising. This fundamental experiment, hypothesized by Einstein, has profound implications for our concepts of the structure of spacetime. The change in frequency is very small for short distances, unless you are near a cosmic black hole.

Fortunately, Rudolf Ludwig Mössbauer had discovered that some nuclear transitions had enormous Q values when the host atoms were held in a crystal, so that most of the recoil momentum was absorbed by the macroscopic crystal instead of a single nucleus. Mössbauer discovered this effect when he was a graduate student, and received the Nobel Prize for this work in 1961.

Using the Mössbauer effect, the Doppler shift is minimized when the photon is emitted and the apparent spectral line width is extraordinarily small. Pound and Rebka used Fe^{57} with a Lorentz line shape that had a FWHM of 1.13×10^{-12}. Air absorption in the tower was reduced by letting the photons propagate in a Mylar bag filled with helium. Temperature differences, which could have been a source of error due to thermal motion in the crystals, were minimized.

The source was 0.4 curie of 270-day Co^{57}; the detector was an absorbing film of Fe^{57} with electroplated Co^{57}on the side away from the source. Because of the narrow line width, there was little absorption in the iron unless the frequency of the photons matched the frequency of the absorbing film, so the resonant condition corresponded to minimum counts at the detector. The experiment measured relative frequency shift by moving the detector at variable speed in both directions until the Doppler shift matched the incoming photons. In other words, a down-shifted photon could be detected by motion toward the source, and an up-shifted photon could be detected by motion away from the source. The original paper of Pound and Rebka is worth reading, not only because of its fundamental result, but also as an example of ingeniuty and careful experimental technique.

5.7 *Q* Multipliers and Dividers

After all energy losses have been reduced as much as possible, an obvious way to increase Q is to use an active feedback system to restore some of the energy that is being lost. This is the basic idea underlying "Q multipliers," which find application in electronics. Stability is a major problem with Q multipliers; you must be careful about the amplitude and phase of the feedback to avoid runaway oscillations.

A mechanical Q multiplier is often used with the Foucault pendulum. In 1851 Jean Lèon Foucalt used a simple pendulum to demonstrate that it was indeed the Earth that was rotating, and not the rest of the universe. The major forces acting on a simple pendulum are gravity and air resistance. A pendulum will swing in a constant plane as the Earth rotates, giving an observer attached to Earth the impression that the plane of the pendulum is changing. To keep the pendulum swinging with fairly constant amplitude throughout the day a sensor is used to activate a mechanical device to add a little kick to the wire support when the amplitude drops below a preset value. Careful tuning makes the Q appear to be infinite, because the amplitude does not decrease with time.

In some situations a high Q is undesirable. For example, a high-Q oscillator is difficult and slow to start and stop. An analytical balance uses an aluminum plate moving in the field of a permanent magnet to decrease the Q and bring the balance to its equilibrium position in a short time. The motion of the conductor in the magnetic field induces eddy currents which remove kinetic energy by $I^2R\Delta t$ losses in the conducting plate. This type of Q reduction uses a true velocity-dependent process, the Lorentz force $q(\mathbf{v} \times \mathbf{B})$, and it avoids the stickiness of friction.

Active feedback is used to reduce Q in optical tables and in some architectural structures. The famous collapse of the Tacoma Narrows suspension bridge could have been averted by Q reduction.

In their version of the Eötvos experiment, P. G. Roll, R. Krotkov, and R. H. Dicke ["The equivalence of inertial and passive gravitational mass," *Annals of Physics* **26**, 442–517 (1964)] used feedback damping to lower the Q of their torsion oscillator. To reduce pressure fluctuations due to Brownian motion, the oscillator was operated in a vacuum of 10^{-8} mm Hg and originally had a time constant of about one year. Active feedback damping certainly speeded up their measurements.

5.8 *Q* Switches

High-Q systems can store large amounts of energy; if this can be dumped in a short time then very high-power pulses can be produced. A "Q switch" or "Q spoiler" is a device that decreases Q very quickly. These find use in generating powerful pulses from lasers. For example, a dye cell can be inserted in a laser cavity; when the cell becomes lossy the previously high-Q laser cavity dumps its energy in a very short time. One Joule dumped in a nanosecond is 10^9 Watts.

Energy stored in rotational motion of a large flywheel can be dumped rapidly with a clutch connected to an electric generator. This is used to provide large amounts of energy for short times for linear electric motors used to accelerate objects such as aircraft at take-off.

5.9 Classical Uncertainty Principle, Part 1

In the time domain, the envelope of energy of a decaying oscillator falls to $1/e$ of any initial value after a time $\Delta t = Q/\omega_o$. Let's use this as a characteristic time interval Δt. The bandwidth $2K$ is an easily-measured characteristic frequency interval, on the energy (and power) resonance curve, which we have already called $\Delta\omega$. What is the $\Delta\omega\Delta t$ product?

$$\Delta\omega\,\Delta t = \frac{\omega_o}{Q}\frac{Q}{\omega_o} = 1. \tag{5-27}$$

Another way of looking at this is,

$$\Delta t = \frac{1}{\Delta\omega}, \tag{5-28}$$

which says that the time interval for a significant change in energy to occur is inversely proportional to the bandwidth. Of course, the numerical value of 1 in Eq. (5-28) depends on how the functions are distributed in time and frequency, and how you define the deltas. Here we have been dealing with an exponential decay and its Lorentz resonance shape. In Chapter 6 we will see that the *best* you can do is,

$$\Delta\omega\,\Delta t \geq 0.5, \tag{5-29}$$

and this is only true for Gaussian functions. We will see something unique: If a physical quantity is distributed like a Gaussian in time, then it is also distributed like a Gaussian in frequency, and conversely.

This is quite a challenge, to say that you cannot measure frequency and time as precisely as you choose. Equations (5-28) and (5-29) say that the longer the time interval (Δt) that you measure a frequency, the more precise will be your knowledge of the frequency. That is, the smaller $\Delta\omega$ will be. But, if you use a long time interval then you will not know what time the system had the frequency that you measure, and if you try to pin down the time then you lose precision in frequency.

This is the way Nature works, and many people have tried to beat the classical uncertainty relation, but no one has yet succeeded. This doesn't mean you shouldn't try, but don't use your own money. We will look more deeply into this question from a different point of view in Chapter 6, and eventually we will get to the quantum uncertainty principles. Actually, Eq. (5-29) is a disguised quantum uncertainty principle for energy and time, in which Planck's constant is hidden because it cancels out. We can multiply both sides of Eq. (5-29) by Planck's constant (actually $h/2\pi$) and use our knowledge of adiabatic invariance and Eq. (2-55) or (2-56), but we don't need to do this now.

We have seen glimpses of another uncertainty relation ($\Delta p\,\Delta x$) in Eqs. (1-62) and (2-29), in the context of phase space. In an uncertainty relation the product of two uncertainties is equal to, or greater than, some constant. The value of the constant is not so important, because it depends on how you define the uncertainties and what system of units you use. The important point is that the product of the uncertainties is not zero. Sharpening one uncertainty dulls the other. *There is a connection between quantities related by uncertainties.*

A high-Q system behaves like a mechanical fly-wheel; neither can change rapidly in time. Generally you have to wait some time interval Δt before you can know with confidence that a significant change has occurred.

5.10 ABC-OF-Q, Worksheet in Time and Frequency

This worksheet lets you experiment with Q in the time and frequency domains. Also explore Q-IN-T-F and Q-OF-LCR. You will find that squeezing a function in one domain expands it in the other domain. The Home screen is shown in Fig. 5-6. If you do not see this, press the Home key or the Ctrl Home keys, depending on your spreadsheet. In Microsoft *Excel*, Ctrl Home takes you to cell A1, whereas the Home key by itself takes you to Column A of the particular row that you are in.

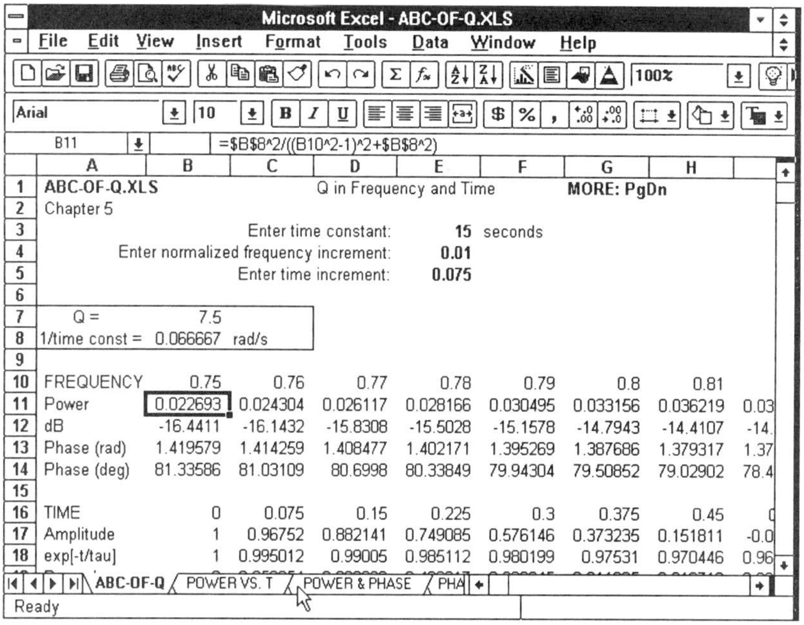

Figure 5-6. Home screen of the ABC-OF-Q worksheet.

5.11 Impedance and Admittance

Impedance and admittance are important because they relate cause and effect, and (like Q) energy storage and loss parameters. Impedance and admittance are general concepts that are applicable to many kinds of systems. We define impedance as

$$Z = \frac{Cause}{Effect} = \frac{Stimulus}{Response} . \tag{5-30}$$

In a mechanical system, we define force F as the cause and velocity v as the effect. In an electrical system, voltage V is the cause and current I is the effect,

$$Z = \frac{V}{I} \quad \text{so} \quad V = IZ. \tag{5-31}$$

Often we are interested in writing an expression for the effect as

$$I = \frac{V}{Z}. \tag{5-32}$$

Since all of these are usually complex quantities, it is convenient to define the admittance Y as the reciprocal of Z to avoid complex division,

$$Y = \frac{1}{Z}, \tag{5-33}$$

so

$$I = YV. \tag{5-34}$$

This says that the effect is proportional to the cause. In general, impedance consists of a real part R (resistance), which transforms ordered energy into disordered energy, and an imaginary part X (reactance) which stores energy,

$$Z = R + jX. \tag{5-35}$$

So, admittance can be written as

$$Y = \frac{1}{Z} = \frac{1}{R + jX}. \tag{5-36}$$

Note that we are using X here to represent reactance, *not* displacement as in $x = |X(\omega)|e^{j\omega t}$. It is useful to separate admittance into its real and imaginary parts,

$$Y = \frac{1}{(R + jX)} \frac{(R - jX)}{(R - jX)} = \frac{R}{R^2 + X^2} - j\frac{X}{R^2 + X^2}, \tag{5-37}$$

where the real part is called G, conductance, and the imaginary part is called B, *susceptance*. In mechanical systems B is often called *compliance*.

$$Y = G + jB. \tag{5-38}$$

Don't confuse the electrical *conductivity* σ with the *conductance* G. Also, don't confuse the electrical *susceptance* χ_e with the *susceptance* B.

- In impedance Z, resistance R and reactance X are completely separated into the real and imaginary parts of Z.
- In admittance Y, resistance and reactance are mixed so that both R and X appear in the real and imaginary parts of Y.
- Conductance G is approximately $1/R$ when $R^2 >> X^2$.
- Susceptance B is approximately $1/X$ when $X^2 >> R^2$.
- The imaginary part of Y has the opposite sign with respect to the imaginary part of Z, so the phase angles have opposite signs.

5.12 Impedance and Admittance: Damped Mass

For simplicity, let's consider a forced, damped mass before we examine the oscillator. We can write Newton's second law for this system as

$$m\frac{dv}{dt} + Rv = F(t), \tag{5-39}$$

and, as we have done before, let's try a phasor solution with $F(t) = F_o e^{j\omega t}$ and $v = |v(\omega)| e^{j\omega t}$. Equation (5-39) now can be written as

$$j\omega m v + R v = F_o, \tag{5-40}$$

and we can solve for F/v (which is Z)

$$Z(\omega) = \frac{F}{v} = R + j\omega m. \tag{5-41}$$

It is also useful to express $Z(\omega)$ in polar form in terms of magnitude and phase, $Z(\omega) = |Z(\omega)| \exp[j\theta(\omega)]$,

$$|Z(\omega)| = \sqrt{R^2 + (\omega m)^2} \quad \text{and} \quad \theta(\omega) = \arctan(\omega m/R). \tag{5-42}$$

There is an exact analogy with an LR circuit, with the input across L and R in series and the output across R. This is like a simple low-pass filter with time constant L/R or m/R seconds. The quantity $j\omega m$ is analogous to inductive reactance $j\omega L$.

For the corresponding admittance in Cartesian form we have

$$Y(\omega) = \frac{1}{Z(\omega)} = \frac{R}{R^2 + (\omega m)^2} - j\frac{\omega m}{R^2 + (\omega m)^2}, \tag{5-43}$$

and in polar form,

$$|Y(\omega)| = \frac{1}{[R^2 + (\omega m)^2]^{1/2}} \quad \text{and} \quad \theta(\omega) = \arctan(-\omega m/R). \tag{5-44}$$

Impedance and admittance for this case are shown in Figs. 5-7 and 5-8.

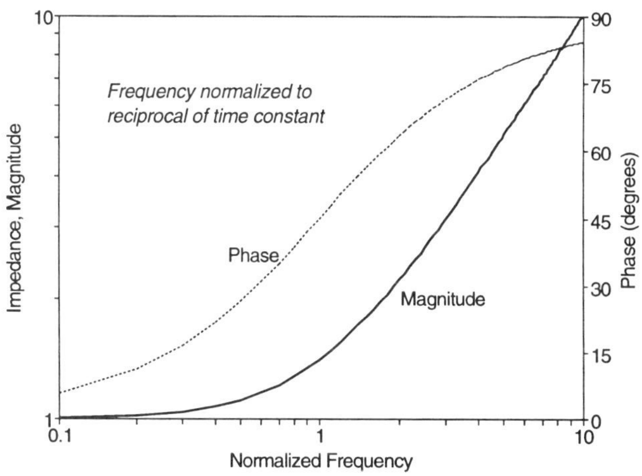

Figure 5-7. Impedance of a forced, damped mass. Note log scales on x-axis and left y-axis. Reciprocal of time constant is the "corner" or "cut-off" frequency.

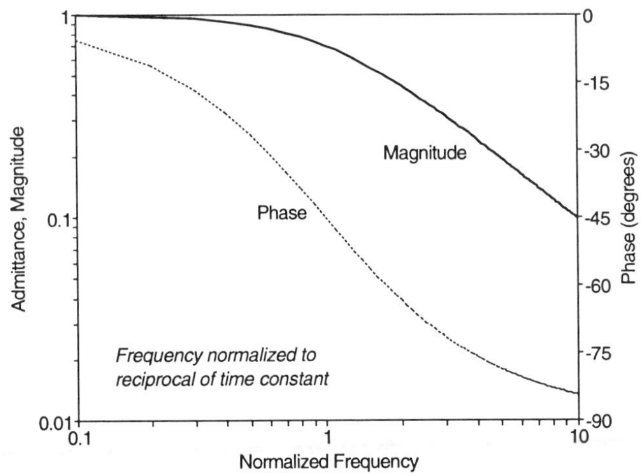

Figure 5-8. *Y* of a forced, damped mass. Note log scales on *x*-axis and left *y*-axis (Bode plot). This is similar to the gain of a single-pole low-pass filter.

5.13 Impedance and Admittance: Harmonic Oscillator

Now that we understand the forced, damped mass it is just a few more steps to do the complete oscillator. Newton's second law can be written conveniently as

$$m\frac{dv}{dt} + Rv + \kappa x = F(t). \tag{5-45}$$

With a phasor solution, $v = |v(\omega)| \exp[j\omega t]$, so $x = \int v\, dt \rightarrow v/j\omega$, and we have

$$j\omega m v + Rv + \kappa\frac{v}{j\omega} = F_o. \tag{5-46}$$

As before, we solve for $Z = F/v$ and after a little algebra we have

$$Z(\omega) = \frac{F}{v} = R + j(\omega m - \kappa/\omega). \tag{5-47}$$

As expected, the imaginary part goes to zero at $\omega^2 = \omega_o^2 = \kappa/m$. In polar form,

$$|Z(\omega)| = \sqrt{R^2 + (\omega m - \kappa/\omega)^2]} \tag{5-48}$$

and

$$\theta(\omega) = \arctan\left(\frac{\omega m - \kappa/\omega}{R}\right). \tag{5-49}$$

In an *LCR* resonator this has the familiar form

$$|Z(\omega)| = \sqrt{R^2 + (X_L - X_C)^2} \tag{5-50}$$

and

$$\theta(\omega) = \arctan\left(\frac{X_L - X_C}{R}\right), \qquad \text{(5-51}$$

where $|X_L| = \omega L$ is the inductive reactance and $|X_C| = 1/\omega C$ is the capacitative reactance. At resonance ($X_L = X_C$) the impedance is purely resistive and its phase angle is zero. Figures 5-9 and 5-10 show impedance in Cartesian and polar forms.

Now we can write admittance as

$$Y(\omega) = \frac{1}{Z(\omega)} = \frac{v}{F} = \frac{1}{R + j(\omega m - \kappa/\omega)}. \tag{5-52}$$

This is more easily understood when the real and imaginary parts are separated,

$$Y(\omega) = \frac{R}{R^2 + (\omega m - \kappa/\omega)^2} - j\frac{(\omega m - \kappa/\omega)^2}{R^2 + (\omega m - \kappa/\omega)^2}. \tag{5-53}$$

Finally, in polar form we have the magnitude and phase angle,

$$|Y(\omega)| = \frac{1}{\sqrt{R^2 + (\omega m - \kappa/\omega)^2}} \tag{5-54}$$

$$\varphi(\omega) = \arctan\left[\frac{-(\omega m - \kappa/\omega)}{R}\right]. \tag{5-55}$$

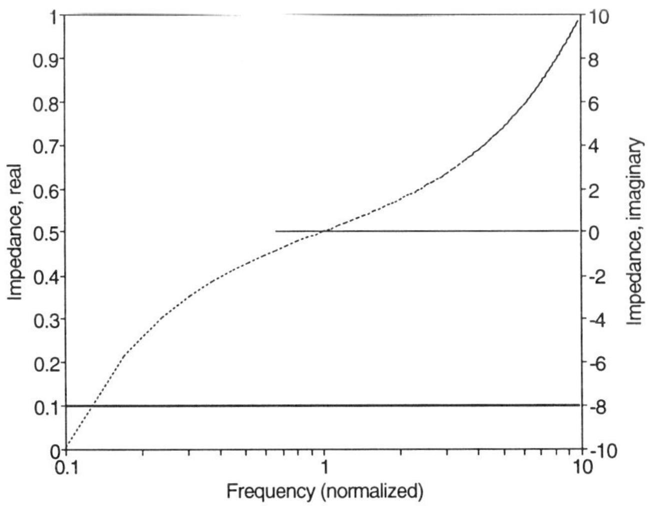

Figure 5-9. Real *(thick line)* and imaginary *(dotted line)* components of harmonic oscillator impedance near resonance. Here, $R = 0.1$ and $Q = 10$.

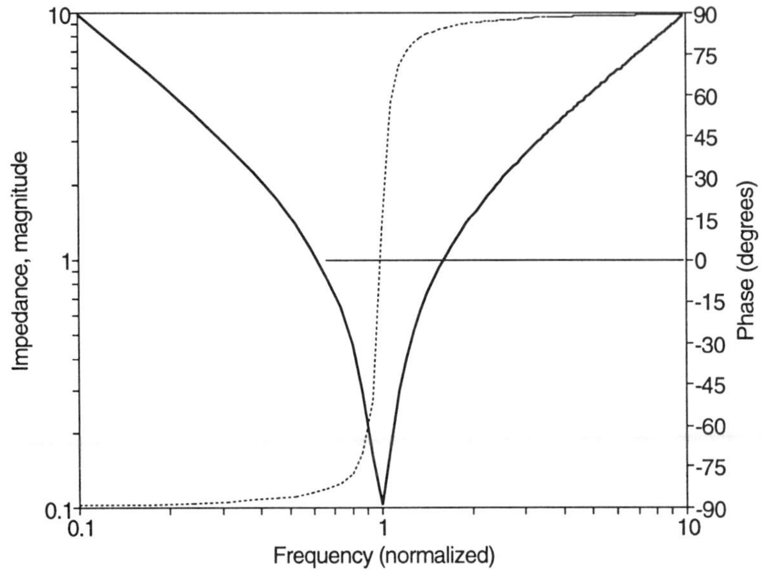

Figure 5-10. Magnitude *(thick line)* and phase *(dotted line)* of harmonic oscillator impedance near resonance, for same parameters used in Fig. 5-9.

Figure 5-11 shows the real and imaginary parts of admittance, Fig. 5-12 shows the magnitude and phase, and Fig. 5-13 shows the polar plot.

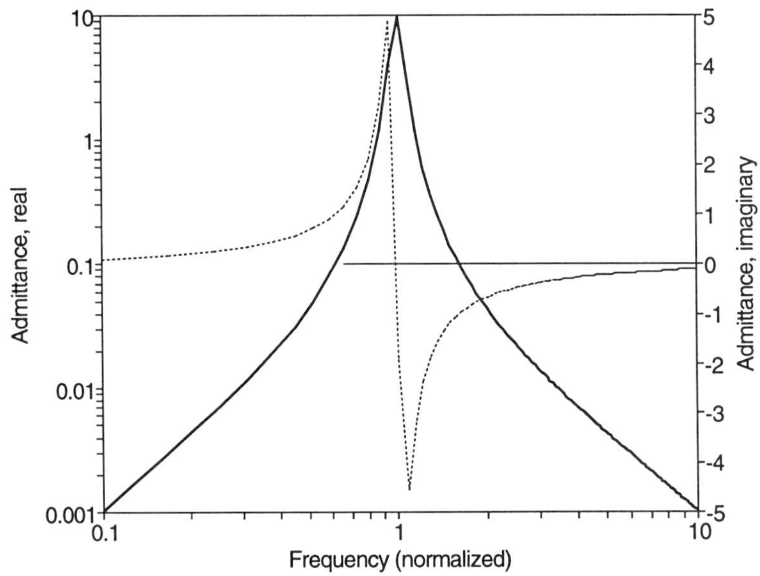

Figure 5-11. Real *(thick line)* and imaginary *(dotted line)* parts of admittance for a harmonic oscillator. In this plot, $Q = 10$.

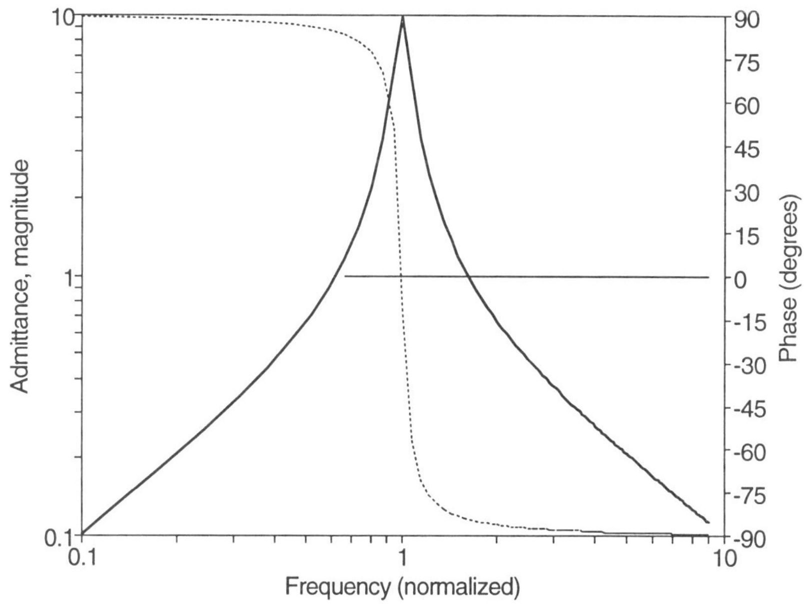

Figure 5-12. Magnitude *(thick line)* and phase *(dotted line)* of admittance for a harmonic oscillator, $Q = 10$. Note similarity to a single-pole band-pass filter.

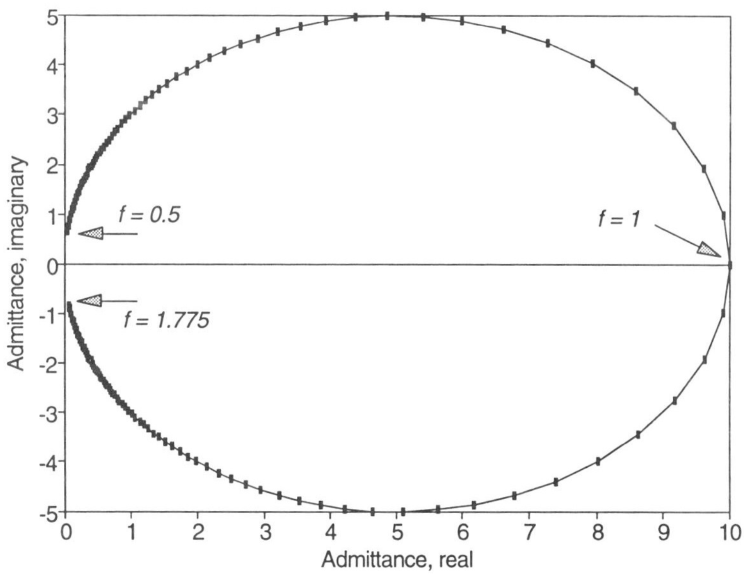

Figure 5-13. Polar plot of admittance for a harmonic oscillator, $Q = 10$. In this plot, $\Delta f = 0.005$ between markers. Normalization is with respect to the resonant frequency. This plot uses the same data shown in Fig. 5-11.

5.14 Y-AND-Z, Admittance and Impedance Worksheet

Using this worksheet you can change system parameters to experiment with admittance and impedance. The Home screen is shown in Fig. 5-14.

5.15 How Henry Cavendish Weighed Planet Earth

Henry Cavendish (1731–1810) was an English physicist and chemist whose name is usually associated with determining the universal gravitational force constant *G*. The Cavendish experiment is an example of use of a high-*Q* oscillator to make very sensitive measurements. In 1996 Dr. Gabriel G. Luther (Los Alamos National Laboratory) was honored by the American Physical Society for his ingenious and precise measurements of *G*.

Cavendish's objective was not to measure *G*. He wanted to determine the mean density of planet Earth. Let's put Cavendish's experiment in modern form and see how he could have measured *G*. Newton (1642–1727) had previously published his law of universal gravitation in 1665, when he was 23 years old,

$$F_g = G\frac{Mm_g}{r^2} \; . \tag{5-56}$$

From Newton's second law of motion the dynamic force F_d and dynamic (inertial) mass m_d are related to the acceleration by,

$$F_d = m_d\, a. \tag{5-57}$$

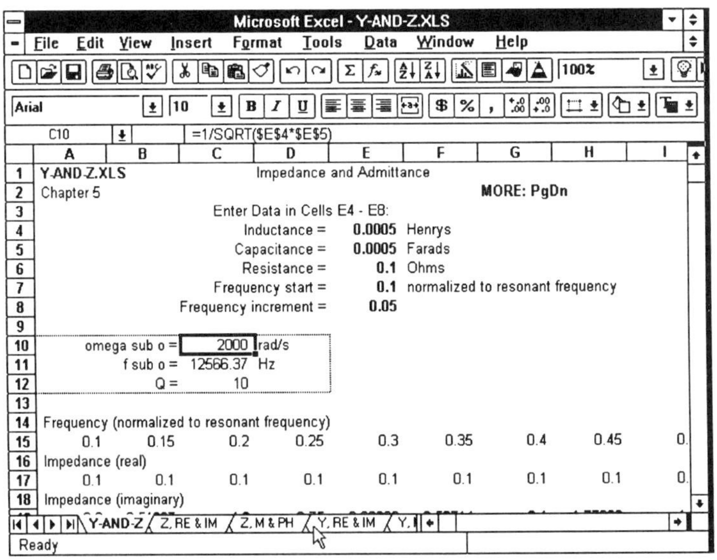

Figure 5-14. Home screen of Y-AND-Z worksheet.

If dynamic mass and gravitational mass are identical, this has profound implications. Einstein, in his principle of equivalence, postulated that a gravitational field is indistinguishable from an accelerating coordinate system. Apparently, Newton and Cavendish believed that inertial mass and gravitational mass were the same. Newton performed experiments with a simple pendulum that showed the masses were equal to within about 10^{-3}. If we equate inertial force and gravitational force we can write

$$m_d\, g = G\frac{Mm_g}{r^2}\,, \tag{5-58}$$

where g is the gravitational acceleration, and the inertial mass m_d on the left-hand side cancels the gravitational mass m_g on the right-hand side.

(The experiments of Baron Roland von Eötvos in 1890, and later those of Roll, Krotkov and Dicke in 1964, show that inertial mass and gravitational mass are the same, to within a few parts in 10^{-12}. These experiments used torsion oscillators.)

Let's assume we are at the surface of planet Earth, $r = R$, where R is the radius of the planet. The radius of planet Earth, assumed spherical, can be easily measured by means of the noon-day shadows of vertical rods stuck in the ground at the same longitude but different lattitudes; the Greek astronomer Eratosthenes did this in Egypt approximately 2200 years ago. (So, why did people think the Earth was flat?) Today we measure Earth's shape using satellites. If dynamic mass is the same as gravitational mass, then we can solve for M,

$$M = \frac{gR^2}{G}\,. \tag{5-59}$$

Therefore, measurements of g, R, and G are necessary and sufficient to measure the mass of planet Earth. Knowing the mass (5.98×10^{24} kg) and the mean radius of the sphere (6.37×10^6 m), you can compute the mean density of planet Earth as 5.5×10^3 kg/m^3. The density of Earth's crust is only about 2.8×10^3 kg/m^3, so Earth must have a very dense core. It is now believed that the outer core density is about 11×10^3 kg/m^3 and the inner core density is about 13×10^3 kg/m^3. Iron, as we know it in the laboratory, has a density of only 7.9×10^3 kg/m^3.

Cavendish's torsion oscillator consisted of a beam 73.3 inches long. At each end of the beam were lead spheres about 2 inches in diameter. This arrangement was suspended by a fiber about 39.25 inches long, and the whole thing was enclosed in a box to reduce effects of air currents. Outside the box were two lead spheres, each weighing 348 pounds. Various suspension fibers were used, giving periods of from 15 minutes to 7 minutes.

A typical modern Cavendish torsion oscillator, designed by Dr. Randall Peters and available from Central Scientific Co., uses two 0.015 kg lead balls at the ends of a light aluminum rod suspended at its center on a 25 μm tungsten wire, about 65 mm long. The oscillator has a period of about 200 seconds, and displacement is sensed with a symmetrical differential capacitance bridge. The gravitational constant is measured by the attraction of the small lead balls to two large lead balls, each of 1 kg mass.

Another modern version, available from Leybold Didactic GMBH, uses a bronze torsion band with a torque constant of 8.5×10^{-9} N·m/rad. There are two 0.015 kg lead masses, each mounted 50 mm from the axis of rotation, and attracted by two 1.5 kg lead masses, each 64 mm in diameter. The period of oscillation is approximately 10 minutes. Both of these torsion oscillators can be interfaced to computers for data acquisition and analysis.

Often, people gasp with astonishment when they see a torsion oscillator begin to move under the gravitational forces between these small masses. The oscillator of Roll, Krotkov, and Dicke could detect the presence of a 100 kg man at a distance of 30 m. Such high sensitivity is needed because the Eötvos experiment seeks a null result. These experiments are best carried out in complete isolation using telemetry to collect data. We don't usually think about the gravitational attraction between small objects because our local gravitational field due to planet Earth, its moon, and the Sun dominate our lives, and much of biology, through the daily tides in sea and land masses.

In 1995 three research groups in various parts of the world produced experimental measurements of *G* that were significantly different from each other and different from the currently accepted value. This caused considerable distress. In 1996 several new experiments were planned in the hope of resolving this problem. R. E. Newman and his colleagues (University of California, Irvine) propose to use a thin quartz plate, suspended by a torsion fiber, between two vertical copper rings. Measurement of *G* is accomplished by measuring the shift in the oscillator frequency when the pair of rings is rotated 90°. The oscillator will be operated at a temperature of 4K to minimize thermal excitation (see Section 10.5). J. H. Gundlach (University of Washington, Seattle) also plans to use a quartz plate in a torsion oscillator, but in his experiments it will be mounted on a rotating platform. Eight spherical masses arranged around the platform will perturb the oscillator as it rotates and *G* can be measured from this perturbation.

Why are people so interested in *G*? First, it is a fundamental constant of the universe that is associated with cosmic structure on a grand scale. The value of *G*, as of today, cannot be derived from any theory. Second, *G* is somehow related to the fundamental micro-structure of spacetime in ways that are not now understood; the so-called Planck length is $(Gh/c^3)^{1/2} \approx 10^{-35}$m and the Planck time is $(G/c^5)^{1/2} \approx 10^{-43}$s. Here, *h* is Planck's constant and *c* is the speed of light in vacuum. Third, according to Mach's principle (which is still open to question) the inertial mass of a body is determined by the total distribution of matter in the universe, so mass could be a tensor instead of a scalar. [See H. G. Robinson, V. W. Hughes, and V. Beltran-Lopez, "Upper limit for the anisotropy of inertial mass from nuclear resonance experiments," *Phys. Rev. Letters* **4**, 342–344 (1960).] It is remarkable that we can investigate such fundamental questions of cosmic importance with simple harmonic oscillators.

In Chapter 9 we will examine coupled oscillators. Coupling may occur by many physical processes, but gravitation is one of the few forces that cannot be shielded or cancelled in some way, and gravity is a long-range force. Even though gravity is weak compared to electromagnetic forces, on a cosmic time scale gravity couples just about everything. Table 5-3 displays some interesting numerical data concerning gravitation in our solar system. It is not generally appreciated that the sun, our moon, and planets Jupiter and Venus have a greater effect on a new-born baby on planet Earth than does the attending obstetrician. The gravitational effect of planet Mars is less than that of the doctor.

Table 5-3. Gravitational Coupling of Selected Objects

Object	*Mass (kg)*	*Distance (m) to 1 kg object on Earth*	*Force (N) on 1kg on Earth*
Sun	1.99×10^{30}	1.5×10^{11}	5.93×10^{-3}
Earth's Moon	7.3×10^{23}	3.63×10^{8}	3.7×10^{-4}
Jupiter	1.9×10^{27}	6.28×10^{11}	3.21×10^{-7}
Venus	4.87×10^{24}	4.14×10^{10}	1.9×10^{-7}
Mars	6.4×10^{23}	2.27×10^{11}	8.28×10^{-10}
Obstetrician	70	1	4.67×10^{-9}

There is a fundamental connection between an inverse square force like gravitation and the harmonic oscillator. This much was clear in Chapter 1, where we mentioned Galileo's observation of Jupiter's moon Callisto and the journey through the center of planet Earth, because a transformation must exist between the orbital motions in the inverse square field and the corresponding oscillatory motions. It is known that the Coulomb-Kepler (or hydrogen-like) problem in quantum mechanics is equivalent to a four-dimensional harmonic oscillator or a pair of coupled two-dimensional oscillators. In classical mechanics, the systems are related by the Kustaanheimo-Stiefel transformation.

What's Next?

Up to this point, everything we have done has come from Newton's laws. The time evolution of a system was always obtained from direct solution of a differential equation. The external forces to which the oscillators were subjected were simple in form, usually sinusoidal. The frequency domain solution was always obtained by a phasor solution of the same differential equation. Clearly, the information content of both time domain and frequency domain solutions must be equivalent, since no information was added or taken away in the solution for either domain.

It may appear paradoxical that, despite our use of an apparently causal, deterministic representation (Newton's laws and idealized deterministic forces), we were led to an uncertainty relation involving frequency and time. In Section 1.9 we found an uncertainty relation involving momentum and displacement. Uncertainty is something that we would like to minimize in an oscillator. An oscillator should be the essence of predictablity; clocks should be accurate and precise. In other words, Q should be as high as possible for maximum predictability.

Part of this apparent paradox is related to the fundamental process of measurement, and it is a tribute to Newton's laws that uncertainty is an intrinsic property of classical mechanics. This apparent paradox will be partially resolved, at least from a mathematical perspective, in terms of the scale-change property of the Fourier transform, which goes to the heart of the measurement process.

Ultimately, however, the fundamentals of measurement seem to be in the jurisdiction of quantum theory. There is still some disagreement concerning interpretations, such as the Einstein-Podolsky-Rosen (EPR) paradox, Bell's theorem, and some recent "delayed decision" experiments involving information in photons in laser beams that travel different paths. Evidently, Yogi Berra anticipated multi-path laser experiments in his famous oracular statement "When you come to a fork in the road, take it."

These experiments involve another continuing paradox: We can make things that work, like quantum cryptographic systems and secure telecommunications, using things that we do not really understand. This is a thread that runs through the fabric of human history; there were medical doctors for thousands of years before people were aware of the existence of bacteria, viruses, cells, DNA, enzymes, and so on.

In the next chapter we are going to take a point of view that is different from anything we have considered before. We are going to discuss Fourier methods that will enable us to apply external forces that are almost as complicated as we choose. These forces can be periodic, non-periodic, and combinations of both types. Eventually we will be able to apply random, noise-like forces to oscillators. This will be more like the real world because all forces (and information) have an element of unpredictability in them.

Chapter 6

Fourier Perspective

Yesterday was my twenty-first birthday, at that age Newton and Pascal had already acquired many claims to immortality.
Baron Jean Joseph Baptiste Fourier

Several times in previous chapters we mentioned the spectra of harmonic oscillators, but we intentionally did not pursue this concept until now. In Section 3.3 we stated that the under-damped oscillator cannot be characterized by a single frequency. In Chapter 4 you saw that a damped oscillator will respond with a characteristic amplitude and phase when it is driven by a force with a frequency that may be different from its resonant frequency. Now we will see there is a direct relationship between the free decay of a damped oscillator and the Lorentz spectrum shape. These ideas can be understood more clearly and more beautifully from the Fourier perspective. Viewing physical phenomena in the frequency domain is a fruitful and often illuminating procedure, yielding both theoretical and numerical results.

It will appear that we can represent almost any periodic function as an infinite set of discrete harmonic oscillators, and a non-periodic function as an infinite set of harmonic oscillators continuously distributed in frequency.

In this chapter we are going to adopt a philosophical point of view that will be reinforced later as we proceed toward the quantum domain. Previously we have regarded displacement, velocity, momentum, acceleration, force, energy, mass, electric charge, and time as if they were quantities that could be measured as precisely as we choose, and which exist independently of our measurements. Now we are going to regard physical quantities as the *results* of measurements, as *signals* that carry *information* that we can collect. We will find in this chapter, and subsequent chapters, that there are some natural restrictions that are imposed on the information that we can obtain, the precision of the information, and the rate at which we can collect information.

6.1 Signal Classification

We can distinguish two broad categories of signals, periodic and non-periodic. It is convenient to define two sub-categories of non-periodic signals, transient and random. Figure 6-1 shows some typical signals of these types.

A periodic signal is characterized by a period p such that the signal repeats or turns into itself after this interval,

$$f(t) = f(t + p). \tag{6-1}$$

Transient and random signals do not have this property. In the real world, many signals are combinations of periodic, transient, and random components. Certainly, if you measure anything carefully enough you will see a random component.

6.2 Periodic Signals

In the nineteenth century, Fourier made a remarkable discovery while he was developing an analytical theory of heat. He found that almost any periodic function can be represented by sines and cosines and an additive constant. In other words, Fourier found that a periodic function could be decomposed into sines and cosines of various amplitudes, frequencies, and phases; this process is called *Fourier analysis*. Conversely, a periodic function can be constructed from sines and cosines of various amplitudes, frequencies, and phases; this is called *Fourier synthesis*.

Today we accept this intuitively, but it was a radical idea when it was introduced. Some eminent mathematicians (including Lagrange) did not believe that the series converged, and others built their reputations proving the conditions of convergence.

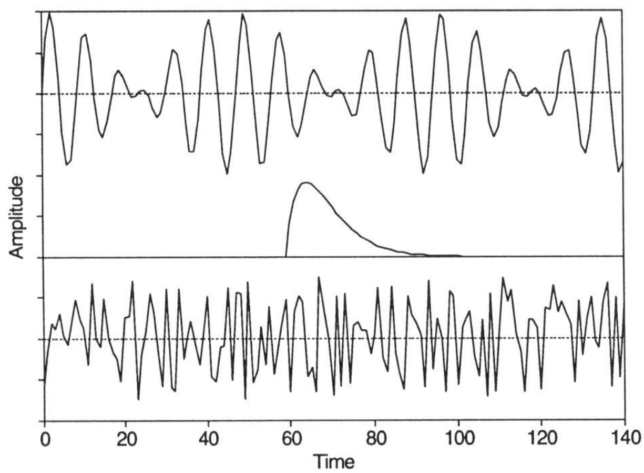

Figure 6-1. Typical Signals. *Top:* Periodic. *Center:* Transient. *Bottom:* Random.

Fourier Series

Fourier series can be expressed in several forms. A form convenient for our present purposes is

$$f(t) = \frac{A_o}{2} + \sum_{n=1}^{\infty}\left[A_n \cos(2\pi nt/p) + B_n \sin(2\pi nt/p)\right]. \tag{6-2}$$

Equation (6-2) looks like a superposition of an infinite collection of discrete harmonic oscillators. We have seen something like this before in Equations (1-5) and (1-14), as a solution for the undamped harmonic oscillator. The term $A_o/2$ is a constant, the "dc" value, which is zero for $[f(t)]_{av} = 0$, that is, zero mean value.

Fourier series can represent a function of any variable. It is not restricted to time. For example, a periodic function in space can be represented as $f(x) = f(x + p)$.

The coefficients A_o, A_n, and B_n are obtained by multiplying Eq. (6-2) by a particular sine or cosine and integrating over the fundamental interval p.

$$A_n = \frac{2}{p}\int_{c}^{c+p} f(t) \cos(2\pi nt/p)\, dt \tag{6-3}$$

$$B_n = \frac{2}{p}\int_{c}^{c+p} f(t) \sin(2\pi nt/p)\, dt, \tag{6-4}$$

where c is a constant. Equation (6-3) is also used to determine A_o for $n = 0$.

This is an example of exploiting the *orthogonality* of the sines and cosines. Here, orthogonality is used in a very general sense. Equations (6-3) and (6-4) select the even and odd parts of $f(t)$ that are parallel to a particular vector labeled n. For example, the dot product (or scalar product) of two vectors $\boldsymbol{P}$ and $\boldsymbol{Q}$ can be written in Cartesian coordinates as

$$\boldsymbol{P}\cdot\boldsymbol{Q} = \sum_{i=1}^{3} P_iQ_i = P_xQ_x + P_yQ_y + P_zQ_z = |\boldsymbol{P}|\,|\boldsymbol{Q}| \cos\theta. \tag{6-5}$$

If $\boldsymbol{P}\cdot\boldsymbol{Q} = 0$ then the vectors are orthogonal. Equation (6-5) selects the part of vector $\boldsymbol{P}$ that is parallel to, or has a projection on, vector $\boldsymbol{Q}$. In this manner, Equations (6-3) and (6-4) can be considered as expressions of orthogonality in time. Fourier series can be thought of as representing a periodic function by means of an infinite set of discrete, mutually orthogonal vectors or harmonic oscillators.

If we represent a vector by its components or projections on some coordinate axes we must include all non-zero components to get an accurate representation of the vector. Similarly, we must include all non-zero Fourier components if we are to get an accurate representation of a periodic function. If we have all components, this is called a *complete set*. You can see incomplete sets in the worksheet FOURIER.

- The Fourier series represents a periodic function in terms of a complete orthogonal set.

It can be shown that, for a given, finite number of terms, the Fourier series is the optimum series representation for a periodic function. Here, "optimum" is used in the sense that no other series has a smaller mean-square error than that of the Fourier series. In other words, the Fourier series is the least-mean-square-error representation of a function, for a given number of terms. This is quite a distinction.

Restrictions on Fourier Series

Fourier series can be used to represent an infinte variety of periodic functions, even discontinuous functions. Lejeune-Dirichlet (who married the sister of the composer Felix Mendelsohn) showed that there are some restrictions on the functions:

- The function must have a finite number of discontinuities in one period.
- The function must have a finite number of maxima and minima in one period.
- The function must be absolutely integrable over a period, that is, the integral must be finite.

For most functions that you are likely to encounter, these restrictions are not very restrictive. Nevertheless, you must be careful to distinguish between a function and its *representation*. One important distinction will show up in the worksheet FOURIER where you will see the so-called Gibbs' phenomenon. This is an overshoot at points of discontinuities in the first derivative of the function. Another distinction occurs at points where the function itself is discontinuous; at these points the series converges to the average value of the function based on points before and after the discontinuity. You can also see this in the worksheet FOURIER, and you can experiment with an incomplete set (missing Fourier components). You can also experiment with the effects of errors in the components, including random errors; this is something that was seldom done before computers but it can be quite informative in analyzing experimental data or generating special functions.

Square Wave

As an example of a square wave, see Fig. 2-10 for the velocity of a classical particle in a box. For specific numerical results, let,

$$f(t) = 2 \quad \text{for} \quad 0 < t < 5 \tag{6-6}$$

$$f(t) = -2 \quad \text{for} \quad 5 < t < 10 \tag{6-7}$$

$$f(t) = f(t + 10). \tag{6-8}$$

The period is 10 and we can now calculate the Fourier coefficients as,

$$A_n = \frac{2}{10}\int_0^5 2\cos(2\pi nt/10)\,dt + \frac{2}{10}\int_5^{10} -2\cos(2\pi nt/10)\,dt \tag{6-9}$$

$$B_n = \frac{2}{10}\int_0^5 2\sin(2\pi nt/10)\,dt + \frac{2}{10}\int_5^{10} -2\sin(2\pi nt/10)\,dt. \tag{6-10}$$

A square wave is an important type of standard test signal. When you use it as a stimulus, the Fourier series tells you that you are actually testing a system simultaneously with the fundamental frequency and odd harmonics of the square wave. So, you are testing a response over a wide bandwidth. Figure 6-2 shows the synthesis of the square wave with 9 components, and Fig. 6-3 shows the spectrum.

In Fig. 6-2 you can count five bumps on the square wave; these correspond to the five non-zero components of the nine harmonics used in this incomplete series. As the number of Fourier components increases, the number of bumps increases. The frequency of the "ripple" becomes higher and the amplitude becomes smaller. The "ears" on the square wave are the Gibbs' pheonomenon. As the number of Fourier components increases, the ears become more narrow but they never go away completely.

Another way of looking at this is to contrast square-wave testing with sinewave testing. With a sinewave you can measure magnitude and phase at a particular frequency, and then you shift the frequency and measure again. This is a *serial* narrow-band operation. With a square wave you are testing with many sinewaves (the Fourier components) simultaneously; this is a *parallel* wide-band operation.

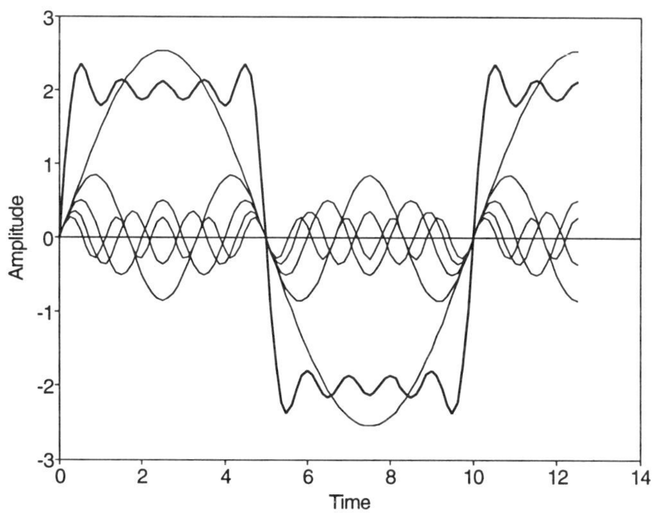

Figure 6-2. Square wave of Eq. (6-6)–(6-8), approximated with 9 harmonics. Even harmonics are zero. Five bumps on square wave indicate 5 non-zero harmonics.

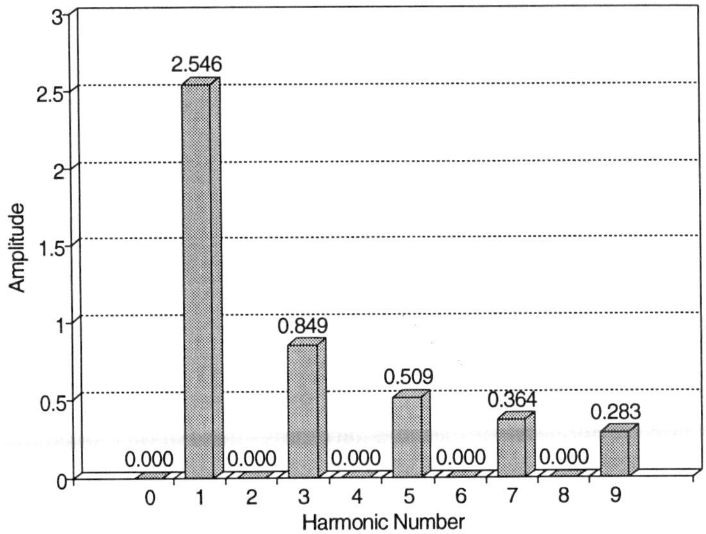

Figure 6-3. Spectrum of the square wave shown in Fig. 6-2, for 9 components.

Triangular Wave

As an example of the triangular wave, see Fig. 2-10 for the displacement of a particle in a box. For the following numerical calculation let,

$$f(t) = 0.8t \quad \text{for} \quad 0 < t < 2.5 \tag{6-11}$$

$$f(t) = -0.8t + 4 \quad \text{for} \quad 2.5 < t < 7.5 \tag{6-12}$$

$$f(t) = 0.8t - 8 \quad \text{for} \quad 7.5 < t < 10 \tag{6-13}$$

$$f(t) = f(t + 10). \tag{6-14}$$

The period is 10 and we can calculate the Fourier coefficients as,

$$A_n = \frac{2}{10}\int_0^{2.5} 0.8t\cos(2\pi nt/10)\,dt + \frac{2}{10}\int_{2.5}^{7.5}(-0.8t+4)\cos(2\pi nt/10)\,dt + \frac{2}{10}\int_{7.5}^{10}(0.8t-8)\cos(2\pi nt/10)\,dt \tag{6-15}$$

$$B_n = \frac{2}{10}\int_0^{2.5} 0.8t\sin(2\pi nt/10)\,dt + \frac{2}{10}\int_{2.5}^{7.5}(-0.8t+4)\sin(2\pi nt/10)\,dt + \frac{2}{10}\int_{7.5}^{10}(0.8t-8)\sin(2\pi nt/10)\,dt. \tag{6-16}$$

Examine the integrals for the A_n . They add to zero; all of the A_n are zero because the triangular wave, as we have chosen it with respect to $t = 0$, is an odd function and the cosine is an even function. I leave it to you to do the integrals for the B_n.

FOURIER, Fourier Series Worksheet

Figure 6-4 shows the Home screen of this worksheet, which lets you experiment with Fourier series by constructing periodic functions with arbitrary amplitudes and frequencies for the sines and cosines. This is equivalent to choosing arbitrary amplitudes, phases, and frequencies for an array of oscillators.

Table 6-1 shows the Fourier coefficients of three common periodic waveforms. In this table, all of the A_n are zero because the waveforms are set to be odd functions with respect to the time axis. Shifting the time axis to make the waveforms symmetric with respect to $t = 0$ will change the coefficients. The constant value is assumed to be zero in this table.

In the FOURIER worksheet you can change the Harmonic Number in Cells A6 through A15. This will enable you to sum more harmonics by eliminating those harmonics that you know to be zero. More harmonics can be added, but only six can be plotted together on a spreadsheet graph.

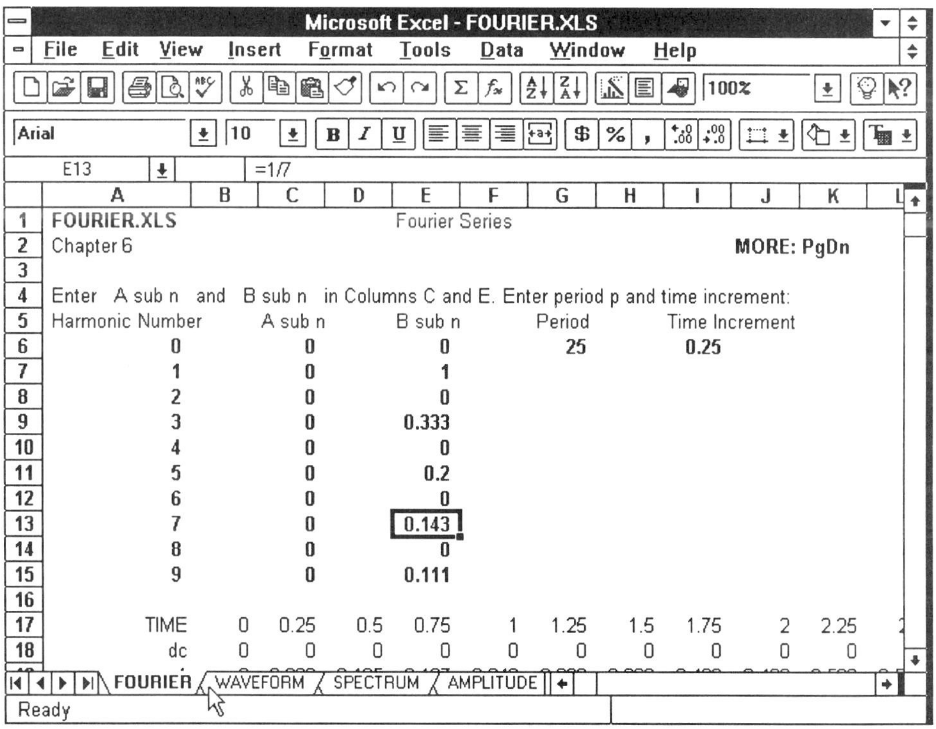

Figure 6-4. Home screen of FOURIER worksheet. Enter coefficients and period to synthesize a periodic function. The ones shown here produce a square wave.

Table 6-1. Fourier Series for Common Periodic Waveforms ($A_o = 0$)

Type	*Harmonic Number*	A_n	B_n
Square Wave	1	0	1/1
	3	0	1/3
	5	0	1/5
	all odd harmonics	0	$1/n$ same sign
Triangular Wave	1	0	$1/1^2$
	3	0	$-1/3^2$
	5	0	$1/5^2$
	all odd harmonics	0	$1/n^2$ alternating sign
Sawtooth Wave	1	0	1/1
	2	0	−1/2
	3	0	1/3
	all harmonics	0	$1/n$ alternating sign

6.3 Oscillator with Periodic Force

Now that we have the Fourier series we can solve the harmonic oscillator for any type of periodic force that can be represented by the series. Because of the linearity of the differential equation we know that the complete solution is the sum of the solutions for individual forces; we have used this idea many times. Now we can apply periodic forces of great variety, decompose each force into a series, and find the solution for each component. It is simply a process of finding solutions for sinusoidal driving forces of different frequencies ω_n and amplitudes A_n and B_n,

$$m\frac{d^2x}{dt^2} + R\frac{dx}{dt} + \kappa x = F(t) = \sum_{n=1}^{\infty} [A_n \cos(\omega_n t) + B_n \sin(\omega_n t)] . \tag{6-17}$$

The oscillator responds independently to each force. This is the essence of a linear system and the principle of superposition. (Also, see Section 6.9 for the Laplace transform, a general method of solving ordinary linear differential equations.)

6.4 Bandwidth and Spectral Density

Now we come to two particularly important subjects that apply to almost everything we do with periodic signals. Let's consider the spectra of periodic pulse trains, such as those shown in Figs. 6-5 and 6-6.

It is convenient to define the *bandwidth* as the frequency interval between the first zero-crossing and the central maximum, because these are two easily identified points in the spectrum. The *pulse repetition frequency* (PRF) is the reciprocal of the time between leading edges of adjacent pulses. The *spectral density* is the number of spectral lines per frequency interval. Fourier series tells us:

- The pulse width controls the bandwidth. The bandwidth is inversely proportional to the pulse width.
  Short pulse ⇔ wide band.
  Long pulse ⇔ narrow band.
- The PRF controls the spectral density. The spectral density is inversely proportional to the PRF.
  Low PRF ⇔ high spectral density.
  High PRF ⇔ low spectral density.

Only a periodic signal has a line spectrum, but if the PRF is very low then the spectral density is very high and it becomes difficult to resolve individual lines; periodicity may be difficult to identify. As PRF $\rightarrow 0$ the spectrum becomes continuous. A system must have a wide bandwidth (low Q) if it is to pass a short pulse.

The preceding is closely related to our interpretation of measurements in general, and, in particular, to sampling and digitizing of data. If we want to determine the value of a measured physical quantity that is changing in time we would want to use a short sampling time to minimize the probability that the physical quantity is changing while we sample it. It might seem best if we could use an approximation to a δ-function (of infinitesimal width). However, the shorter the sampling time the more the frequency spread of the measurement. A better procedure would be to use a sampling time that is matched to the measurement in such a way that the physical quantity is practically constant over the sampling time interval but the required bandwidth is not excessive (the frequency spread is minimal).

In sampling data, we would like to sample often so we know what is happening; we don't know what is happening while we're not observing. However, if we know what is happening at closely-spaced times the spectral density decreases; we only have information at the spectral lines in the frequency domain. Nevertheless, we must sample at twice the frequency of the maximum frequency that we want to know with confidence. This is the Nyquist theorem and the maximum frequency that we know with confidence is called the Nyquist frequency. For example, if we want to measure up to 100 kHz we must collect at least 200,000 samples per second. These concepts apply to all measurements, not only to analog-to-digital conversion.

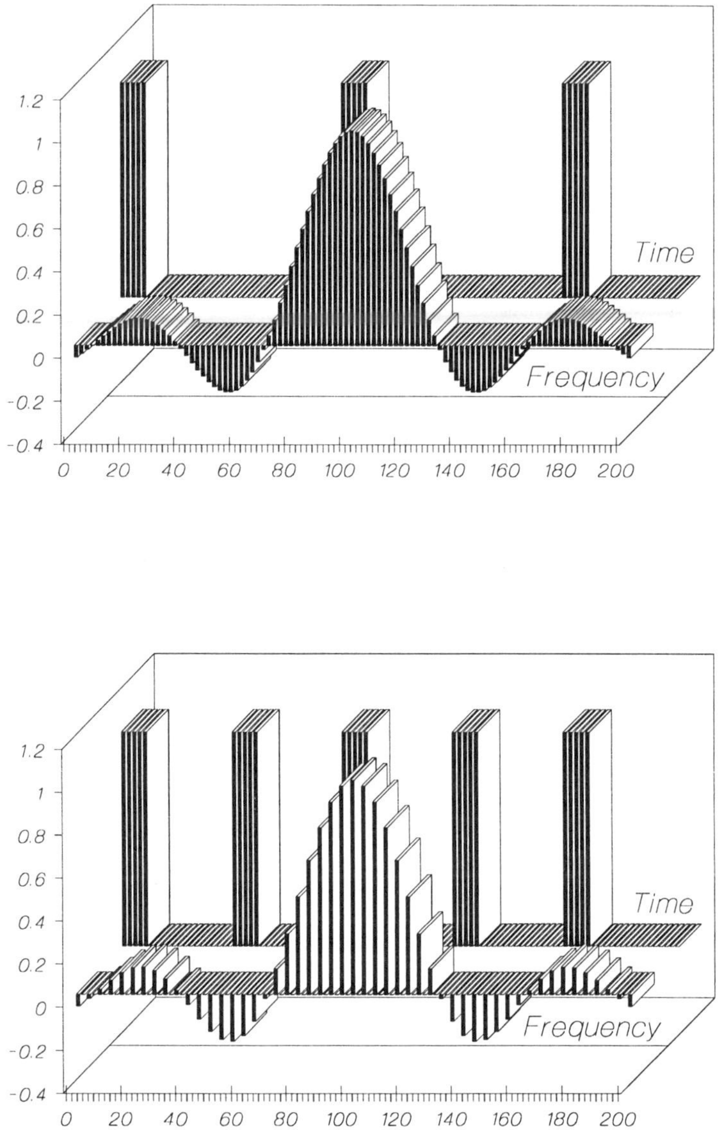

Figure 6-5. *Top:* Short pulse with low PRF. Compare Fig. 6-6 (top). *Bottom:* Same pulse width with two times the PRF. Compare Fig. 6-6 (bottom). Increasing PRF decreases spectral density but bandwidth is constant because pulse width is constant. PRF controls spectral density; pulse width controls bandwidth.

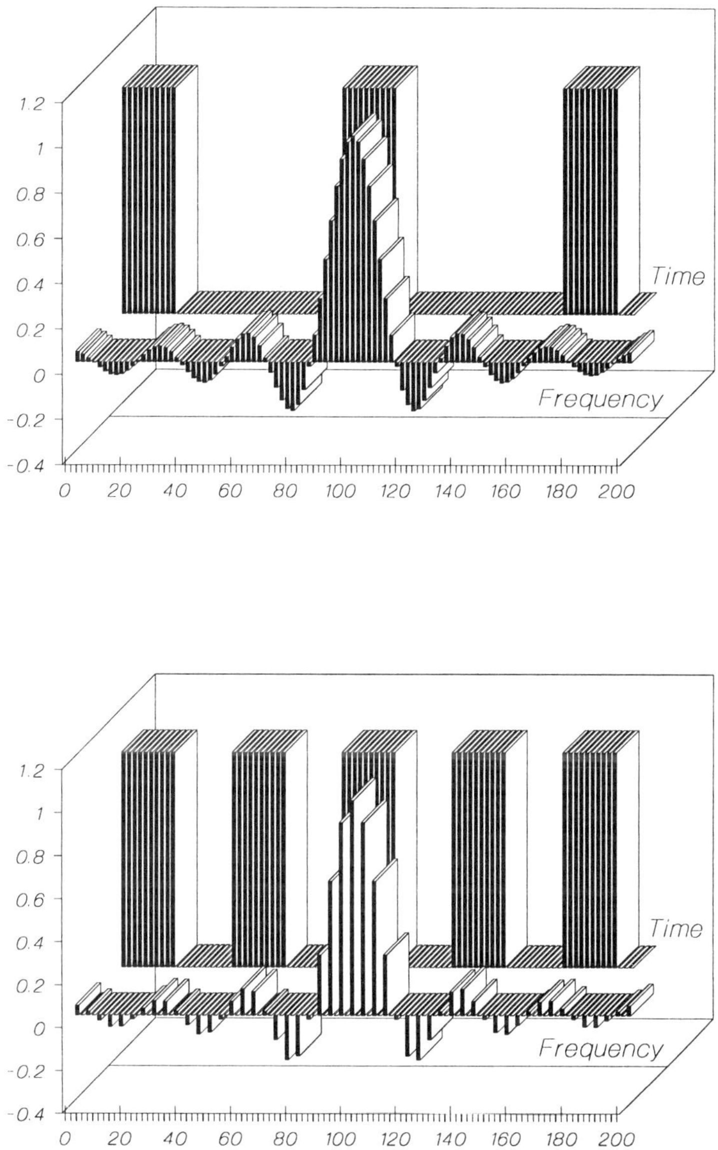

Figure 6-6. *Top:* Pulses two times longer than pulses in Fig. 6-5, but same PRF, has half the bandwidth. Compare Fig. 6-5 (top). *Bottom:* Same pulse width at top but two times the PRF. Bandwidth is same as at top, but spectral density is half. Compare Fig. 6-5 (bottom).

6.5 Laplace's Equation

Laplace's equation plays a central role in electromagnetics, mechanics, fluid mechanics, heat flow, and mathematics. Solutions of Laplace's equation are known as *harmonic functions*. Methods of solution have engaged mathematicians for almost 200 years. Here, we are only going to look at the tip of this iceberg. We are not going to produce a complete solution, but you can finish it; we are going to see how the Fourier series arises in this ubiquitous equation. For simplicity, let's write Laplace's equation in two dimensions, using Cartesian coordinates,

$$\nabla^2 U(x, y) = \frac{\partial^2 U}{\partial x^2} + \frac{\partial^2 U}{\partial y^2} = 0. \tag{6-18}$$

Let's try a solution by separation of variables,

$$U(x, y) = X(x)Y(y) \tag{6-19}$$

and substitute this in Laplace's equation to obtain,

$$Y\frac{\partial^2 X}{\partial x^2} + X\frac{\partial^2 Y}{\partial y^2} = 0. \tag{6-20}$$

Now divide by XY,

$$\frac{1}{X}\frac{\partial^2 X}{\partial x^2} + \frac{1}{Y}\frac{\partial^2 Y}{\partial y^2} = 0 \tag{6-21}$$

so,

$$\frac{1}{X}\frac{\partial^2 X}{\partial x^2} = -\frac{1}{Y}\frac{\partial^2 Y}{\partial y^2}. \tag{6-22}$$

The left side is a function only of x and the right side is a function only of y. How can a function of one variable equal a function of another independent variable? This can be true only if both functions are equal to a constant (the separation constant). Let's call the constant $-k^2$ for reasons that will soon be apparent. Then,

$$\frac{1}{X}\frac{d^2 X}{dx^2} = -\frac{1}{Y}\frac{d^2 Y}{dy^2} = -k^2. \tag{6-23}$$

Now we have two ordinary second-order differential equations that can be solved separately. The constant was free to choose, but only for $-k^2$ can we write,

$$\frac{d^2 X}{dx^2} = -k^2 X. \tag{6-24}$$

This is an ordinary differential equation now because we have separated the variables. This ordinary differential equation is just our old harmonic oscillator, but in space instead of time. We know a solution,

$$X(x) = A\cos(kx) + B\sin(kx). \tag{6-25}$$

Let's put on specific boundary conditions and observe how this produces an infinite discrete set. Let $X(0) = 0$ and $X(L) = 0$. The first boundary condition sets $A = 0$. The second boundary condition sets $\sin(kL) = 0$. This requires,

$$k_n L = n\pi, \quad n = 1, 2, 3, \ldots \tag{6-26}$$

We have put a subscript on k because each one belongs to a specific n. So,

$$k_n = \frac{n\pi}{L} \tag{6-27}$$

and the solution for these specific boundary conditions now becomes,

$$X(x) = \sum_{n=1}^{\infty} B_n \sin(n\pi x/L). \tag{6-28}$$

For general boundary conditions we would have,

$$X(x) = \sum_{n=1}^{\infty} [A_n \cos(n\pi x/L) + B_n \sin(n\pi x/L)]. \tag{6-29}$$

This is just the Fourier series for space instead of time. It is straightforward to obtain solutions for other dimensions and other coordinate systems. For example, in plane polar coordinates the sines and cosines are replaced by Bessel functions.

Laplace's equation has many interesting interpretations. Basically it implies that the solution is as smooth as it can be, consistent with the boundary conditions.

- It is not difficult to show by means of a Taylor series expansion that Laplace's equation says that the average value of its solution around a point is equal to the value of the solution at that point.

This is the starting point for powerful computer-based finite element methods, but we will not pursue this here. (A finite elephant method is shown in Fig. 6-7.) For our purposes, it is enough to see how Fourier series and the harmonic oscillator equation arise naturally in situations where you might not expect to see them.

Cauchy-Riemann Equations

Functions that satisfy these two partial differential equations are *analytic functions*. Analytic functions play important roles in engineering, the sciences, and mathematics; we have already used analytic functions many times in our solutions for undamped and damped oscillators. We will run into them again in the quantum harmonic oscillator. Now we will see that the real and imaginary parts of an analytic function are conjugate functions, and they are solutions of Laplace's equation. Finally, we will see another very deep connection between Laplace's equation and harmonic oscillators.

Suppose we have a function,

$$U(x,y) = u(x,y) + jv(x,y) . \tag{6-30}$$

If $u(x,y)$ and $v(x,y)$ obey the Cauchy-Riemann equations,

$$\frac{\partial u}{\partial x} = \frac{\partial v}{\partial y} \quad \text{and} \quad \frac{\partial v}{\partial x} = -\frac{\partial u}{\partial y} \tag{6-31}$$

then, differentiating again, we obtain,

$$\frac{\partial^2 u}{\partial x^2} = \frac{\partial^2 v}{\partial x \partial y} \quad \text{and} \quad \frac{\partial^2 v}{\partial y \partial x} = -\frac{\partial^2 u}{\partial y^2} . \tag{6-32}$$

Interchanging the order of differentiation and subtracting gives Laplace's equation,

$$\nabla^2 u(x, y) = \frac{\partial^2 u}{\partial x^2} + \frac{\partial^2 u}{\partial y^2} = 0 \tag{6-33}$$

and similarly for $v(x,y)$. In a simply connected region, $u + jv$ is an analytic function.

The physical importance of all of this is that, for a field that can be described by Laplace's equation, the field is connected in such a way that all parts of the field influence all other parts. The real and imaginary parts are conjugate functions.

The classical source-free wave equation for electromagnetic waves, mechanical waves, and sound waves can be written as Laplace's equation in four dimensions by expressing time as jct, where c is the speed of the wave. The wave in space-time is a disturbance that is as smooth as it can be, consistent with the boundary conditions. When sources such as electric charges, currents, or masses are present we have Poisson's equation. For example, if an electric charge density $\rho(x,y,z)$ exists in a region of space which is a vacuum except for the charges, Poisson's equation is

$$\nabla^2 U(x,y,z) = \frac{\rho(x,y,z)}{\varepsilon_o} . \tag{6-34}$$

This makes sense because a source (or sink) makes the value of the field at a point differ from the average value of the field around that point.

As a graphic example, the baby elephant in Fig. 6-7 constitutes a quadruped quadrupole (all with the same sign). If the elephant were not there, the average value around any point would be equal to the value at that point. The field is distorted where the elephant's feet contact the pool cover.

Figure 6-7. Physical example of Poisson's equation. Notice how the presence of mass distorts the field (rectangular grid lines). When Bubbles, the elephant, is removed the field is as smooth as possible, consistent with the boundary conditions set by the turnbuckles on the circumference, and Poisson's equation reduces to Laplace's equation. Don't try this at home. (Courtesy of LOOP-LOK Safety Swimming Pool Covers, Hauppauge, NY.)

6.6 Non-Periodic Signals

Fourier Transform

A common form of the Fourier transform pair can be written as,

$$f(t) = \frac{1}{2\pi}\int_{-\infty}^{+\infty} F(\omega)e^{+j\omega t}d\omega, \tag{6-35}$$

where, if the integral exists, the spectrum $F(\omega)$ is given by,

$$F(\omega) = \int_{-\infty}^{+\infty} f(t)e^{-j\omega t}dt\,. \tag{6-36}$$

The first thing you notice about the Fourier transform pair is that, to know the waveform completely, you must know its spectrum over an infinite bandwidth. To know the waveform's spectrum completely, you must know the waveform over an infinite time interval *including the future*. Because we can never know these things, our knowledge is doomed to be incomplete in an essential, fundamental sense. This causes some analysts to use the term *spectral estimation* for all of their work.

- Look again at Fig. 6-1. If this is all the information you have about the three signals, it is possible that all three are periodic, perhaps with a period of 140 time units. With these time-limited data it is not possible to know that the "transient" signal and the "random" signal are not periodic.
- In Fig. 6-1 the "periodic" signal at the top may not be periodic. Perhaps it is only a transient; perhaps there is nothing before $t = 0$ and nothing after $t = 140$. If it is not in our set of measurements, we do not know it.

It is tempting to extrapolate from what we know to what we do not know, and from what we measure to what we do not measure. This can lead to trouble. When we focus our attention only on what we measure in the real world, we are on safe ground and this has important consequences, as we shall see.

Centered Rectangular Pulse

Consider a single rectangular pulse, symmetric with respect to $t = 0$,

$$f(t) = A \quad \text{for} \quad \frac{-T}{2} \le t \le \frac{+T}{2} \tag{6-37}$$

$$f(t) = 0 \quad \text{otherwise.} \tag{6-38}$$

When this pulse is inserted into the Fourier integral for $F(\omega)$ we have

$$F(\omega) = \int_{-T/2}^{+T/2} Ae^{-j\omega t} dt \tag{6-39}$$

and we can immediately integrate to obtain

$$F(\omega) = \frac{A}{-j\omega}(e^{-j\omega T/2} - e^{+j\omega T/2}), \tag{6-40}$$

or,

$$F(\omega) = \frac{A}{j\omega}(e^{+j\omega T/2} - e^{-j\omega T/2}) . \tag{6-41}$$

This can be simplified by recalling the exponential representation of the sine function,

$$\sin(\theta) = \frac{e^{+j\theta} - e^{-j\theta}}{2j} \tag{6-42}$$

so,

$$F(\omega) = \frac{2A}{\omega}\sin(\omega T/2) = \frac{AT\sin(\omega T/2)}{\omega T/2} . \tag{6-43}$$

This is a very important result. It says that a constant value A measured for a finite time interval T is distributed or spread out in the frequency domain. The symmetric pulse, centered at $t = 0$, and its Fourier transform are shown in Fig. 6-8. The form of the spectrum is related to the sinc function, $\text{sinc}(\theta) = \sin(\pi\theta)/\pi\theta$.

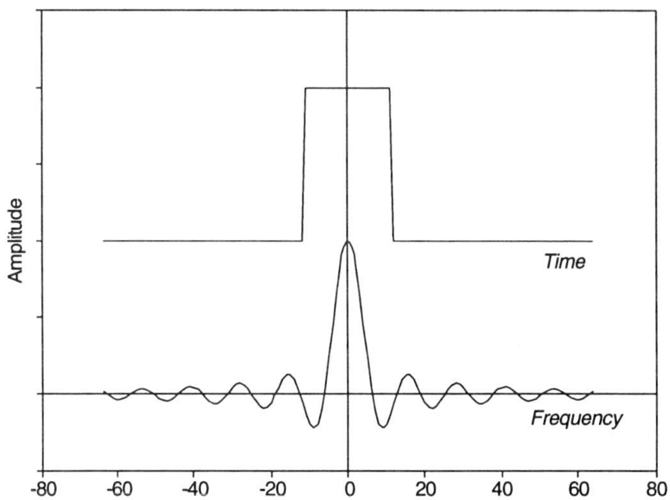

Figure 6-8. Symmetric pulse (centered in time) and its Fourier transform.

Because of the symmetry in the Fourier integral, a function distributed in time like $f(t) = \sin(\omega_o t)/\omega_o t$ would appear as a rectangular, band-limited spectrum.

The power spectrum, often called the power spectral density or PSD, is given by

$$|F(\omega)|^2 = \frac{(AT)^2 \sin^2(\omega T/2)}{(\omega T/2)^2} . \tag{6-44}$$

An important interpretation of this result is related to the fundamental measurement process. As a special case, suppose you measure the voltage of a battery using a very high impedance voltmeter so the battery is disturbed an insignificant amount (an insignificant amount of current flows so internal resistance and internal heating can be neglected). Surely, a battery voltage in this situation is the epitome of a dc measurement. Nevertheless, Eq. (6-44) implies that a broad spectrum of frequencies will be present in your measurements, and the shorter the measurement ($\Delta t = T$), the broader the spectrum.

This transform pair is extremely important because of its generality, not only as a fundamental measurement paradigm, but also as an example of *generalized diffraction*. The pulse can be an aperture or slit in space and its transform is then the single-slit diffraction pattern. The power spectrum, Eq. (6-44) is shown in Fig. 6-9.

A second reason why this transform pair is important is that pulses are ubiquitous signals; oscillators are subjected to them 24 hours a day, every day.

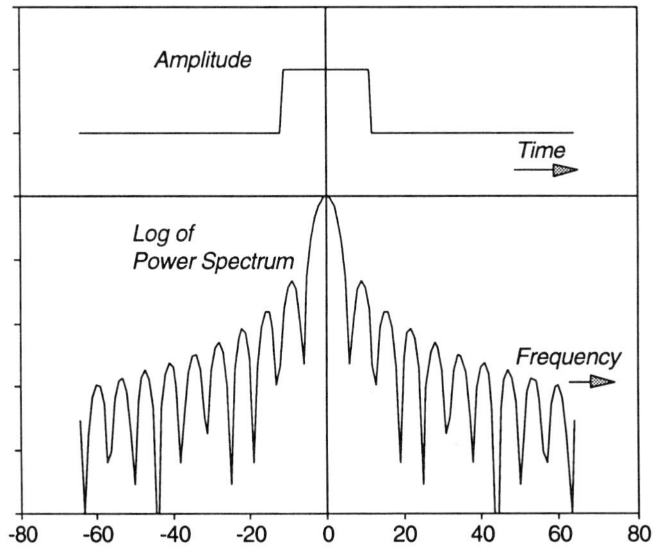

Figure 6-9. Single pulse (centered in time) and its power spectrum. The power spectrum is shown on a logarithmic scale for greater clarity of the structure away from the central maximum. This is the same information shown in Fig. 6-8.

Pulse Bandwidth

According to Eq. (6-43), the spectrum has a zero-crossing when,

$$\sin(\omega T/2) = 0. \tag{6-45}$$

This has zeros at

$$\frac{\omega_n T}{2} = n\pi. \tag{6-46}$$

The first zero occurs on each side of the central maximum when $n = 1$ so the frequency of the first zero is

$$\omega_1 = \frac{2\pi}{T}, \tag{6-47}$$

or

$$f_1 = \frac{1}{T} \text{ Hz}. \tag{6-48}$$

As with periodic signals (Section 6.4), the frequency interval between the central maximum and the first zero-crossing of the spectrum of the rectangular pulse (Fig. 6-5) is a convenient, standard measure of bandwidth. *The bandwidth is the reciprocal of the pulse width.*

$$\text{Bandwidth (Hz)} = \frac{1}{T\,\text{(seconds)}} \tag{6-49}$$

Of course, a bandwidth definition like this only applies to spectra with zero-crossings. Previously we have used the full-width at half-maximum power as the bandwidth definition for a spectrum of the Lorentz type, which has no zeros.

Uncentered Rectangular Pulse

Now let's shift the rectangular pulse later in time so that it starts at $t = 0$, like something you would see on an oscilloscope screen,

$$f(t) = A \text{ for } 0 \le t \le +T \tag{6-50}$$

$$f(t) = 0 \text{ otherwise}. \tag{6-51}$$

The only effect of this time shift is to change the limits on the Fourier integral,

$$F(\omega) = \int_0^{+T} Ae^{-j\omega t}dt \tag{6-52}$$

so the result is

$$F(\omega) = \frac{A\,(e^{-j\omega T} - 1)}{-j\omega}. \tag{6-53}$$

Eq. (6-53) looks complicated but it isn't. Let's separate it into its real and imaginary parts, and then express it in polar form with magnitude and phase. Separating the real and imaginary parts we find

$$F(\omega) = \frac{A\cos(\omega T) - 1}{-j\omega} - j\frac{A\sin(\omega T)}{-j\omega} = \frac{A\sin(\omega T)}{\omega} + j\frac{A\,[\cos(\omega T) - 1]}{\omega}. \tag{6-54}$$

We can bring this into a more familiar form,

$$F(\omega) = \frac{AT\sin(\omega T)}{\omega T} + j\frac{AT\,[\cos(\omega T) - 1]}{\omega T}. \tag{6-55}$$

With some algebra and $\sin^2(\theta/2) = [1 - \cos(\theta)]/2$, we find the magnitude,

$$|F(\omega)| = \frac{AT\sin(\omega T/2)}{\omega T/2} \tag{6-56}$$

which is, of course, the same as the centered pulse, but now the phase is,

$$\theta(\omega) = \arctan\left\{\frac{[\cos(\omega T) - 1]/\omega T}{\sin(\omega T)/\omega T}\right\}. \tag{6-57}$$

Figure 6-10 shows the real and imaginary parts of this Fourier transform, Fig. 6-11 shows the polar plot, and Fig. 6-12 shows the magnitude and phase.

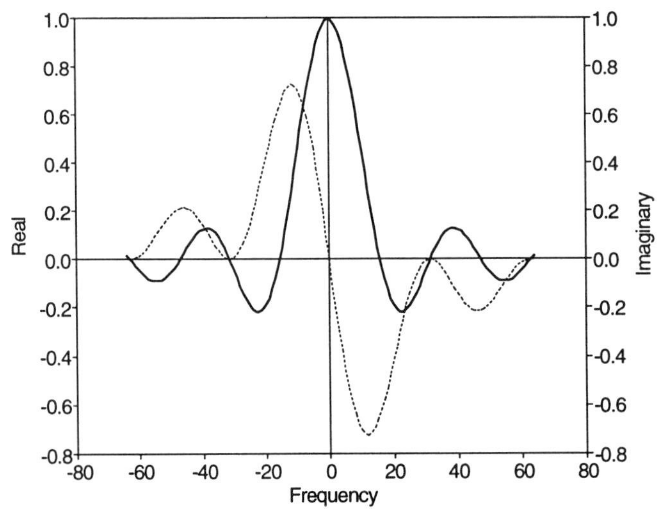

Figure 6-10. Spectrum of an uncentered rectangular pulse. *Solid line:* Real. *Dotted line:* Imaginary . Note even and odd symmetries of real and imaginary.

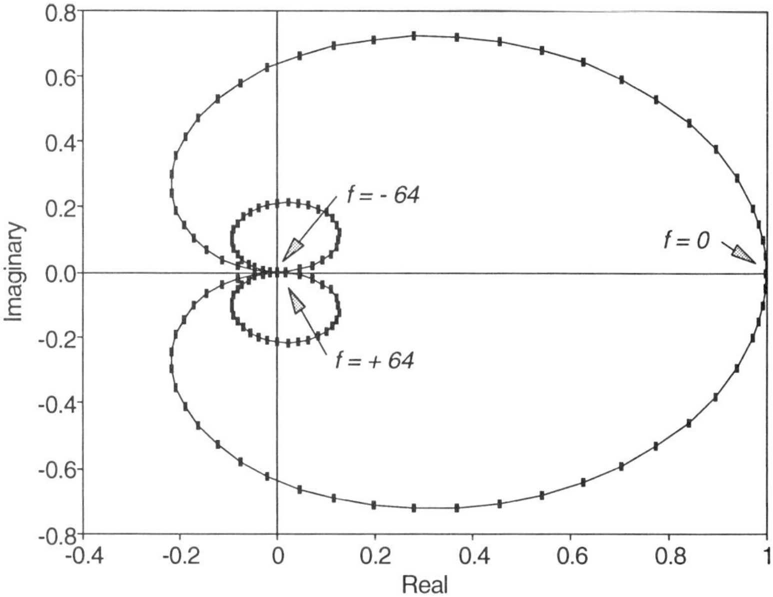

Figure 6-11. Polar plot of an uncentered rectangular pulse, starting at $t = 0$. This is the same information plotted in Fig. 6-9. $\Delta f = 1$ between markers.

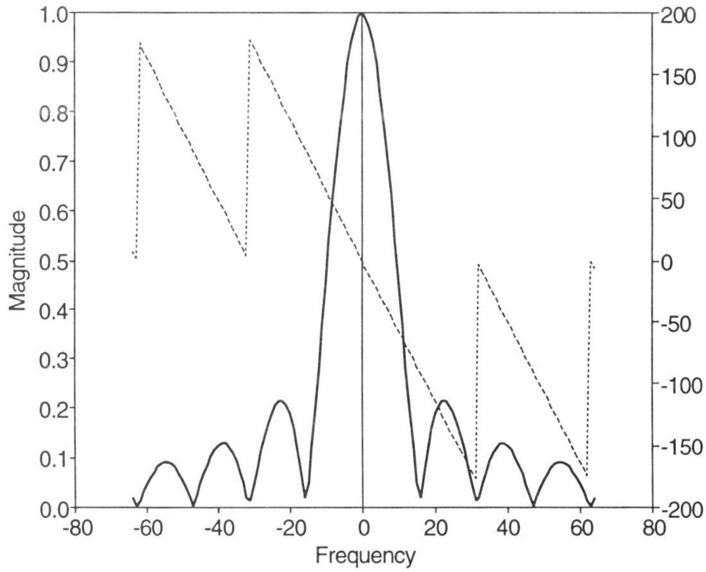

Figure 6-12. Magnitude *(solid line)* and phase *(dotted line)* of an uncentered pulse. Pulse starts at $t = 0$ and has a pulse width of T. Note the linearity of the phase shift through the center of the spectrum.

The only effect of shifting the pulse in time is to make the imaginary part of the spectrum non-zero, and therefore to introduce a non-zero phase in the spectrum. Obviously, the magnitude and power spectrum cannot change just because we look at the pulse at another time on our personal clock.

Exponential Decay

The exponential decay is associated with many physical processes. In particular we have seen it as the envelope of the displacement of a free, damped, harmonic oscillator, and as the decay of energy and momentum of a damped mass. Let's look at the Fourier transform of a typical exponential such as

$$f(t) = Ae^{-Kt} \quad \text{for} \quad 0 \le t \le +\infty \tag{6-58}$$

$$f(t) = 0 \quad \text{otherwise.} \tag{6-59}$$

When we insert this into the Fourier integral we obtain

$$F(\omega) = \int_0^{+\infty} Ae^{-Kt}e^{-j\omega t}dt\,. \tag{6-60}$$

Now collect the exponents,

$$F(\omega) = \int_0^{+\infty} Ae^{-(K+j\omega)t}dt. \tag{6-61}$$

This is a simple integral and the result is

$$F(\omega) = \frac{A\,e^{-(K+j\omega)t}\big|_0^{\infty}}{-(K+j\omega)} = \frac{A\,(0-1)}{-(K+j\omega)} = \frac{A}{(K+j\omega)}\,. \tag{6-62}$$

Multiply the numerator and denominator by the complex conjugate of the denominator to separate the real and imaginary parts,

$$F(\omega) = A\left(\frac{K}{K^2+\omega^2} - j\frac{\omega}{K^2+\omega^2}\right). \tag{6-63}$$

And, as usual, the magnitude and phase are obtained as

$$|F(\omega)| = \frac{A}{\sqrt{\omega^2+K^2}} \tag{6-64}$$

$$\theta(\omega) = \arctan(\mathrm{Im}/\mathrm{Re}) = \arctan(-\omega/K)\,. \tag{6-65}$$

Finally, the power spectral density (PSD) is

$$|F(\omega)|^2 = \frac{A}{K+j\omega} \cdot \frac{A}{K-j\omega} = \frac{A^2}{\omega^2+K^2} . \qquad (6\text{-}66)$$

Figure 6-13 shows the exponential decay and its power spectrum. The real and imaginary parts are shown in Fig. 6-14, the polar plot of imaginary versus real is shown in Fig. 6-15, and the magnitude and phase are shown in Fig. 6-16.

Oscillatory Exponential Decay

This example is important because of its fundamental nature and its applications.

- We will see that the spectrum of the complex exponential decay is the Lorentz shape of Fig. 4-14.
- This is a specific case of the *complex modulation* property, which we will explain later.
- This is an example of the spectrum of an *analytic function*, a pervasive concept in mathematics, engineering, and physics. This is the first time that we will have the pleasure of calculating the Fourier transform of a complex function.
- We will compare the spectrum of the real part of the complex function with the spectrum of the complex function, which will show one reason why complex functions are necessary in quantum mechanics.

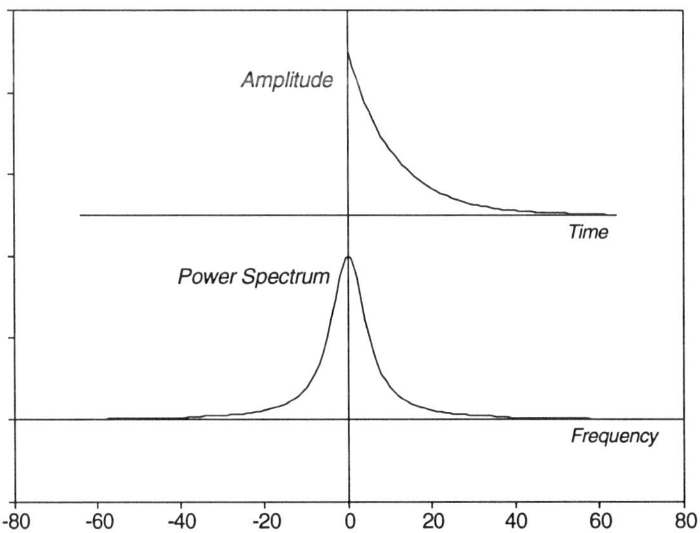

Figure 6-13. *Upper:* Non-oscillatory exponential decay, Eqs. (6-58) and (6-59), starts at $t = 0$. *Lower:* Power spectrum of the exponential decay, described by Eq. (6-66). Note that this power spectrum has no zeros.

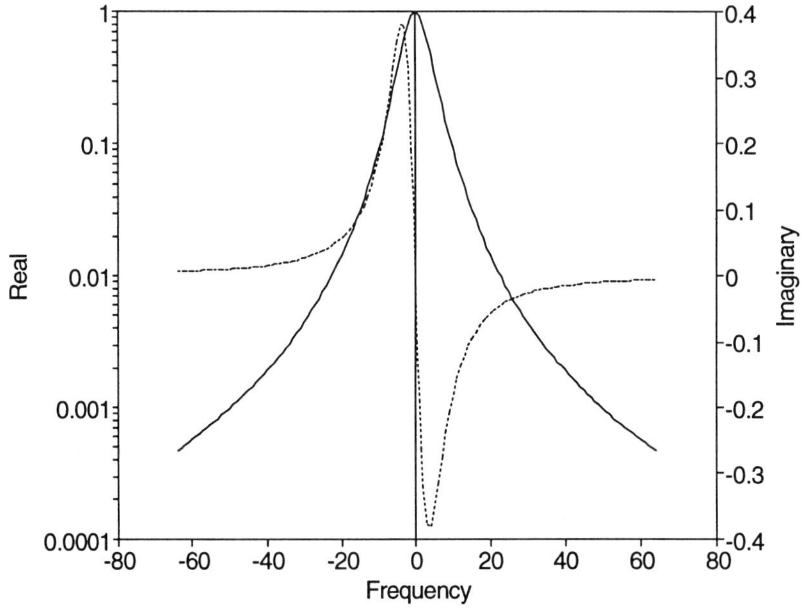

Figure 6-14. Real and imaginary parts of the non-oscillatory exponential decay. The real part has even symmetry and the imaginary part has odd symmetry.

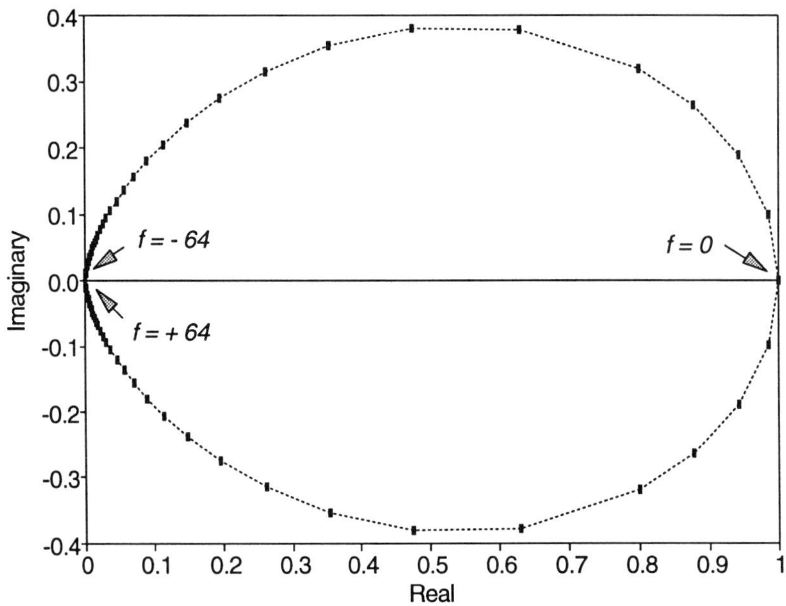

Figure 6-15. Polar plot of the non-oscillatory exponential decay spectrum. $\Delta f = 0.5$ for the first 4 markers on each side of $f = 0$. For other markers, $\Delta f = 1$.

Look again at Fig. 6-13. Do you recognize the PSD of the non-oscillatory exponential decay? It is the Lorentz shape from Chapter 4, down-shifted to $\omega_o = 0$ and un-normalized to $A = K$. The magnitude and phase shown in Fig. 6-16 should look familiar.

For our complex oscillatory exponential decay let's choose,

$$f(t) = Ae^{-Kt}e^{j\omega_o t} \quad \text{for} \quad 0 \le t \le +\infty$$
$$f(t) = 0 \quad \text{otherwise} \tag{6-67}$$

so the spectrum can be written as

$$F(\omega) = \int_0^{+\infty} Ae^{-Kt}e^{+j\omega_o t}e^{-j\omega t}dt\,. \tag{6-68}$$

Let's collect the exponentials and arrange them conveniently as

$$F(\omega) = \int_0^{+\infty} Ae^{-[K-j(\omega-\omega_o)]t}dt. \tag{6-69}$$

This can be integrated immediately as

$$F(\omega) = \frac{A\,e^{-[K-j(\omega-\omega_o)]t}\Big|_0^{\infty}}{-[K-j(\omega-\omega_o)]}\;. \tag{6-70}$$

Equation (6-70) is zero at the upper limit $t = \infty$, and at the lower limit $t = 0$ we have,

$$F(\omega) = \frac{A}{K-j(\omega-\omega_o)}\;. \tag{6-71}$$

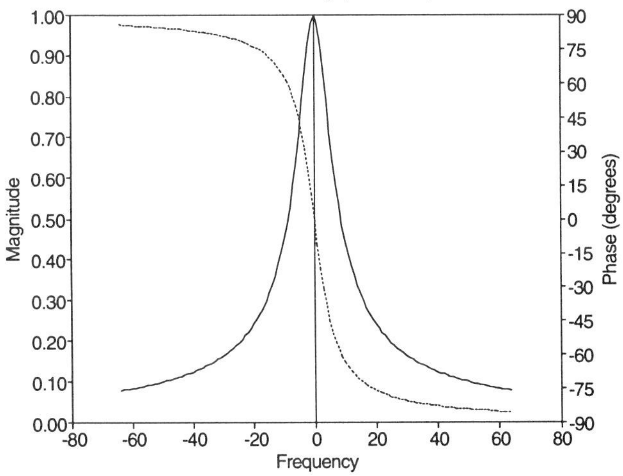

Figure 6-16. Magnitude *(solid line)* and phase of the non-oscillatory exponential decay spectrum. Compare with Fig. 4-11 for the forced, damped oscillator.

This has a magnitude-squared of

$$|F(\omega)|^2 = \frac{A}{K - j(\omega - \omega_o)} \cdot \frac{A}{K + j(\omega - \omega_o)} = \frac{A^2}{(\omega_o - \omega)^2 + K^2} \,. \tag{6-72}$$

This power spectrum, shown in Fig. 6-17, is the Lorentz shape if we normalize it with $A^2 = K^2$. Compare Eq. (6-72) with Eq. (4-91).

6.7 Classical Uncertainty Principle, Part 2

Throughout this chapter we have used deterministic functions, many of which are identical to those found in earlier chapters as solutions of simple harmonic oscillators. And yet, we have seen something of a paradox when we found an uncertainty relation for $\Delta p \Delta x$ in Chapter 1, Section 1.9, and Chapter 2, Section 2.5. Now we will see that the paradox deepens and will continue to do so until we come to the quantum domain (where some people say it gets deeper and wider). There is inevitably some overlap between what we are going to do here and what we will do in Chapter 11 with the statistical perspective, but here we will continue to use idealized, completely deterministic signals (no random component).

We have examined the spectra of a rectangular pulse, an exponential decay, and an oscillatory exponential decay, and in each case we found values of $\Delta\omega$ and Δt that were convenient for measurements. These are summarized below, but we need to find some common measurement so we can compare the $\Delta\omega\,\Delta t$ products. The half-power points in time and frequency are useful for this common measure.

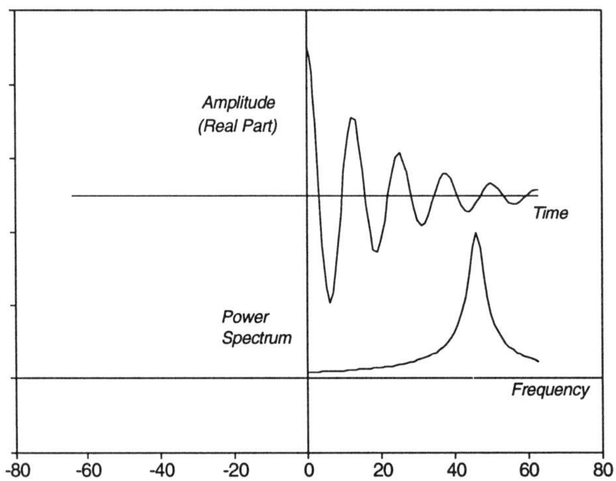

Figure 6-17. *Upper:* Oscillatory exponential decay (real part). *Lower:* PSD of the analytic signal. Note only positive frequencies are present even though the decay is uncentered in time. This is characteristic of an analytic signal.

Rectangular Pulse

For this pulse we defined the bandwidth as the full frequency interval between the first zeros adjacent to the central maximum of the spectrum (see Fig. 6-8),

$$\Delta\omega = \Delta\omega_1 = \frac{2\pi}{T},$$

because this was easy to measure. Similarly, the pulse width is a convenient measure of Δt,

$$\Delta t = T$$

so,

$$\Delta\omega\,\Delta t = 1.$$

Non-Oscillatory Exponential Decay

The reciprocal of the decay constant K is a convenient Δt for this signal,

$$\Delta t = 1/K.$$

Alternatively, the half-life could be used. This spectrum has no zeros; the full width at half-maximum power is a standard and convenient measure of $\Delta\omega$ for this signal,

$$\Delta\omega = 2K$$

so,

$$\Delta\omega\,\Delta t = 2.$$

However, these measures are inconsistent with those used for the rectangular pulse.

Oscillatory Exponential Decay

As we have seen in Section 6.6, the oscillatory exponential decay is just the non-oscillatory exponential decay shifted from the central frequency $\omega_o = 0$ to the central frequency $\omega_o \neq 0$. Therefore the uncertainty relation is the same. This shift is a property of the Fourier transform known as *complex modulation* (see Table 6-3).

Gaussian Function

In Chapter 10 we will consider the Gaussian function,

$$f(x) = \frac{1}{\sqrt{2\pi}\,\sigma} e^{-x^2/(2\sigma^2)}, \tag{6-73}$$

where σ is the standard deviation, availablc in spreadsheets as the STD function. For simplicity here let's consider a spectrum of the (un-normalized) Gaussian form,

$$F(\omega) = \exp[-(\omega - \omega_o)^2 t_o^2]\,. \tag{6-74}$$

The amplitude falls to $1/e$ at $\omega = \omega_o \pm 1/t_o$, where $t_o^2 = 1/(2\sigma^2)$ is a characteristic time, somewhat analogous to a time constant. (The square of the standard deviation is called the *variance*, and is available in spreadsheets as the VAR function.)

If all of the Fourier components are in phase at $t = 0$ the Gaussian will have minimum width and we can express the signal by the Fourier inversion,

$$f(t) = \frac{1}{\sqrt{2\pi}} \int_{-\infty}^{+\infty} \exp[-(\omega - \omega_o)^2 t_o^2] \exp[+j\omega t]\, d\omega\,. \tag{6-75}$$

We can simplify this by a rearrangement,

$$(\omega - \omega_o)^2 t_o^2 + j\omega t = [(\omega - \omega_o)t_o + j\frac{t}{2t_o}]^2 + j\omega_o t + \frac{t^2}{(2t_o)^2} \tag{6-76}$$

and change of variable,

$$z = (\omega - \omega_o)t_o + j\frac{t}{2t_o}\,. \tag{6-77}$$

Now we can write Eq. (6-75) as

$$f(t) = \exp[-t^2/(2t_o)^2] \exp[+j\omega_o t] \frac{1}{\sqrt{2\pi}} \int_{-\infty}^{+\infty} \exp[-z^2] \frac{dz}{t_o}\,, \tag{6-78}$$

and the result is,

$$f(t) = \frac{1}{t_o} \exp[-t^2/(2t_o)^2] \exp[-j\omega_o t]. \tag{6-79}$$

Notice the extraordinary sensitivity of $f(t)$ and $F(\omega)$ to the characteristic time t_o. The characteristic time *appears in the exponential* and it is *squared.*

- A large characteristic time (standard deviation) expands the Gaussian in time and compresses it in frequency. A small characteristic time compresses the Gaussian in time and expands it in frequency.
- Both $f(t)$ and $F(\omega)$ are Gaussian in form. *The Gaussian is the only function that is its own Fourier transform.* Only the width changes but the form is *invariant* under the Fourier transformation.

This is a special case of the scale-change property of the Fourier transform, but it is a very important case because it can be proved that the Gaussian has the smallest $\Delta\omega\, \Delta t$ product of any function. For the Gaussian, $\Delta\omega\, \Delta t \geq 0.5$.

Recall, the mean value or first moment of a distribution is defined as

$$\omega_{av} = \frac{\int_{-\infty}^{+\infty} \omega\, F^*(\omega)F(\omega)\, d\omega}{\int_{-\infty}^{+\infty} F^*(\omega)F(\omega)\, d\omega}. \tag{6-80}$$

Let's define Δt = r.m.s. value of t over the distribution $|f(t)|^2$ and $\Delta\omega$ = r.m.s. value of ω over the distribution $|F(\omega)|^2$. Then the mean square deviation from the mean (the *variance*) is

$$(\Delta\omega)^2 = \frac{\int_{-\infty}^{+\infty} (\omega - \omega_{av})^2\, F^*(\omega)F(\omega)\, d\omega}{\int_{-\infty}^{+\infty} F^*(\omega)F(\omega)\, d\omega}. \tag{6-81}$$

Notice that the mean value ω_{av} is analogous to the first moment (the center of mass) of a mass distribution, and the variance $(\Delta\omega)^2$ (the second moment) is analogous to the square of the radius of gyration, referred to the center of mass.

Table 6.2 Some Uncertainty Relations Based on Half-Power Points

Waveform	Δt	$\Delta\omega$	$\Delta\omega\, \Delta t$
Rectangular Pulse	$\frac{T}{2}$	$\frac{2}{T}(1.3915574\ldots)$	$1.3915574\ldots$
Exponential Decay	$\frac{\ln(2)}{2K}$	$2K$	$\ln(2) \approx 0.693149$
Gaussian	$2\left[\frac{\ln(2)}{2K}\right]^{1/2}$	$\frac{1}{4}\left[\frac{\ln(2)}{2K}\right]^{-1/2}$	0.5

In general we find,

$$\Delta\omega\, \Delta t \geq 0.5, \tag{6-82}$$

where the equal sign holds for any two Gaussian-distributed variables related by the Fourier transform. All other distributions have greater uncertainty products. A small window into the classical uncertainty relation in frequency and time is shown in Fig. 6-18.

We can see how the moments of a spectral distribution are suggestive of a statistical interpretation (see Chapter 12) of the quantum wavefunction product $\Psi^*\Psi$ (like F^*F), but in the early days of quantum theory this was not so obvious.

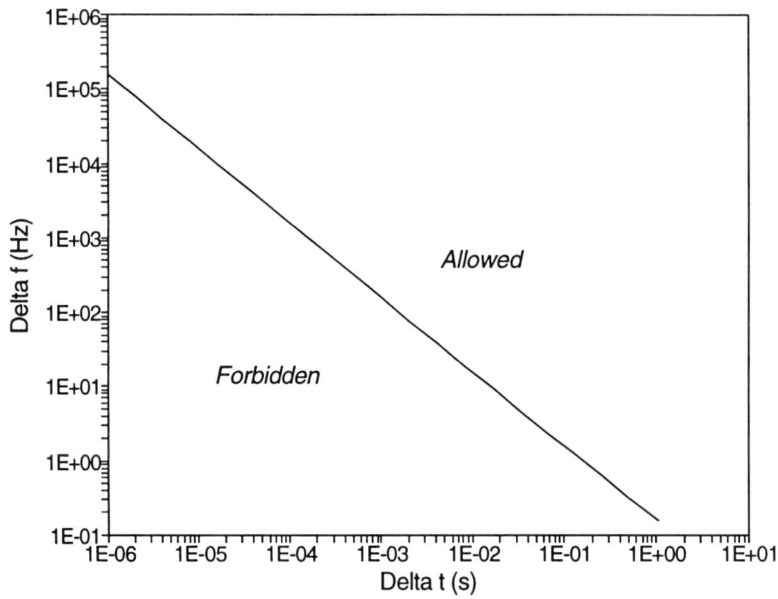

Figure 6-18. Classical uncertainty $\Delta f \Delta t$, over a small range of frequency and time intervals. The line is the minimum uncertainty product (Gaussian).

Max Born was awarded the Nobel prize in 1954 for the statistical interpretation of the quantum wavefunction, but his path to this revelation was through an investigation of the scattering of particles, like the Rutherford experiment.

Similarly, the uncertainty relations of quantum theory follow from the scale change property of the Fourier transform. This was also not so obvious and Werner Heisenberg received the Nobel prize in 1932 for his formulation of quantum mechanics which included the uncertainty principles. The famous mathematician Klein said "All mathematics is obvious once you've seen it."

The scale change property of the Fourier transform is only one of many properties. A few of them are listed in Table 6-3, which shows the basis for an extraordinary range of applications. We have already proved the delay and the complex modulation properties when we examined the uncentered pulse and the oscillatory exponential decay.

6.8 Alternative Formats

The most mystical relation in the annals of mathematical mythology is

$$e^{j\omega t} = \cos(\omega t) + j\sin(\omega t). \tag{6-83}$$

Table 6-3. Some Useful Properties of the Fourier Transform		
Name of Property	*Time*	*Frequency*
Linearity	$Af_1(t) + Bf_2(t)$	$AF_1(\omega) + BF_2(\omega)$
Delay	$f(t - t_o)$	$F(\omega)\, e^{-j\omega t_o}$
Complex modulation	$f(t)\, e^{-j\omega_o t}$	$F(\omega - \omega_o)$
Scale change	$f(kt)$	$\frac{1}{\lvert k\rvert} F(\omega/k)$
Multiplication	$f(t)\, h(t)$	$\int_{-\infty}^{+\infty} F(\omega')\, H(\omega - \omega')\, d\omega'$
Convolution	$\int_{-\infty}^{+\infty} f(\tau)\, h(t-\tau)\, d\tau$	$F(\omega)\, H(\omega)$

Feynman said of Eq. (6-83), "... this most amazing jewel ... the most remarkable formula in mathematics." Equation (6-83) was discovered by Abraham de Moivre, a French mathematician who escaped to England with his parents, and Leonhard Euler, the Swiss mathematical genius who often worked with his noisy children in the room and a baby on his knee. When asked a question, sometimes Newton said "I don't know, ask de Moivre." Late in his life de Moivre spent his time in English taverns discussing probability. Using Eq. (6-83) we can write the Fourier transform pair as,

$$f(t) = \frac{1}{2\pi}\int_{-\infty}^{+\infty} F(\omega)\,[\cos(\omega t) + j\sin(\omega t)]\, d\omega, \tag{6-84}$$

and

$$F(\omega) = \int_{-\infty}^{+\infty} f(t)\,[\cos(\omega t) - j\sin(\omega t)]\, dt, \tag{6-85}$$

or

$$F(\omega) = \int_{-\infty}^{+\infty} f(t)\cos(\omega t)\, dt - j\int_{-\infty}^{+\infty} f(t)\sin(\omega t)\, dt. \tag{6-86}$$

These are called the *Fourier cosine and sine integrals*, and they explicitly select the even and odd parts of the signal.

Discrete Fourier Transform

The discrete Fourier transform (DFT) can be defined in several ways. Bracewell's definition has been widely accepted in engineering and the sciences. With k and τ as discrete indices, the forward DFT is usually written as

$$F(k) = \frac{1}{N}\sum_{\tau=1}^{N} f(\tau)\exp[-j2\pi(k/N)\tau] \tag{6-87}$$

and the inverse DFT is

$$f(\tau) = \sum_{k=1}^{N} F(k)\exp[+j2\pi(\tau/N)k]\,. \tag{6-88}$$

In some works in mathematics the signs in the exponentials are reversed and the normalizing factor $\sqrt{N}$ appears before each summation. The Fast Fourier Transform (FFT) and its applications will be discussed in Chapter 7.

6.9 Laplace Transform

The Laplace transform, useful for solving linear ordinary differential equations with constant coefficients (and some with variable coefficients), resembles the Fourier transform but the transforms are quite different. Part of the beauty of the Laplace transform is that *initial conditions are automatically included in the solution*. The other part of the beauty is that the Laplace transform reduces the solution of an ordinary differential equation to an algebraic operation. The transform is defined as

$$L\{f(t)\} = F(s) = \int_0^{\infty} f(t)\,e^{-st}\,dt\,, \tag{6-89}$$

where s is a complex variable. Table 6-4 lists some common Laplace transforms.

An example will illustrate these points. Let's consider a simple harmonic oscillator with mass $m = 2$ kg, resistive force constant $R = 16$ N m/s, and restoring force constant $\kappa = 50$ N/m. The initial conditions are $x(0) = 0$ and $v(0) = 0$, and the driving force is $100\sin(3t)$ for $t > 0$. In other words, the oscillator is at rest until the driving force is applied. We expect that the complete solution will consist of a transient solution and a steady-state solution with amplitudes and phases determined by the oscillator parameters and the force. The system is described by

$$2\frac{d^2x}{dt^2} + 16\frac{dx}{dt} + 50x = 100\sin(3t), \tag{6-90}$$

or

$$\frac{d^2x}{dt^2} + 8\frac{dx}{dt} + 25x = 50\sin(3t). \tag{6-91}$$

Using the third entry in Table 6-4 for the force, the Laplace transform is

$$(s^2 + 8s + 25)X = \frac{150}{s^2+9}. \tag{6-92}$$

Solving for X we have,

$$X = \frac{150}{(s^2 + 8s + 25)(s^2+9)} \tag{6-93}$$

and after some algebra to put this in a usable format for the inverse Laplace transform we find,

$$X = \frac{75}{26}\frac{1}{s^2+9} - \frac{75}{52}\frac{s}{s^2+9} + \frac{75}{26}\frac{1}{(s+4)^2+9} + \frac{75}{52}\frac{s+4}{(s+4)^2+9}. \tag{6-94}$$

Using pairs 3, 4, 5, and 6 in Table 6-4 we have the complete solution,

$$x(t) = \frac{25}{26}\sin(3t) - \frac{75}{52}\cos(3t) + \frac{25}{26}e^{-4t}\sin(3t) + \frac{75}{52}e^{-4t}\cos(3t). \tag{6-95}$$

The first two terms are the steady-state solution and the last two terms are the transient solution. Figure 6-19 shows the displacement as a function of the force.

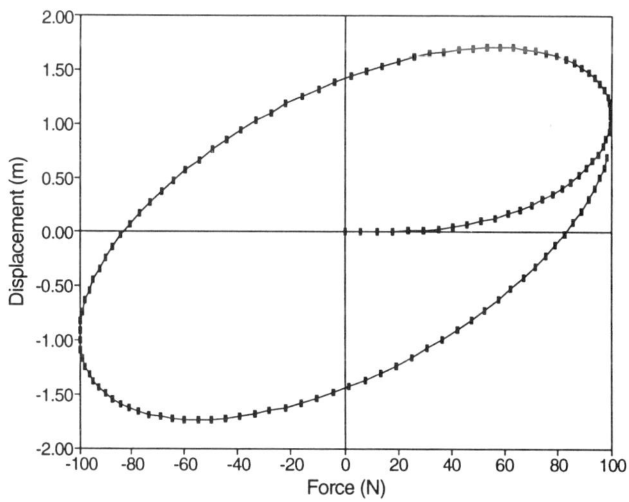

Figure 6-19. Transition to the steady state for the oscillator described in Eq. (6-95). Oscillator starts from rest at $x = 0$. Time interval is 20 ms between markers.

Table 6-4. Short Table of Laplace Transforms	
$F(s)$	$f(t)$
$\frac{1}{s}$	1
$\frac{1}{s-b}$	e^{bt}
$\frac{1}{s^2+a^2}$	$\frac{\sin(at)}{a}$
$\frac{s}{s^2+a^2}$	$\cos(at)$
$\frac{1}{(s-b)^2+a^2}$	$\frac{e^{bt}\sin(at)}{a}$
$\frac{s-b}{(s-b)^2+a^2}$	$e^{bt}\cos(at)$
e^{-bs}	$\delta(t-b)$
$\frac{1}{s}\tanh\left(\frac{bs}{2}\right)$	Square wave, amplitude = 1, period = $2b$

Some mathematicians prefer to use the Laplace transform instead of the Fourier transform, which avoids some difficulties by using a complex variable s and integrating along an appropriate path in the s plane. However, the same procedure can be used with a Fourier transform, considering ω as a complex variable. Engineers and scientists often prefer to use the Fourier transform because it directly involves frequency, a quantity of immediate physical interest. Laplace transforms are more abstract and less in touch with the physical situation. The Fourier transform keeps you in touch with the problem at intermediate points in calculations, and numerical computation with the FFT is a major reason for the popularity of Fourier analysis.

6.10 Hilbert Transform

Examine the real and imaginary parts of Equations (6-52) and (6-64). Causality imposes a connection between the real and imaginary parts of causal response functions, such as permittivity, susceptibility, and impedance. This connection is the Hilbert transform. For a causal response function, the Hilbert transform lets you calculate the real part if you know the imaginary part, and conversely. In other words, you cannot arbitrarily choose both the real and imaginary parts. If you know one part, the other part is determined. The Hilbert transform is completely general for a causal system. For a causal response function such as $Z(\omega) = Z_r(\omega) + jZ_i(\omega)$, the Hilbert transform pair can be written as,

$$Z_r(\omega) = \frac{1}{\pi} P\int_{-\infty}^{+\infty} \frac{Z_i(\omega')}{\omega - \omega'} d\omega' \tag{6-96}$$

$$Z_i(\omega) = -\frac{1}{\pi} P\int_{-\infty}^{+\infty} \frac{Z_r(\omega')}{\omega - \omega'} d\omega', \tag{6-97}$$

where P indicates the Cauchy principal value of the integral is to be taken. A few Hilbert transform pairs are given in Table 6-5. The first two pairs are Equations (6-63) and (6-55) which we obtained earlier by direct calculation using the Fourier transform. See Appendix 2 for a derivation and an application of this transform.

Table 6-5. Some Hilbert Transform Pairs

Real Part	*Imaginary Part*	*Signal*
$\frac{K}{\omega^2 + K^2}$ for $K > 0$	$-\frac{\omega}{\omega^2 + K^2}$	Exponential decay, $\exp[-Kt]$ for $t > 0$
$\frac{\sin(\omega T)}{\omega}$ for $T > 0$	$\frac{\cos(\omega T) - 1}{\omega}$	Rectangular pulse, duration T, $t > 0$
$\cos(\omega t)$ for $\omega > 0$	$\sin(\omega t)$	
$\cos(\omega t)$ for $\omega < 0$	$-\sin(\omega t)$	
$\sin(\omega t)$ for $\omega > 0$	$-\cos(\omega t)$	
$\sin(\omega t)$ for $\omega < 0$	$\cos(\omega t)$	

Unlike the Fourier and Laplace transforms, the Hilbert transform does not relate functions in different domains; it relates parts of a function in the same domain. In x-ray analysis the Hilbert transform pairs are often called the *Kramers-Kronig relations*, and in fundamental particle physics they are called the *dispersion relations*.

Analytical calculation of the Hilbert transform can be a challenging task except for the most simple cases because it requires contour integration. For experimental data it frequently happens that the real part of the response function is measured and the imaginary part is to be computed. Computer programs are available for this but you must be careful to avoid nonphysical results due to finite data sets.

6.11 Parseval's Theorem

This theorem, a respectable mathematical statement by itself, provides another connection between the time domain and the frequency domain, or any two quantities related by the Fourier transform. In physical terms it expresses the fact that the energy of a waveform is the same as the energy of its spectrum. For a discrete waveform and its spectrum, Parseval's theorem takes the form,

$$\sum_{n=0}^{N-1} |f_n(t)|^2 = \frac{1}{N}\sum_{n=0}^{N-1} |F_n(\omega)|^2. \tag{6-98}$$

This tells you how to normalize a spectrum, and it gives you a way to check on the accuracy of the spectrum by computing and comparing both sums. For the continuous case Parseval's theorem is,

$$\int_{-\infty}^{+\infty} |f(t)|^2\, dt = \frac{1}{2\pi}\int_{-\infty}^{+\infty} |F(\omega)|^2\, d\omega. \tag{6-99}$$

For periodic functions with period p, described by Fourier series, Parseval's theorem is,

$$\frac{1}{p}\int_{-p/2}^{+p/2} |f(t)|^2\, dt = \frac{1}{4}A_o^2 + \frac{1}{2}\sum_{n=1}^{\infty} (A_n^2 + B_n^2). \tag{6-100}$$

6.12 Joint Time-Frequency Analysis

Using the pure Fourier transform, it is possible to get an estimate of the spectrum of a signal, but it is not possible to determine exactly when the signal had that spectrum. To overcome this deficiency, the short-time Fourier transform (STFT) performs the FFT on finite blocks of data with selectable time-overlap of the blocks, so it can provide an estimate of a relatively slowly time-varying spectrum. Joint time-frequency analysis (JTFA) plots amplitude or power as a function of both time and frequency. Figure 6-20 shows a complicated time-varying spectrum represented by the STFT with a Hamming window and 50 percent overlap.

The Gabor spectrogram, named for Dennis Gabor (1971 Nobel prize for his discovery of the principles of holography), is a particularly powerful method of performing JTFA.

Wavelet analysis, an even more recent invention, and a subject of intense current research, uses special orthogonal functions with compact support to provide spectral analysis on different scales. "Compact support" means that these functions have the unique property of orthogonality over a finite (usually very small) interval.

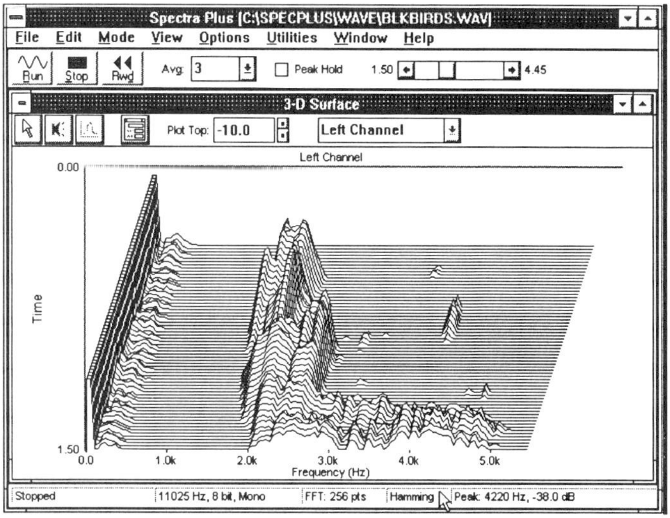

Fig. 6-20. A time-varying spectrum represented by a short-time Fourier transform with a block size of 256 data points and 50 percent overlap, seen through a Hamming window.

6.13 Inductive Leap

The inductive operation referred to here is related to deductive-inductive logical processes, not a quantum leap in an inductance. In 1883 Mark Twain (real name, Samuel Clemens) wrote in *Life on the Mississippi*, "There is something fascinating about science. One gets such wholesale returns of conjecture out of such a trifling investment of fact." Let's collect our information about classical uncertainty relations, from Chapter 1 through this chapter, and add some trifling investments. For Gaussian distributions, our two uncertainty relations are,

$$\Delta p_x \, \Delta x \ \geq \ 0.5 \, \frac{E}{\omega} \tag{6-101}$$

$$\Delta\omega \, \Delta t \ \geq \ 0.5 \, . \tag{6-102}$$

Recall that Eq. (6-101) came from Newton's laws and Eq. (6-102) came from the Fourier transform. Now let's add a trifling investment of quantum information, $E = hf = (h/2\pi)\omega$. If we put this in Eqns. (6-101) and (6-102) we see a remarkable similarity,

$$\Delta p_x \, \Delta x \ \geq \ 0.5 \, (h/2\pi) \tag{6-103}$$

$$\Delta E \, \Delta t \ \geq \ 0.5 \, (h/2\pi) \, . \tag{6-104}$$

Momentum and its associated displacement go together; they are dynamically conjugate variables (Chapter 8). The quantity $h/2\pi$ occurs so often that it is usually written as $\hbar$.

Next let's add a trifling investment of relativity information: Space and time should be treated on the same basis. We can change the time scale in Eq. (6-104) to a space scale by multiplying time by the speed of light, $t \rightarrow ct$. Now the time scale is measured in meters. To retain the form and correct units of Eq. (6-104) we must divide E by c,

$$\Delta(E/c)\,\Delta(ct) \geq 0.5\,(h/2\pi)\,. \tag{6-105}$$

But, E/c is the momentum of a photon, or the momentum carried by an electromagnetic wave in the classical picture. So, Eqs. (6-104) and (6-105) are just other expressions of Eq. (6-103). Energy is the "momentum" associated with the fourth coordinate, time. Relativistic electrodynamics confirms this.

It would probably fascinate Mark Twain that there is even more. Since ω and t, and therefore E and t, comprise a Fourier transform pair, Equations (6-103) and (6-104) suggest that p_x and x are related by the Fourier transform. Later we will see that the quantum wavefunction in x, $\psi(x)$, and in p_x, $\varphi(p_x)$, comprise a Fourier transform pair.

Emma Noether (Göttingen and Bryn Mawr College), possibly the world's greatest woman mathematician to date, proved Noether's theorem, which tells us that there is a conserved quantity corresponding to every symmetry. For example, symmetry in space implies conservation of momentum. Conversely, every conserved quantity implies a symmetry. Noether's theorem is a general statement of something that we have used implicitly, *shift invariance*. Shift invariance (symmetry) in space is associated with conservation of momentum, and shift invariance in time is associated with conservation of energy. Shift invariance in angle leads to conservation of angular momentum. Apparently Newton, Lagrange, and Hamilton understood this. Shift invariance also led to success in the Fourier transform and Laplace's equation.

Would Mark Twain be surprised that the conservation laws can appear to be "violated" a little bit? Equation (6-103) says that momentum conservation appears to be violated a bit if we demand high precision in displacement. Equation (6-104) says that energy conservation appears to be violatable for short times, and the shorter the time, the greater the apparent violation. Equation (6-104) leads to a plausible photon argument for the inverse square nature of the Coulomb force.

What's Next?

In Figures 6-8 through 6-16 we saw that the Fourier transform of a real function is redundant to the extent that, if we know the part for positive frequencies, then we know the part for negative frequencies. When we calculate the discrete Fourier transform using Equations (6-87) and (6-88) we find additional redundancies because there is a pattern in which the same numbers appear.

In the next chapter we will use the fast Fourier transform (FFT), which exploits redundancies, to reduce the task of computation. The amount of reduction is astounding. Transforms that would require years on the fastest mainframe computers can be done in seconds (often milliseconds) with the FFT on your personal computer. The Fourier transform is like the key to the front door of Fort Knox, and the FFT is like the Lear jet that you can use to carry off the gold. Chapter 7 will show you where the gold is hidden, and how to get it.

The Fourier transform generally has a phase, or an imaginary part, except for the special case of a perfectly centered function. There is a question concerning the relative amount of information contained in the magnitude and the phase (or the real and the imaginary). One might think that the magnitude or the real part would convey the most information but there is some experimental evidence that, at least in the case of very asymmetric two-dimensional data, the phase is most important ["The role of amplitude and phase of the Fourier transform in the digital image processing," I. Juvells, S. Vallmitjana, A. Carnicer, and J. Campos, *Am. J. Phys.* **59**, 744–748 (1990)]. This somewhat counterintuitive result requires further investigation.

In Chapter 7 we will explore the harmonic oscillator in the general context of linear systems. There is a considerable body of knowledge now about linear systems and one chapter, or even one book, could not cover everything, but we will see how harmonic oscillators fit into the general scheme of things. The use of the FFT will enable you to perform feats of computation that people could only dream about in earlier years. You will be able to experiment with random noise in addition to well-behaved, integrable functions, and you will be able to study harmonic oscillators in ways you probably never imagined.

Physical quantities that are related by the Fourier transform play a central role in our description of Nature. In quantum mechanics, the position probability amplitude $\psi(x)$ and the momentum probability amplitude $\varphi(p)$ comprise a Fourier transform pair. Consequently, both probability densities contain the same information, but the information is encoded in different ways. The two representations are complementary. In Chapter 10 we will examine probability distributions for the classical harmonic oscillator. Chapter 12 will introduce probability distributions for the quantum oscillator.

Certainly, the position probability distribution in the most familiar but the momentum probability distribution is becoming increasingly important in experimental research because it corresponds to directly measured quantities. For example, we obtain momentum distributions in nuclear scattering experiments, neutron scattering in solids and liquids, and high-energy collisions in elementary particle physics.

A vital feature of the Fourier transform and the FFT is that they provide *numerical results*. It is impossible to overestimate the practical importance of this fact for the experimental engineer and scientist. Bless you, Baron Fourier.

Fourier's theorem is not only one of the most beautiful results of modern analysis, but it may be said to furnish an indispensable instrument in the treatment of nearly every recondite question . . .

Lord Kelvin

I worked on the theory of Fourier integrals . . . for a good many years before I discovered that this theory has applications . . .

E. C. Titchmarsh
Journal of the London Mathematical Society **25**, 85 (1950)

Professor Titchmarsh is the author of *Introduction to the Theory of Fourier Integrals* (Oxford, 1937). Titchmarsh's statement is understandable; he worked with the legendary Godfrey Harold Hardy at Oxford. Professor Hardy (of Oxford, Princeton, and California Institute of Technology) was a genius at age two. He considered himself a *pure* mathematician and once said "I have never done anything useful."

This difficult investigation requires a special calculus based on new theorems . . . it leads us to the point where we can obtain numerical values, a necessary property of any such method, without which we obtain nothing but useless formal transformations.

Baron Jean Joseph Baptiste Fourier
Théorie Analytique de la Chaleur (1822)

Chapter 7

Elements of Linear Systems

The essence of knowledge is, having it, to apply it.
Confucius

Now we are going to consider the classical simple harmonic oscillator in the context of the general theory of linear systems. In fact, it is the quintessential linear system. There are many books devoted entirely to the subject of linear systems; one chapter can only scratch the surface. Our objective is to go light on theory and, following Confucius, concentrate on applications, using numerical methods for real-world situations to make your knowledge more concrete. Our most important tool in these practical applications will be the Fourier transform and one of its discrete industrial-strength algorithms, the FFT.

7.1 What Is a Linear System?

Physically, a linear system is one in which the output (response) is directly proportional to the input (stimulus). If this sounds familiar, it's because we discussed impedance and admittance in Chapter 5. Mathematically, a linear system is described by linear equations, so we are assured that the principle of superposition holds. There is a slightly fuzzy border between linear and nonlinear systems, in a mathematical sense, because some physically nonlinear systems are described by nonlinear equations that can be transformed into linear equations with time-varying coefficients. In the physical world many systems are intrinsically nonlinear, even at low power levels, and most linear systems become nonlinear at high power levels.

Causality in linear systems takes the form of a deceptively simple statement that the stimulus must precede the response; there can be no output before the input arrives. This is an added assumption that strictly limits the types of functions and operators that can be used to characterize linear systems. Note that causality is an additional requirement that we impose because it seems to be a sensible thing to do. However, this opens a difficult and subtle set of questions, including determinism and the energy-time uncertainty relation, which makes "before and after" and simultaneity lose their sharp meanings. Here we will disregard these questions and assume that we are in the classical domain with measurements on a scale that cannot detect quantum uncertainties. This puts the lid on Pandora's box, for now.

One practical measure of linearity is that the frequencies in the output are the same as the frequencies of the input; no new frequencies appear in the output. The reason that this is so useful is that very small amounts of energy can be measured in the frequency domain, even in the presence of very large amounts of energy at other frequencies. Identification and elimination of harmonic and subharmonic energy is an important subject of current experimental and theoretical interest. The waveform in electrical power systems is maintained sinusoidal to very high precision, because even small amounts of current at harmonic frequencies can have serious economic consequences and may interfere with control circuits and power carried by the neutral line.

7.2 Ode to the Sinusoid

Sinusoids (and, therefore, complex exponentials) appear to be Nature's building blocks. Why do sinusoids play such a dominant role? One reason is that, when a linear system is excited by a sinusoidal time function, the steady-state response throughout the system is sinusoidal, differing at most by amplitude and phase from one part to another. The linear system may be of unlimited complexity; the response remains sinusoidal. *No other periodic function has this claim to fame.*

Also, *no other periodic function remains unchanged when it is added to a similar function,* but of different amplitude and phase. A sinusoid, when added to any other sinusoid of the same period and different amplitude and phase, produces a function having the same sinusoidal form as the constituents. Ernst Guillemin (MIT) called this "an uncanny property." What we regard as commonplace is actually unusual.

The shape of a sinusoid is unique, and, in a sense, it is the smoothest of all functions. Psychoacoustically, a pure tone sounds beautiful. The derivative of a sinusoid is also a sinusoid, and therefore the integral of a sinusoid is also a sinusoid. Only the amplitude and phase are changed due to differentiation or integration; the sinusoidal shape is invariant to differentiation or integration. In general, you know that differentiation tends to accentuate irregularities of a function, and integration tends to smooth out irregularities. *The sinusoid is unique in that its shape is invariant to integration or differentiation, no matter how many times the operations are repeated.*

Fourier series, composed of sines and cosines, enable you to approximate an arbitrary function (subject to Dirichlet conditions) with least-mean-square error. By going to the limits of arbitrarily long and short periods you can represent practically any function of time or space with a series of sinusoids. Thus the sinusoid is the basic building block from which almost anything else can be constructed, and *we can find the response to any excitation if we know the response to sinusoids.*

It is impossible to over-emphasize the importance of the sinusoid and Fourier methods in science and technology. If sinusoids and these methods did not exist, surely they would have to be invented.

7.3 Fast Fourier Transform

Before we discuss linear systems, let's introduce the spreadsheet version of the principal tool that we will be using. The FFT is a discrete Fourier transform algorithm that is unusually fast. Actually, there are several versions of the FFT. The one in spreadsheets requires that the number of data points in the original and transform domains be an integer power of two, such as 128, 256, 1024, and so on. The spreadsheet contains the forward transform, which we will call FFT, and the inverse FFT, which we will call IFFT. We will test the accuracy of the transformations by first going forward and then inverse, and see how close we come to the original data.

In the spreadsheet, FFT and IFFT do not recalculate automatically. So, every time the input data change you must call up the FFT and perform the operation again. You can write a macro to automate this task. There are several tricks to using the FFT, and we will take them up as we go.

Be sure to assign the same number of cells occupied by input data to the row or column where you want the transformed data set to reside. In the on-screen dialog box be sure that the "inverse" box *is not checked* if you want the FFT, and be sure that it *is checked* if you want the IFFT.

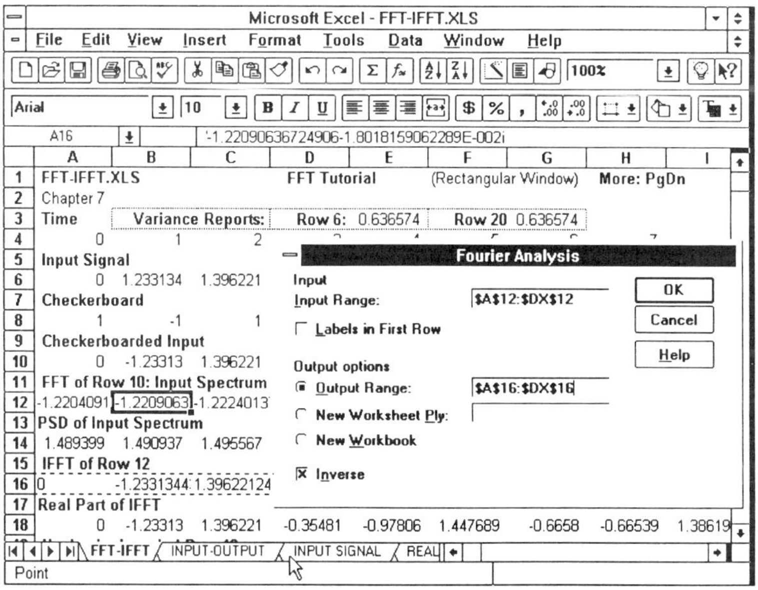

Figure 7-1. Home screen of FFT-IFFT. The Fourier Analysis dialog box is active. Note complex contents of Cell B12, shown in editing line below icons.

The transformed data in your worksheet (row 12, Fig. 7-1) are not in numerical form; each entry is preceded by an apostrophe (') that indicates it is non-numerical. The reason for this is that most transformed data are complex, even when the original data are purely real. Notice that the transformed data are in *Excel* 5.0 complex format '$x + yi$ in the formula editing box in Fig. 7-1, which shows the Home screen of the tutorial file FFT-IFFT with the Fourier Analysis dialog box in view.

We will use spreadsheet complex functions IMREAL and IMAGINARY to separate transformed data so we will have the real and imaginary parts, and then we will use IMABS and IMARGUMENT for magnitude and phase. These functions, and more, are in Data Analysis which should be visible in the menu box when you click on Tools at the top of the screen. If Data Analysis is not visible, run Setup to install these analysis options. Figure 7-2 shows the very small errors involved in transforming from the original data domain with the FFT and inverse-transforming the spectrum back to the original using the IFFT. You will notice in FFT-IFFT that the input data are purely real but the results of the IFFT produce a non-numerical data set back in the original domain. You need to convert the IFFT output into numerical data using the spreadsheet function IMREAL as shown on your screen in row 18.

Later we will perform some analytical calculations so you can compare the exact results with the FFT results. Variance reports are in Cells E3 and G3. Press PgDn to see the non-checkerboarded variances, identical to six decimal places. The FFT provides you with astounding power, but it has some peculiarities that you should be aware of. A major source of errors for nonperiodic data is that the FFT always assumes that data are periodic. As a result, the FFT may be inaccurate for non-periodic data as you approach the ends of the positive and negative frequency axis.

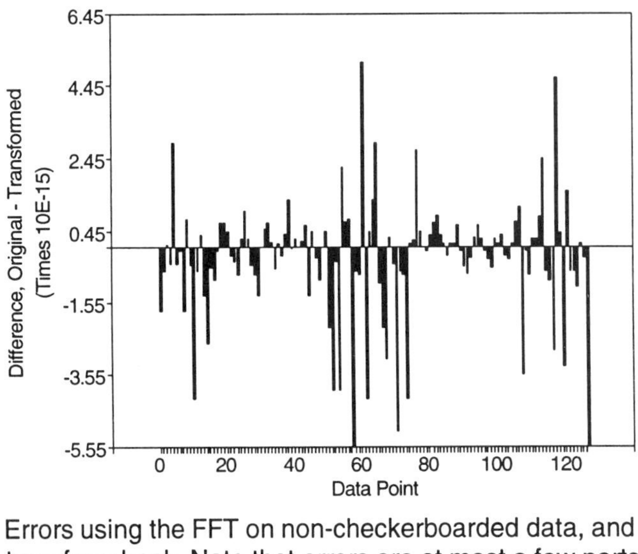

Figure 7-2. Errors using the FFT on non-checkerboarded data, and then using the IFFT to transform back. Note that errors are at most a few parts in 10^{15}.

This *imposed periodicity* is clear when you compare FFT results with the exact results of real and imaginary, magnitude and phase, and the polar plot of imaginary versus real. The polar plot always closes because the FFT always expects to join smoothly with an identical data set on the right and left ends. Let's summarize some more tips and tricks that you'll need to use for successful operation with the FFT.

Nyquist shift, "checkerboarding," and centering spectra. Application of the FFT to a raw data set may shock you at first because it results in a funny-looking spectrum that has the central part at the ends of the graph and the ends at the center. Multiplying every odd-numbered data point by –1 before the FFT operation gives a spectrum that looks like one you would calculate by hand. If you are not interested in examining the spectrum directly, and only want to use it in computations, you can eliminate the Nyquist shift which takes time and computing power. To see how the Nyquist shift is applied, use your mouse to explore rows 8 and 10 in the FFT-IFFT worksheet, shown in Fig. 7-1.

Zero padding. If your original data set does not have a number of points that is an integer power of 2, you can add zeros to the data to bring the number up to the next power of 2. For example, if you have 118 data points, add 10 zeros to bring the number to 128. Why not add 138 zeros to bring the number to 256? Experiment with this and see what happens. The first thing you'll notice is that the spectrum *appears* to be much smoother, but you have not added any more information. The downside of zero-padding is that, statistically speaking, the variance is artificially decreased and the bias is increased. Experiment with a small data set, say 14 points. Add 2 points to get 16, and look at the FFT. Compare with adding 18, 50, and 114 zeros.

Frequency scaling. Because the FFT is a discrete transform, there is a connection to the Nyquist sampling theorem. If the sampling interval is Δt seconds then the associated sampling frequency is

$$f_{sample} = \frac{1}{\Delta t} \text{ Hz.}$$

Many data acquisition boards can collect 10^5 samples per second, so Δt is 10^{-5}s. If the total number of samples is N, then the frequency interval between data points in the spectrum after FFT processing is

$$\Delta f = \frac{1}{N \Delta t} \text{ Hz.}$$

For example, if N is 512 samples and Δt is 10 μs then the frequency scale is 195.3125 Hz between data points in the spectrum. Information is located only at these data points; there is nothing between them. Increasing the number of samples makes the scale smaller. Collecting more samples is always a good idea because it increases information, but it also requires more time and computing power.

Sampling and Aliasing. Harry Nyquist showed that it is possible to reconstruct a continuous-time signal completely from discrete equally spaced samples, under the condition that the highest frequency that you want to know with confidence is less than half of the sampling frequency. Sampling at a lower frequency results in a phenomenon called "aliasing," which occurs when a high frequency masquerades as a low frequency. This strobe effect can be seen in movies (a sampled data system with 24-Hz sampling rate) when wheels appear to rotate slowly, or backwards.

When collecting data you should stay far below the Nyquist frequency. In a data acquisition system with 100 kHz sampling, the Nyquist frequency is 50 kHz. Most sound cards in computers have a sampling rate of 44.1 kHz with an associated Nyquist frequency of 22.05 kHz, which is just above the upper limit of human hearing. Some digital sound systems may have a specification of "8 times over sampling," which means the Nyquist frequency is eight times the highest frequency amplified by the system. Oversampling enables use of inexpensive, simple filters after A/D conversion.

"Anti-aliasing" filters are low-pass filters used ahead of A/D converters to reduce energy above the Nyquist frequency prior to analog-to-digital conversion. It is important that an anti-aliasing filter have a linear phase characteristic in the passband to minimize phase distortion.

Data Windowing. The Nyquist sampling theorem provides the connection between continuous time signals and digitized (discrete) information, but it does not tell the entire story. The width of the sampling pulse and the start and stop of sampling are important in terms of spectrum "leakage." An abrupt start and stop of sampling a continuous-time signal smears frequency resolution; leakage involves spreading energy into adjacent frequencies. For an example of this problem, see Figs. 6-5 and 6-6, which show the influence of pulse width and pulse spacing on spectrum width and spectral density.

"Data windows" have been invented to address this problem by softening the start and stop of the data block. Many types of windows are in use; each one has special attributes and there is at present no way to decide which one is best. In fact, asking which one is best is an ill-posed question; there is no way to decide what "best" means because of the multiplicity of parameters that are associated with windows, in addition to the primary purpose of enhancing resolution (see Section 7.17). For simplicity in the tutorial, Fig. 7-1 uses no window, but this is not really true because this means that a rectangular window is automatically used. A rectangular window has an abrupt start and stop; it is usually the worst choice.

Common windows that will be used in later worksheets are the Hamming, Kaiser-Bessel, von Hann, and Blackman-Harris. These windows provide amazing improvements over a rectangular window; sometimes the improvement can be as great as a factor of 10^9 in reduction of spectrum leakage.

Excel for Windows 95 is considerably faster and easier to use than *Excel* Version 5.0 when it computes the FFT.

7.4 Test Signals

There is an infinite variety of possible signals that you can use to probe a system. Of this infinite set some are more useful than others because of our ability to generate approximations to specific signals. The sine, the unit step, the impulse (delta) function, and random noise have been found to be particularly useful. Sines are very narrow-spectrum signals; steps, impulses, and noise are wide-spectrum signals.

7.5 Step Function Response

A step function is sometimes called a Heaviside function after Oliver Heaviside. The response to a step function completely characterizes a linear system. To illustrate the method in an easy way let's examine a familiar non-oscillatory system. Consider a series RC circuit with the capacitor fully discharged at $t = 0$.

For our present purposes let $V = V_o$ be a constant voltage. At $t = 0$ we impose this voltage across R and C in series and add up the voltages,

$$V_o - IR - \frac{q}{C} = 0. \tag{7-1}$$

At $t = 0$, $q = 0$ and $V = V_o$. We can include the initial conditions by means of a unit step function $u(t)$, which has the properties that $u(t)$ is zero until $t = 0$ and after that $u(t) = 1$ forever. You can start the step function at any time t_o by using $u(t - t_o)$, and you can form a pulse of length T by using $u(t) - u(t - T)$. Since $I = dq/dt$ Eq. (7-1) can be written as

$$R\frac{dq}{dt} + \frac{q}{C} = u(t)V_o\,, \tag{7-2}$$

which has a particular solution $q(t) = u(t)CV_o$. The reduced equation can be written as

$$\frac{dq}{q} = -\frac{1}{RC}dt, \tag{7-3}$$

which has a solution,

$$q(t) = u(t)\,CV_o\,e^{-t/RC}. \tag{7-4}$$

So the complete solution is

$$q(t) = u(t)CV_o - u(t)CV_o\,e^{-t/RC}, \tag{7-5}$$

or

$$q(t) = u(t)CV_o\,(1 - e^{-t/RC}). \tag{7-6}$$

It is convenient to normalize this to the charge $q_n = CV_o$, and when we do this we obtain the normalized response to a unit step function input,

$$q_n(t) = \frac{q(t)}{CV_o} = u(t)(1 - e^{-t/RC}). \tag{7-7}$$

7.6 Impulse Response

The time derivative of $q_n(t)$ is the normalized current $I_n(t)$,

$$I_n(t) = u(t)\frac{1}{RC}e^{-t/RC} \tag{7-8}$$

and this is the normalized impulse response (the response to a delta function input). The standard notation for the normalized impulse response is $h(t)$ and we will use this starting now,

$$h(t) = u(t)\frac{1}{RC}e^{-t/RC}. \tag{7-9}$$

We have explicitly included $u(t)$ as a reminder that the impulse response is a causal function: There is no output before there is input. Because the delta function $\delta(t)$ is the derivative of the unit step function, the impulse response is the derivative of the response to a unit step function. (We're omitting some subtle mathematical questions here.) Both are standard ideal test signals with the useful property that each completely describes the system in the time domain. They can only be approximated using hardware because nothing has a zero rise-time and nothing goes on forever.

The RC circuit we have been discussing is a single-pole low-pass filter (integrator), because the input is across R and C in series and the output is across C. Its response to a unit step function and impulse response are shown in Fig. 7-3.

7.7 Frequency Domain Response Using Phasors

Now that we've taken care of the time domain let's examine the frequency domain. We start with a general input V_{in} and sum the voltages around the loop after V_{in} has been applied for a time much greater than RC.

$$V_{in} - IR - \frac{q}{C} = 0, \tag{7-10}$$

so

$$V_{in} = IR + \frac{q}{C}. \tag{7-11}$$

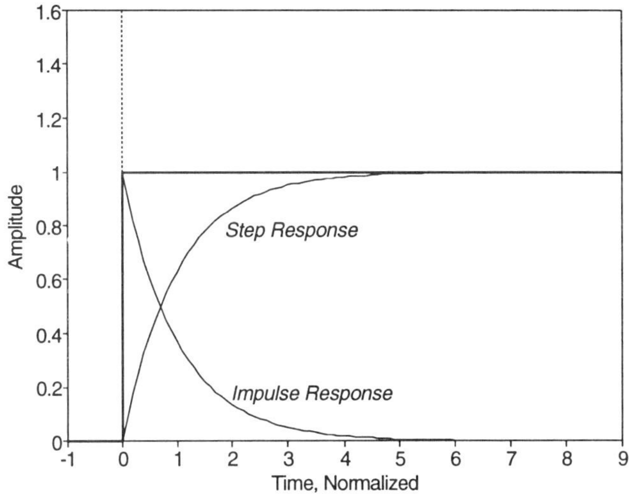

Figure 7-3. Unit step function response and impulse response of the *RC* low-pass filter. Time is normalized in units of *RC* seconds. Note causality.

Now let $q(t) = q_o \exp[j\omega t]$, so $I(t) = j\omega q(t)$ and we have

$$V_{in} = j\omega qR + \frac{q}{C} = (j\omega RC + 1)\frac{q}{C}. \tag{7-12}$$

But q/C is just the voltage across the capacitor, which is the output voltage V_{out} so we can write,

$$\frac{V_{in}}{V_{out}} = 1 + j\omega RC \tag{7-13}$$

Invert Eq. (7-13) to obtain the usual expression for the Gain, $G(j\omega)$,

$$G(j\omega) = \frac{V_{out}}{V_{in}} = \frac{1}{1 + j\omega RC} = \frac{1}{1 + (\omega RC)^2} - j\frac{\omega RC}{1 + (\omega RC)^2}. \tag{7-14}$$

This is the frequency domain response of the system in Cartesian coordinates, that is, in terms of the real and imaginary parts. From now on we are going to call the frequency domain response $H(\omega)$,which is standard notation for the system function or transfer function,

$$H(\omega) = \frac{1}{1 + (\omega RC)^2} - j\frac{\omega RC}{1 + (\omega RC)^2}. \tag{7-15}$$

$H(\omega)$ completely describes the system in the frequency domain, just as $h(t)$ or the step-function response completely describes the system in the time domain. The same information is contained in all of these functions.

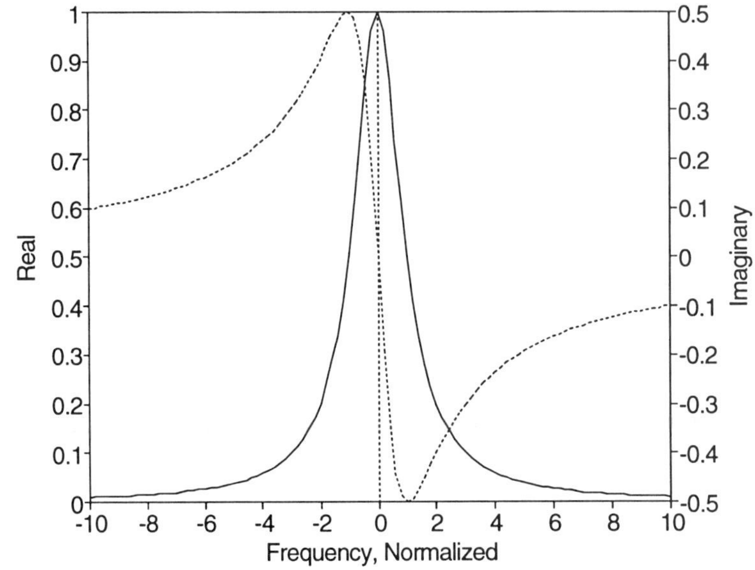

Figure 7-4. Real and imaginary parts of $H(\omega)$ for the *RC* integrator. Compare Fig. 6-14. *Solid line:* Real part. *Dotted line:* Imaginary part. Frequency is normalized in units of 1/*RC*. Compare FFT phase results in Fig. 7-14.

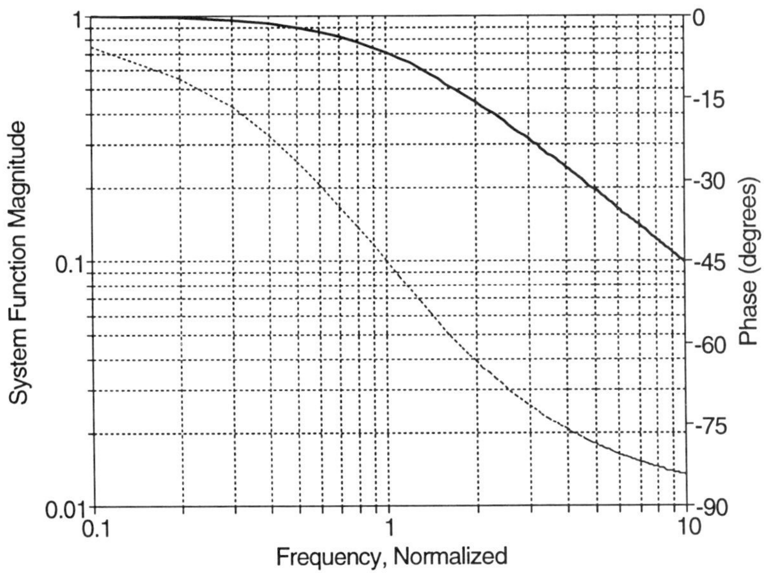

Figure 7-5. Magnitude and phase of $H(\omega)$ for the *RC* integrator. *Solid curve:* magnitude. *Dotted curve:* phase. Frequency is normalized in units of 1/*RC*.

In Sections 7.9 and 7.10 we will see how to use the Fourier transform to find $H(\omega)$ if you know $h(t)$, and how to find $h(t)$ if you know $H(\omega)$. In other words, the impulse response and the system function comprise a Fourier transform pair.

The real and imaginary parts of $H(\omega)$ in Eq. (7-14) are shown in Fig. 7-4 and the magnitude and phase in Eqns. (7-19) and (7-20) are shown in Fig. 7-5. These exact analytical results will be compared with FFT results using the spreadsheet.

It is often more useful to express this in polar coordinates, that is, in terms of the magnitude $|H(\omega)|$ and phase $\theta(\omega)$,

$$|H(\omega)| = \sqrt{\mathrm{Re}^2 + \mathrm{Im}^2} \tag{7-17}$$

$$\theta(\omega) = \arctan(\mathrm{Im}/\mathrm{Re}). \tag{7-18}$$

So, the frequency domain response becomes

$$|H(\omega)| = \frac{1}{\sqrt{1 + (\omega RC)^2}} \tag{7-19}$$

$$\theta(\omega) = \arctan(-\omega RC). \tag{7-20}$$

Because of the wide range of the magnitude it is convenient to scale the magnitude in decibels. In terms of volts, $\mathrm{dB} = 20 \log_{10}(V_{out}/V_{in})$

$$\mathrm{dB} = 20 \log_{10}(|H(\omega)|) = -10 \log_{10}[1 + (\omega RC)^2] . \tag{7-21}$$

Note in Fig. 7-5 and Eq. (7-21) that the magnitude falls to 0.707, and the phase shift is –45 degrees at the corner frequency $1/RC$ rad/s according to Eq. (7-20).

Figure 7-5 is a standard plot format with the magnitude on a log scale, phase linear in degrees, and a logarithmic frequency scale. For a single-pole low-pass filter the magnitude always falls to –3 dB (half power) and the phase shift is always –45 degrees at a frequency that is the reciprocal of the time constant, which is equal to a frequency of 1 on this normalized frequency scale.

Now that we have $H(\omega)$ in polar form we can express it conveniently as

$$H(\omega) = |H(\omega)|\, e^{j\theta(\omega)}, \tag{7-22}$$

and we can do several interesting things:

- If we know $H(\omega)$ and the input spectrum $F_{in}(\omega)$ we can find the output spectrum. This is *linear filtering* in the frequency domain, and we can then find the filtered waveform by using the inverse Fourier transform.

$$F_{out}(\omega) = H(\omega)F_{in}(\omega). \tag{7-23}$$

- If we know the output $F_{out}(\omega)$ and $H(\omega)$ we can find the input. This is called *deconvolution* or inverse filtering; we shall return to this.

$$F_{in}(\omega) = \frac{F_{out}(\omega)}{H(\omega)} . \tag{7-24}$$

- We can determine the system function by sending in a known signal $F_{in}(\omega)$ and measuring the output $F_{out}(\omega)$. This means that we can measure $H(\omega)$ using any signal, even random noise. Finding $H(\omega)$ is called *system identification.*

$$H(\omega) = \frac{F_{out}(\omega)}{F_{in}(\omega)} . \tag{7-25}$$

7.8 Impulse Response and Causality

A classical system is not physically realizable if its impulse response is not causal. One way to check on causality is by means of the Paley-Wiener condition. If $h(t)$ is square integrable, a necessary (but not sufficient) condition for $h(t)$ to be causal is given in terms of its transfer function $H(\omega)$. The Paley-Wiener condition is

$$\int_{-\infty}^{+\infty} \frac{|\ln|H(\omega)||}{1+\omega^2}\, d\omega < \infty.$$

Causality in $h(t)$ is not a simple problem (see R. E. A. C. Paley and N. Wiener, *Fourier Transforms in the Complex Domain*, American Mathematical Society Colloquium Publications, Vol. XIX, New York, 1934). Paley died at age 26 in a skiing accident.

If $h(t)$ is non-causal it is often possible to introduce a time delay such that $h(t)$ is negligible before $t = 0$. A long wait is the price you pay for approximating a non-causal function.

Causality and the Hilbert Transform

Just as causality is manifested in the time domain by the impulse response, the Hilbert transform (Section 6.10) expresses causality in the frequency domain. Causality imposes relations between the real and imaginary parts of linear response functions such as the system function, impedance, admittance, and complex permittivity and permeability. Because of these relations it appears that you do not have to measure both the real and imaginary parts; if you know one, then you know the other. This is commonly done, and computer programs exist just for this purpose. There are some subtle difficulties with implementation of the Hilbert transform and we will not pursue it further at this point (see Appendix 2 for a derivation and application). You should try it with a classical harmonic oscillator.

7.9 System Function via Fourier Transform

As a simple example let's consider $h(t)$ for the RC circuit, given by Eq. (7-9). This is a special case, but it is more general than it appears because of the complex modulation-spectrum shift property of the Fourier transform. The Fourier transform of the impulse response is given by

$$H(\omega) = \frac{1}{RC}\int_{-\infty}^{+\infty} u(t)e^{-t/RC}e^{-j\omega t}\,dt. \tag{7-27}$$

Collect exponentials and set the lower limit with $u(t)$,

$$H(\omega) = \frac{1}{RC}\int_{0}^{\infty} \exp[-(1/RC + j\omega)t]\,dt \tag{7-28}$$

and this integrates to,

$$H(\omega) = \frac{1}{RC}\frac{\exp[-(1/RC + j\omega)t]|_0^\infty}{-(1/RC + j\omega)} = \frac{1}{RC}\frac{(0-1)}{-(1/RC + j\omega)}. \tag{7-29}$$

so,

$$H(\omega) = \frac{1}{RC}\frac{1}{(1/RC + j\omega)} = \frac{1}{1 + j\omega RC}. \tag{7-30}$$

As usual let's separate the real and imaginary parts and simplify,

$$H(\omega) = \frac{1}{1 + (\omega RC)^2} - j\frac{\omega}{1 + (\omega RC)^2}. \tag{7-31}$$

This is exactly what we found for $H(\omega)$ in Eq. (7-16) using phasors. Now you know how to find the frequency response of a system if you measure the system's response in the time domain. It is often much faster to measure something in the time domain and then use the FFT to obtain the frequency response; this is the secret of Fourier-transform spectroscopy, used in the infrared and in pulsed NMR. On the other hand, it is usually more accurate to measure spectra slowly in the frequency domain.

7.10 Impulse Response via Fourier Transform

Next, let's see how you can compute the impulse response of a system if you measure the frequency response, the system function. Suppose you have measured $H(\omega)$ and now you need $h(t)$. In general, $h(t)$ is the inverse Fourier transform of $H(\omega)$,

$$h(t) = \frac{1}{2\pi}\int_{-\infty}^{+\infty} H(\omega)\, e^{+j\omega t}\, d\omega \tag{7-32}$$

so for the special case of the *RC* filter we have

$$h(t) = \frac{1}{2\pi}\int_{-\infty}^{+\infty} \frac{e^{+j\omega t}}{1 + j\omega RC}\, d\omega. \tag{7-33}$$

This integral looks tricky, but it isn't. There are several ways to do this integral. First, you could look it up in a table of Fourier transforms. Second, if you write the exponential in terms of $\cos(\omega t) + j\sin(\omega t)$ and rewrite the integral in terms of the real and imaginary parts, then the odd part integrates to zero because it is an odd function and you are left with the even function.

If you are alone on a desert island you can do it without any tables, and this may be the best way. Consider $j\omega$ as a complex variable and rewrite Eq. (7-33) as,

$$h(t) = \frac{1}{2\pi jRC}\int_{-\infty}^{+\infty} \frac{e^{+j\omega t}}{1/RC + j\omega}\, d(j\omega). \tag{7-34}$$

This can be evaluated immediately with the Cauchy residue theorem. The integrand has a pole at $j\omega = -1/RC$ so the integral is $2\pi j$ times the value of the residue, $\exp[-t/RC]$, at the pole and the impulse response is

$$h(t) = u(t)2\pi j \frac{1}{2\pi j}\frac{1}{RC} e^{-t/RC} = u(t)\frac{1}{RC} e^{-t/RC}. \tag{7-35}$$

This is just what we found in Eq. (7-9) by the solution of a differential equation. Now you know how to find $h(t)$ and $H(\omega)$ by several methods.

We have spent a lot of time with the *RC* integrator instead of an oscillator because the *RC* circuit is simple, easy to understand, and it saves printer's ink in equations. It should be obvious that what we did translates into harmonic oscillators because of the *complex modulation* property of the Fourier transform, Table 6-3. Multiplication of a function of time by $\exp[-j\omega_o t]$ shifts its spectrum by $F(\omega - \omega_o)$.

7.11 Convolution

Convolution is a special way of multiplying two sequences of numbers. The process of convolution is everywhere. Human speech involves convolution of the pulses from the vocal chords with the cavity resonances of the throat and head; the sound is further convolved before reception. Vision and hearing are convolutional processes with restricted bandwidths, complicated by nonlinearities. Transmission of electromagnetic signals through the atmosphere, the ionosphere, and fiber optics, signals in circuits, and other dispersive, absorptive media are similar convolutions. When two probability densities are joined, they must be joined by convolution instead of simple addition.

Convolution in space is also common. For example, the slit function $h(x)$ of a spectrometer corresponds to $h(t)$. The slit convolves the light passing through it and smears the spectrum. Another example, involving both space and time, is when a laser beam sweeps over a detector. Suppose the laser beam has a Gaussian radial intensity distribution and the detector is a narrow strip. The detector output, a function of time, is the convolution of the detector width, the laser beam intensity distribution in space and the sweep speed, and the detector response time.

Convolution can, of course, be performed with any variable but to be specific we will discuss functions of time. The convolution of two functions $f_1(t)$ and $f_2(t)$ is defined as

$$f_1(t) \otimes f_2(t) = \int_L^U f_1(t)\, f_2(t-\tau)\, d\tau\,, \tag{7-36}$$

where the symbol $\otimes$ indicates convolution. Convolution is just another kind of multiplication in which one function is multiplied by a time-reversed version of another function, and their overlap products are summed for various time displacements τ.

Setting the lower and upper limits in Eq. (7-36) is sometimes a source of difficulty. For two functions with lower bounds L_1 and L_2 and upper bounds U_1 and U_2, choose the largest of the lower bounds as the lower limit, and choose the smallest of the two upper bounds as the upper limit.

The filter aspect of convolution can be visualized most vividly when $f_2(t)$ is the impulse response of a filter, $h(t)$. Because a delta function contains all frequencies with equal amplitudes, $h(t)$ completely describes the system. As we have seen, $h(t)$ comprises a Fourier transform pair with the system function $H(\omega)$ which also completely describes the system, in the frequency domain. In the time domain,

$$f_{out}(t) = f_{in}(t) \otimes h(t) = \int_L^U f_{in}(t)\, h(t-\tau)\, d\tau\,. \tag{7-37}$$

Basically, the process of convolution in Eqs. (7-36) and (7-37) is shift, multiply, and add. Note that the convolution integral results in a function of t, not τ; it is important to be aware of this distinction when comparing convolution with the correlation function (Section 7.14).

The integral definition of convolution is intuitively satisfying, but the discrete implementation is the most useful and practical with actual data, The operation is particularly easy to implement for small discrete data sets on a spreadsheet. The shift-multiply-and-add property makes convolution a filter by a sort of weighted moving average. Sometimes this is called a *sliding window*. Equations (7-36) and (7-37) and their discrete versions are relatively slow to execute for large data sets.

Fortunately, the convolution theorem enables us to perform convolution swiftly and easily in the frequency domain using the FFT.

Convolution Theorem

If you have two probability density functions of two continuous independent random variables then the convolution integral gives the probability density of their sum. This fact, discussed in most works on statistics, would be a sufficient claim to fame for the convolution integral but there is more. Of the several convolution theorems for the Fourier, Laplace, Mellin, and other transforms, what is often known as "the" convolution theorem says that convolution of the functions $f_1(t)$ and $f_2(t)$ in the time domain is equivalent to simple multiplication of the Fourier transforms $F_1(\omega)F_2(\omega)$ in the frequency domain. The product of the two spectra is called the cross-power spectral density or cross-PSD. This interpretation expresses the fact that the output spectrum has only the frequencies that the two spectra share in common.

This is important because it is the key to the use of the FFT in convolution computations,

$$f_1(t) \otimes f_2(t) = \frac{1}{2\pi}\int_{-\infty}^{+\infty} F_1(\omega)\, F_2(\omega) e^{+j\omega t}\, d\omega \tag{7-38}$$

where, as usual,

$$F_i(\omega) = \int_{-\infty}^{+\infty} f_i(t) e^{-j\omega t}\, dt.$$

In view of Eq. (7-38), $H(\omega)F_{in}(\omega)$ is the cross-PSD of the output spectrum $F_{out}(\omega)$,

$$F_{out}(\omega) = H(\omega)F_{in}(\omega)\,. \tag{7-39}$$

- The output spectrum is the cross-PSD composed of the system function and the input spectrum. The output spectrum only has frequencies shared in common by the system function and the input spectrum.
- The output waveform is the inverse Fourier transform of the output spectrum.

This shows the simplicity of convolution in the frequency domain. Compute the input spectrum, multiply it by the system function to form the cross-PSD, and compute the inverse Fourier transform of the cross-PSD. When the FFT is used, this is called *circular convolution* because of the imposed periodicity of the FFT.

The inverse theorem says that multiplication of the original functions in the time domain is convolution of the Fourier transforms in the frequency domain.

The complex operations outlined above can be analytically formidable, but the discrete version can be performed with ease, grace, and speed in a spreadsheet.

Let's summarize:

- Convolution in time is equivalent to multiplication in frequency.
- Convolution in frequency is equivalent to multiplication in time.
- Convolution, in time, of a signal with a time-reversed impulse response is equivalent to the inverse Fourier transform of the product of the signal spectrum with the system function.

The system function and the input spectrum are generally complex. Expressed in complex polar form Eq. (7-39) says that the output spectrum, a product of two complex functions, is the result of a "shrink and turn" vector operation (in Feynman's words).

This is a useful aid for visualization of what appears to be a rather abstract process. Recall that, for the *RC* circuit, the output is always equal to or less than the input (because of losses in *R*) and the phase shift goes from 0 to –90 degrees. Equation (7-39) in polar form says that the amplitude of the output is the product of the amplitudes of the system function and the input, so this is a "shrink." The phases add, so this is a "turn."

Because the amplitudes multiply and the phases add, it is possible to think of the *RC* circuit in the frequency domain as something that shrinks and turns (backward) an infinite number of vectors (corresponding to a continuous frequency variable) applied to the input. The amount of shrink-and-turn depends on frequency, as shown in Fig. 7-5. If a system adds energy to the signal at any frequency we might call it a "stretch and turn" process. Causality in the convolution process is somewhat like Feynman's path integral description of quantum mechanics in which all of the past, up to the present, influences the present.

CONVOLVE, Convolution Worksheet

Figure 7-6 shows the Home screen of this worksheet which contains two methods of performing discrete convolution. The first method uses the FFT and IFFT, and the second method is based on the discrete version of Eq. (7-37). Press PgDn to see the second method. Use your mouse to explore the cell formulas so you can see how they work.

Random noise can be added to the input signal, and the worksheet reports the input and output variances (the squares of the standard deviations). Note the variance reduction in the output. You can modify the impulse response in Row 9 to adjust the time constant or to change to an oscillator impulse response which you can obtain from Table 4-3. In this table, replace the force $F(t)$ by $F_o\delta(t)$.

With random noise added to the input signal, press Function Key F9 to change the noise sample and recalculate. This Function Key retains the same noise statistics but it uses a different noise sample. Change the noise statistics using RAND.

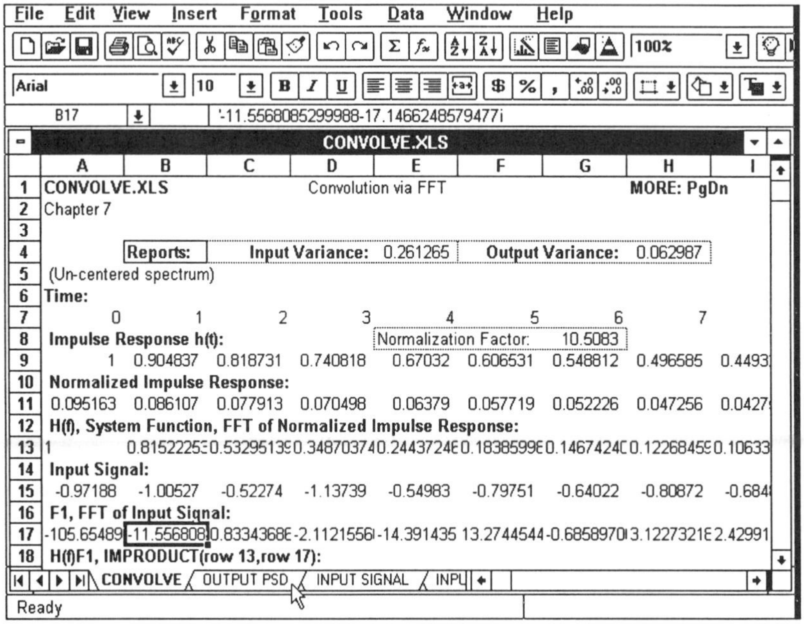

Figure 7-6. Home screen of CONVOLVE. Note reports of input and output variances. Modify input signal in Row 15 and impulse response in Row 9.

Don't forget to run the FFT on the input signal and the IFFT on the convolved signal every time the signal changes. If you change the impulse response, you will have to run the FFT on this to generate the new system function. Notice that no checkerboarding is used on the input signal because we are not particularly interested in the spectra, only the input and output waveforms. Graphs of the uncentered spectra are available by clicking the appropriate tab on the bottom, as in Fig. 7-6.

While viewing the Home screen, press PgDn until you come to the section that performs convolution with a finite-length $h(t)$ in the original data domain, without using the FFT. Move your mouse pointer to explore the cells and examine the formulas to see how easy this is with a small data set. Just below this section there is an embedded graph that is "live." Press Function Key F9 repeatedly and watch the input and output waveforms change as the noise sample changes. You can experiment by modifying the input signal, noise, and impulse response.

This section has a built-in test signal consisting of two pulses of different lengths and a segment of a sine wave, Fig. 7-7. The pulses have different spectral widths, but their spectra are broad-band. The sine wave is a relatively narrow-band signal.

Notice the blurring of the pulses and the phase and amplitude changes in the sine wave. Change the pulse widths and pulse spacing. For visual clarity, set the noise to zero. Look in the on-screen Help to see the different types of noise you can add, such as uniform noise or Gaussian noise. All of the effects you see are consistent with the "shrink-and-turn" frequency response shown in Fig. 7-5.

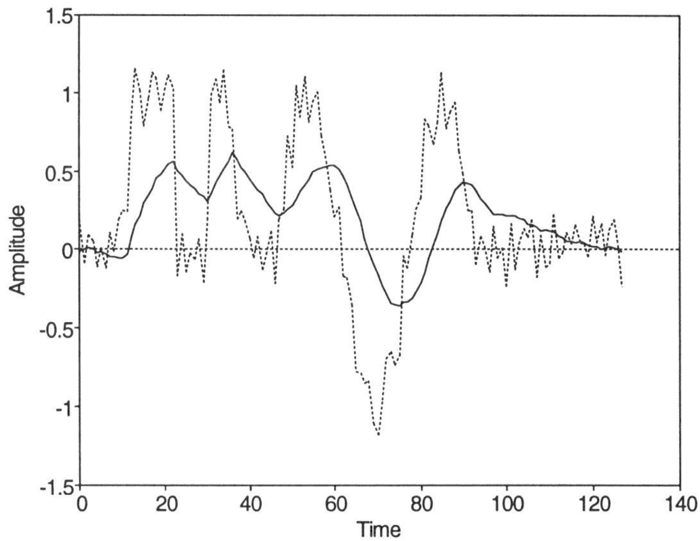

Figure 7-7. Original signal *(dotted line)* and convolved output *(solid line).* Original signal: Two noisy rectangular pulses and a segment of a noisy sine wave. Note blurring of pulses, and phase and amplitude shifts in sine wave. Variance reduction (see Chapter 10) is reported in Cells E4 and H4, Fig. 7-6.

7.12 Power Spectral Density

Measurement and computation of the PSD is an extremely important process for itself, and also because the PSD and cross-PSD are often used in advanced analysis involving other functions. The energy of a signal may be expressed as

$$E = \int_{-\infty}^{+\infty} |f(t)|^2 dt \qquad (7\text{-}40)$$

and according to Parseval's theorem, Eq. (6-99), this is also given by

$$E = \frac{1}{2\pi}\int_{-\infty}^{+\infty} |F(\omega)|^2 d\omega. \qquad (7\text{-}41)$$

If $f(t)$ is measured in volts and $|f(t)|^2$ is measured in volts^2/ohm, which is power in Watts, then integrating over time gives energy. Correspondingly, if $|F(\omega)|$ is measured in volt·s or volt/Hz then $|F(\omega)|^2$ is measured in (volt/Hz)^2/ohm, J·s, or J/Hz and is called the power spectral density, PSD, or (less precisely) the power spectrum. We will return to the concept of power spectra and generalize to cross-power spectra, which are important in cross-correlation functions, system identification, and coherence functions. Measurement of power spectra is often called "PSD estimation" for reasons that will become clear later.

7.13 Wiener-Khintchine Theorem

This theorem states a remarkable and useful connection between the spectral densities of two functions and their cross-correlation function $R_{1,2}(\tau)$. The cross-PSD and the cross-correlation function comprise a Fourier transform pair.

$$R_{1,2}(\tau) = \frac{1}{2\pi}\int_{-\infty}^{+\infty} F_1(\omega)\, F_2^*(\omega)\, e^{+j\omega t}\, d\omega \tag{7-42}$$

$$F_1(\omega)\, F_2^*(\omega) = \int_{-\infty}^{+\infty} R_{1,2}(\tau)\, e^{-j\omega t}\, d\tau \tag{7-43}$$

If the two spectra are the same, then this operation produces the auto-correlation function. As we have just seen, the product of two identical spectra is called the power spectral density, PSD, or just the power spectrum; if the spectra are not the same the product is called the cross-PSD. The cross-PSD is a measure of the frequencies that both spectra have in common, and therefore it is clearly a measure of the similarity of two signals in the frequency domain. Note the complex conjugate used in Eq. (7-42) is not used in convolution, Eq. (7-38). Some people find comparison in the frequency domain to be somewhat more abstract than comparison in time or space domains but it's very practical using spectrum analyzers.

Similarly, the power spectral density $F_1^*(\omega)F_1(\omega)$ and the auto-correlation function $R_{1,1}(\tau)$ comprise a Fourier transform pair. In the discussion that follows we will represent the power spectral density as $S_{1,1}(\omega)$ and the cross-power spectral density as $S_{1,2}(\omega)$. The auto-correlation function is useful in finding hidden periodicities and power spectral densities, among other things like system functions.

7.14 Auto- and Cross-Correlation Functions

It is important to distinguish between the correlation function, which we are discussing here, and the correlation coefficient familiar in statistics. The correlation function is a function of time delay τ. The correlation coefficient is a single number, Eq. (10-67), related to covariance and standard deviations of two data sets.

Correlation Function without FFT

We introduced the cross-correlation in the frequency domain in terms of the Wiener-Khintchine theorem. Now let's go back to the time domain and compare correlation with convolution. Correlation looks much like convolution except that:

- There is no time reversal of one of the functions,
- The integral (or sum) is over time so the correlation function is a function of the time delay τ, not the time t.

The same considerations apply when the correlation function is computed with other variables, such as spatial displacement and frequency.

The cross-correlation function is often normalized to unity, so that the maximum varies from zero (no similarity) to 1 (perfect match). The idea of correlation originated in antiquity when people compared patterns by sliding one pattern over another, which is exactly what the correlation integral does. In the limit as T goes to infinity, the cross-correlation function is defined as,

$$R_{1,2}(\tau) = \frac{1}{T}\int_0^T f_1(t)\, f_2(t+\tau)\, dt\,, \tag{7-44}$$

where $R_{1,2}(\tau)$ is the cross-correlation function. When $f_1(t) = f_2(t)$, the integral is called the auto-correlation function $R_{1,1}(\tau)$.

Performing the correlation integrals is often tedious and time consuming, even on a computer; once again the process is shift, multiply, and add. Thanks to the Wiener-Khintchine theorem, the FFT can be used to speed things up.

As in convolution, the integral form is easy to understand and satisfying to people because we are educated using continuous functions. However, for actual data the discrete definition is more useful in practice. We let τ take on only discrete, integer values and we shift data by integer values. In other words, we are dealing with sampled data and the product of the two functions $f_1(t)$ and $f_2(t+\tau)$ has a non-zero value only when the shift is equal to an integer multiple of the sampling interval.

As the stored replica of the transmitted signal moves across the received signal, the overlap product is calculated and plotted for each shift. Note that the maximum of the cross-correlation function occurs at the time delay corresponding to the time of arrival of the leading edge of the received signal. The maximum value of the cross-correlation function is a measure of the similarity of the received signal and the stored replica.

This stored replica may not actually be that of the original signal if you do not have the original signal; it may be an estimate of what you think the original signal is, or a replica of a signal that you are looking for. Perhaps you have a library of signal replicas in which you have an interest. This is the way some astronomers, cryptographers, and electronic surveillance wizards work.

Applications of Auto- and Cross-Correlation

Signal recovery by ensemble averaging is not possible if the signal is not repetitive, but, for a single event, cross-correlation of the signal-plus-noise with a stored copy of an expected signal often produces astonishingly good results in detecting the existence of the known signal. Some people call this an *Elvis detector.*

The cross-correlation function can be very useful in investigating characteristics of time delays and signal propagation paths. The correlation velocity v_{cor} is defined as

$$v_{corr} = \frac{d}{\tau_{\max}} ,$$

where d is the propagation distance between the source and detector and the time delay parameter $\tau_{\max}$ is that at which the cross-correlation function $R_{rt}(\tau)$ achieves its maximum value, A change in the dispersive properties of the system can be detected through a change in the delay time parameter τ and the maximum value of the cross-correlation function.

The Wiener-Khintchine theorem shows explicitly that the information conveyed by the cross-correlation function in the time domain is the same as that conveyed by the cross-PSD function in the frequency domain.

The noise reduction property of cross-correlation is of considerable importance. Consider a received signal $f_r(t)$ with additive random noise $f_n(t)$, so that the composite signal at the detector is $f_r(t) + f_n(t)$. Then the cross-correlation function of the noisy received signal is given by

$$R_{r,n,o}(\tau) = \frac{1}{2\pi} \int_{-\infty}^{+\infty} [F_r(\omega) + F_n(\omega)]\, F_o^*(\omega)\, e^{+j\omega t}\, d\omega, \qquad (7\text{-}45)$$

where $F_r(\omega)$, $F_n(\omega)$, and $F_o(\omega)$ are the Fourier transforms of the received signal, the noise, and the original signal, respectively. Equation (7-45) is simply,

$$R_{r,n,o}(\tau) = R_{r,o}(\tau) + R_{n,o}(\tau) \approx R_{r,o}(\tau). \qquad (7\text{-}46)$$

The cross-correlation function of the original signal with the random noise, $R_{n,o}(\tau)$ will ordinarily be vanishingly small compared to $R_{r,o}(\tau)$, so that the cross-correlation function of the noisy received signal is very close to that of the noise-free signal.

This noise rejection, when enhanced by ensemble signal averaging with repetitive coded signals, has the potential for remarkable feats of signal detection.

CROSSCOR, Cross-Correlation Worksheet

The Home screen of this worksheet and typical results are shown in Figs. 7-8 and 7-9. As in the CONVOLVE worksheet, both FFT and non-FFT methods of computing the cross-correlation function are available. The Home screen shows the FFT method. Note the reports of time delay for maximum cross-correlation, and the peak value of the cross-correlation function, normalized to the noiseless stored replica.

The peak value is determined using the MAX spreadsheet function on the cross-correlation function. The time delay is more subtle; it uses the IF spreadsheet function, a logical function similar to an exclusive-or gate used in digital electronics. Use your mouse and on-screen Help to learn more about how this function works.

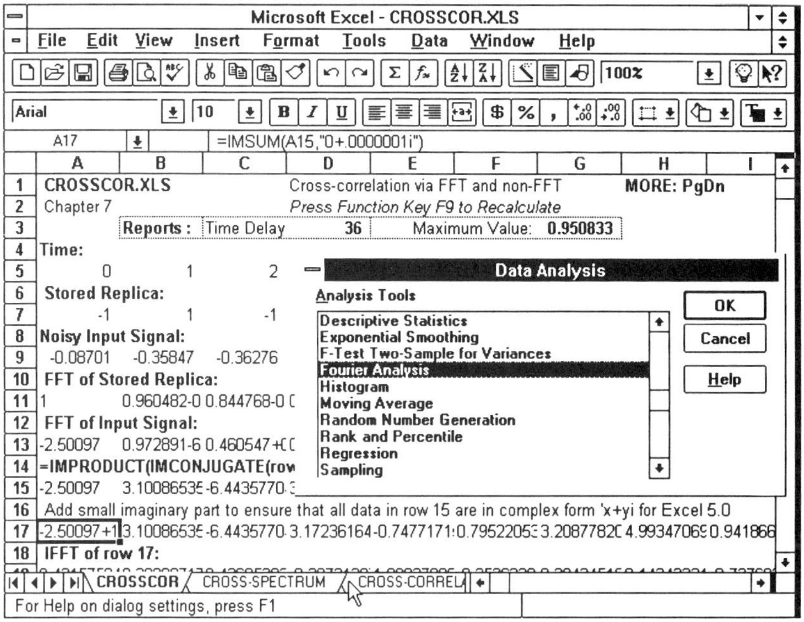

Figure 7-8. Home screen of CROSSCOR. Note reports in cells D3 and G3.

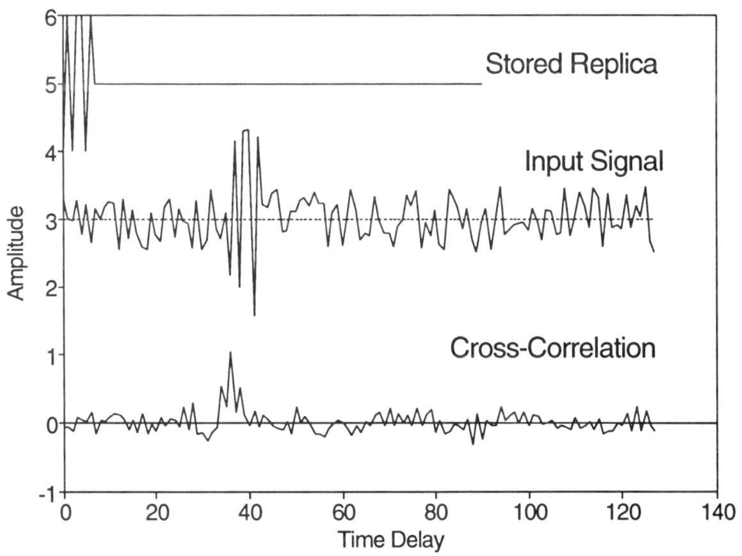

Figure 7-9. Stored replica, noisy received signal, and cross-correlation function. Note that the time of arrival of signal is marked by strong, stable peak in cross-correlation function even when the actual signal is buried in noise.

7.15 Coherence Function

The coherence function is related to the cross-correlation function and can be obtained as a natural by-product of the same computations. The magnitude-squared of the coherence function $\gamma^2(\omega)$ is defined as

$$\gamma^2(\omega) = \frac{|S_{1,2}(\omega)|^2}{S_{1,1}(\omega)\, S_{2,2}(\omega)} , \tag{7-47}$$

where $S_{1,2}(\omega)$ is the cross-PSD of two signals, and $S_{1,1}(\omega)$ and $S_{2,2}(\omega)$ are the auto- or self-PSDs of the signals. In terms of the PSDs, the coherence function is readily computed along with the cross-correlation function. Equation (7-47) is often called the MSC function, for "magnitude-squared coherence".

The MSC function has a value that ranges from 0 to 1. If the output of a linear system is due only to the input, then MSC = 1. If the output of the system has no causal relation with the input, then MSC = 0. A coherence function with a value less than one and greater than zero indicates one or more of the following:

- Noise from an unknown source, or jamming, is present.
- The system relating the input to the output involves nonlinearity.
- The output is due to one or more inputs in addition to the known input.

The SNR (signal-to-noise ratio) function is a simple extension of the MSC function; the SNR function gives the signal/noise power ratio *at each frequency*. This can be a very useful parameter. The SNR function is a function of frequency, not the usual S/N expressed as a single number, a ratio, or in dB. The SNR function is defined as

$$\mathrm{SNR}(\omega) = \frac{\gamma^2(\omega)}{1 - \gamma^2(\omega)} \tag{7-48}$$

Notice that the SNR function can blow up when $\gamma^2(\omega)$ is close to 1. Usually this is not a problem with real-world signals because of random noise that accompanies measurements. Figure 7-10 shows the Home screen for the COHERENC worksheet, which contains the MSC and SNR functions.

Figures 7-11 and 7-12 show how useful this can be. To enhance frequency resolution, this worksheet uses a von Hann data window. Experiment with raw signals without the data window, and compare the frequency resolution with the window in use. Change the noise sample, retaining the same statistics, by pressing Function Key F9. If you want to retain the identical noise sample for all calculations, convert random noise samples from the formula using RAND to Values. To accomplish this, use Copy, then Paste Special, and choose Values for the entire sequence. Always recalculate the FFT with the dialog box, or write a macro to do this.

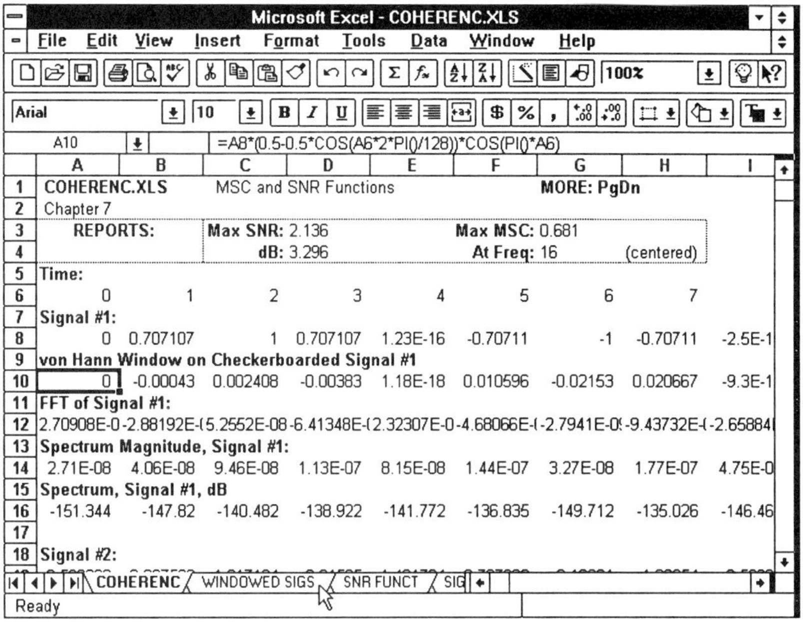

Figure 7-10. Home screen for COHERENC worksheet. Note reports of SNR and MSC functions for signals shown in Fig. 7-11. Window formula with checkerboard is in Cell A10. Checkerboard COS(PI()*A6) is tacked on to window.

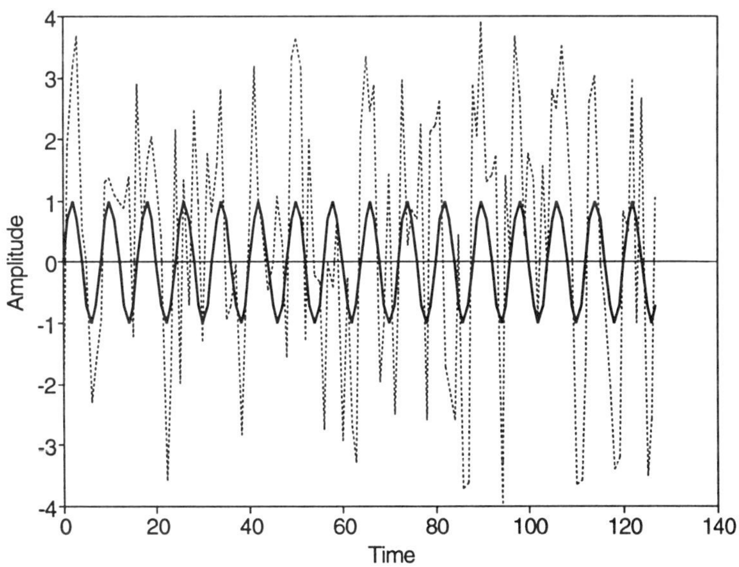

Figure 7-11. These two signals appear to have little or no coherence. *Solid line:* Signal 1. *Dotted line:* Signal 2. See Fig. 7-12 for associated SNR function.

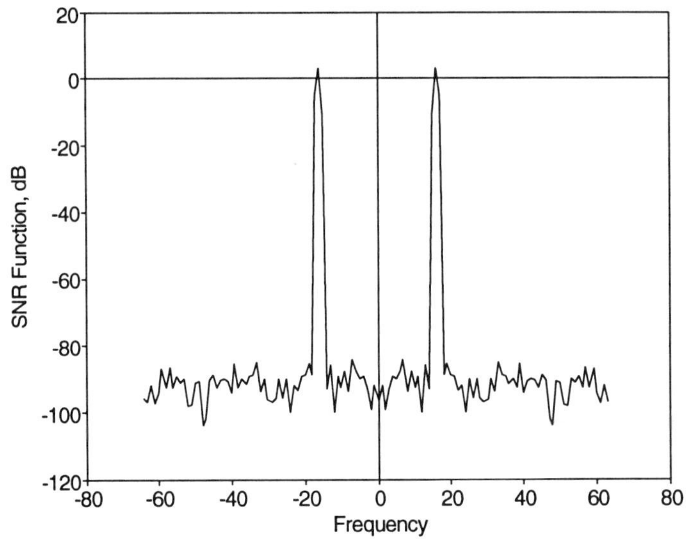

Figure 7-12. SNR function for the two signals shown in Fig. 7-11. There is strong narrow-band coherence where the SNR function is large.

7.16 System Identification

We introduced system identification earlier in Eq. (7-25),

$$H(\omega) = \frac{F_{out}(\omega)}{F_{in}(\omega)} \qquad (7\text{-}25)$$

and now we are going to examine it in more detail. This is a way to determine the heart and soul of a linear system, that is, its system function and impulse response. Equation (7-25) looks simple enough, and it works in many cases. Send in a test signal of your own choosing, so you know its complex spectrum $F_{in}(\omega)$, Then you measure the complex spectrum of the system's response $F_{out}(\omega)$, and perform the indicated complex division,

There is a potential difficulty with this, because any zero in the input spectrum will make Eq. (7-25) blow up. This would seem to eliminate the use of pulses as test signals because they have many zeros. Even though this looks dangerous, the chance of hitting a perfect zero in the real world is very small, considering discrete signals and the inevitable presence of noise. To be on the safe side, you can add a small amount of noise with a spectrum $N(\omega)$ to the test signal,

$$H(\omega) = \frac{F_{out}(\omega)}{F_{in}(\omega) + N(\omega)} \ . \qquad (7\text{-}49)$$

This is basically the idea of the Wiener inverse filter, derived by Norbert Wiener as the optimum deconvolution or inverse filter. Here, "optimum" is used in the least-mean-square sense. Equation (7-49) seems to be a reasonable thing to do, but there is no indication here that this is optimum in any way. It's good to know that such a simple idea is the best idea, at least in the LMS sense.

So, if Equation (7-49) is so good, why do we need $F_{in}(\omega)$? Doesn't this just look like a nonrandom addition to noise? The answer is, we do not need $F_{in}(\omega)$ and we can use random noise as a perfectly good test signal. It's a good idea to use many noise samples, make many measurements, and average the ensemble. If no other problems are present, then the error in $H(\omega)$ will be proportional to the square root of the number of measurements. The noise spectrum should have a bandwidth at least as great as the expected system function bandwidth.

An alternative to Eq. (7-49) uses the power spectral density and the cross-PSD. Multiply numerator and denominator of Eq. (7-49) by the complex conjugate of the output spectrum, obtaining

$$H(\omega) = \frac{F_{out}(\omega)\ F^*_{out}(\omega)}{F_{in}(\omega)\ F^*_{out}(\omega)} = \frac{|F_{out}(\omega)|^2}{F_{in}(\omega)F^*_{out}(\omega)}\ . \tag{7-50}$$

A noise power spectrum can be added to the denominator, or the denominator could simply be the cross-PSD of the input noise and the output noise. Remember, the numerator is real and the denominator is complex.

Figure 7-13 shows the Home screen of the worksheet SYST-I-D so you can experiment with system indentification using the FFT.

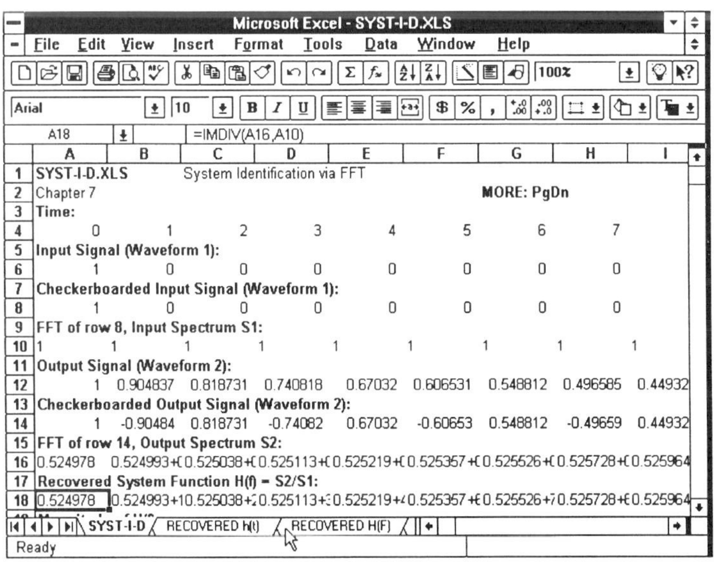

Figure 7-13. Home screen of SYST-I-D. Input test signal in row 6 is a unit impulse.

The unknown system in the worksheet has an exponential impulse response, similar to the RC integrator. The computed system function based on the unit impulse test signal is shown in Fig. 7-14 and its associated impulse response is shown in Fig. 7-15. The difference between the actual impulse response and the computed impulse response is shown in Fig. 7-16. Notice the phase at the ends of the computed system function imply periodicity; compare with the exact phase shown in Fig. 7-4.

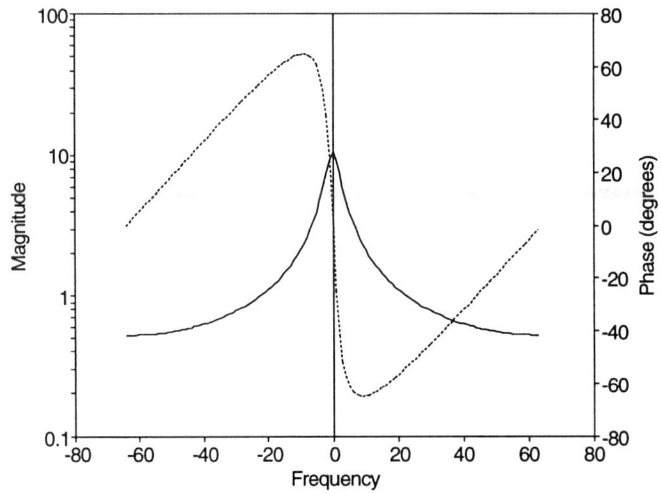

Figure 7-14. Computed $H(\omega)$ using unit impulse input to unknown system.

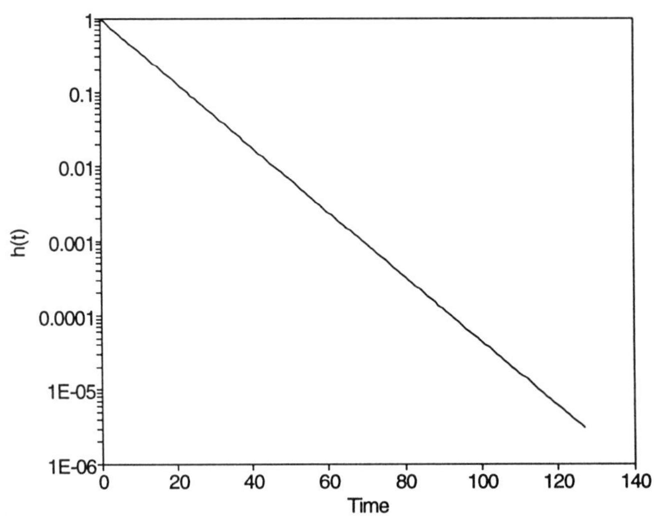

Figure 7-15. Computed $h(t)$ from data in Fig. 7-14; note excellent exponential characteristic. See Fig. 7-16 for errors in this computation.

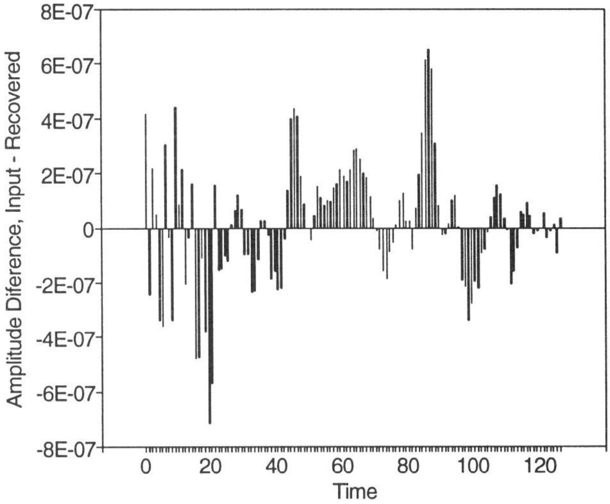

Figure 7-16. Difference between actual impulse response and computed impulse response, Fig. 7-15. With errors of a few parts in 10^7, these results are quite good, considering all of the processing involved.

System identification is very much like safe-cracking, especially when you use random noise as a test signal. A safe-cracker twiddles the dial randomly (test signal) and listens and feels the tumblers in the lock (measuring the output). From these input/output data the safe-cracker determines the combination (system function). This is also analogous to code-breaking. It is well known that, if a code-breaker can get his or her own signal into a cryptographic system and get it out again, the code can usually be broken.

This is a fundamental method in science and technology. For example, Lord Rutherford sent alpha particles (input test signals) through a gold foil (unknown system) and measured their scattering (output signal) to discover the nuclear structure of atoms (system function).

Now that we know how to compute the details of an unknown system using our own input and its output, let's see how we can use this knowledge to determine what an unknown input was when we measure its output. This is analogous to decyphering a coded signal.

7.17 Deconvolution

Equation (7-24) expressed the elementary idea of deconvolution, which is using the known system function and a measured output to determine what the original signal was. This technique has many applications. In telecommunications systems it is called "equalization," which means that the dispersive properties of the system are partially cancelled over a part of the spectrum. Deconvolution is particularly easy to do in the frequency domain, as was shown in Eq. (7-24),

$$F_{in}(\omega) = \frac{F_{out}(\omega)}{H(\omega)} . \tag{7-24}$$

The only problem that might arise involves possible zeros in $H(\omega)$, which would make the process blow up. A little noise would be helpful here. This is related to the original problem addressed by Norbert Wiener. The Wiener inverse filter is

$$F_{in}(\omega) = \frac{F_{out}(\omega)}{H(\omega) + N(\omega)} . \tag{7-51}$$

This works best if $N(\omega)$ is the actual noise spectrum in the system at the time that the output is measured. If this noise spectrum is updated for each measurement, then Eq. (7-51) behaves like an adaptive inverse filter. When $H(\omega) > N(\omega)$ ordinary deconvolution is performed, but when $H(\omega) < N(\omega)$ the output is uniformly low for broadband noise. The Home screen of the DECONVOL worksheet is shown in Fig. 7-17. Figure 7-18 shows the original test signal, the convolved signal, and the deconvolved signal. Figure 7-19 shows the errors in Fig. 7-18.

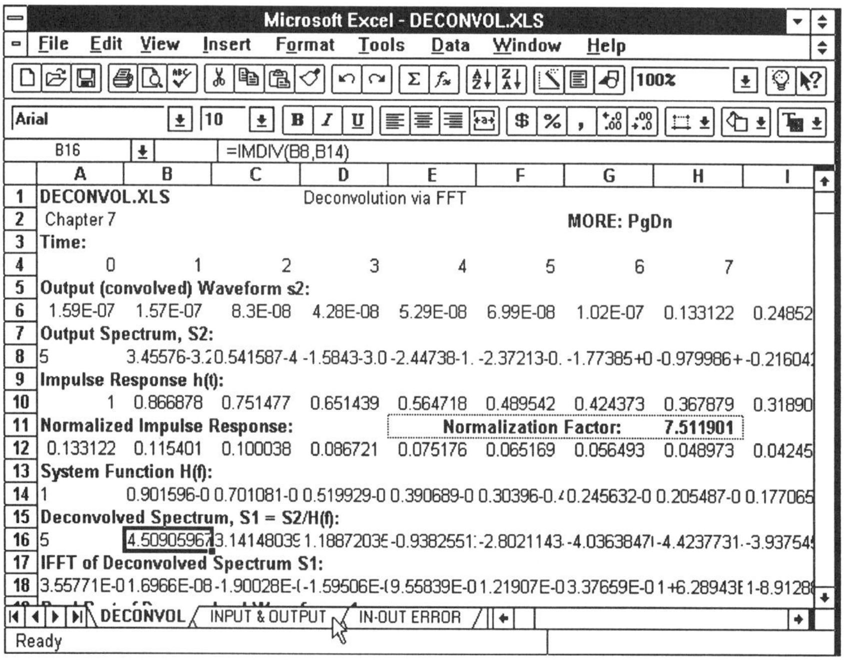

Figure 7-17. Home screen of DECONVOL. This produces a user-defined convolved test signal with a user-defined impulse response, and then deconvolves the signal. Note use of IMDIV for spectral deconvolution in Cell B16.

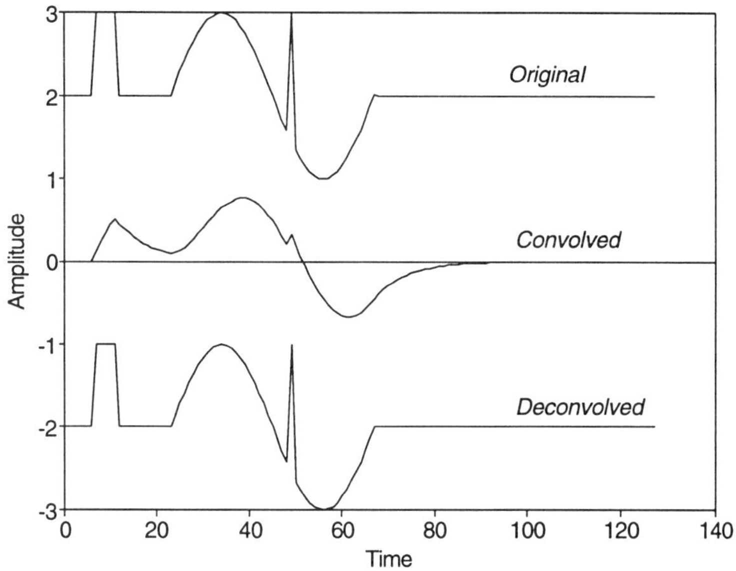

Figure 7-18. Comparison of original test signal, convolved signal, and deconvolved signal. Note blurring of pulses, with most severe distortion on the shorter pulse. Convolved sine wave has reduced amplitude and shifted phase. (Traces are displaced vertically for graphic clarity.)

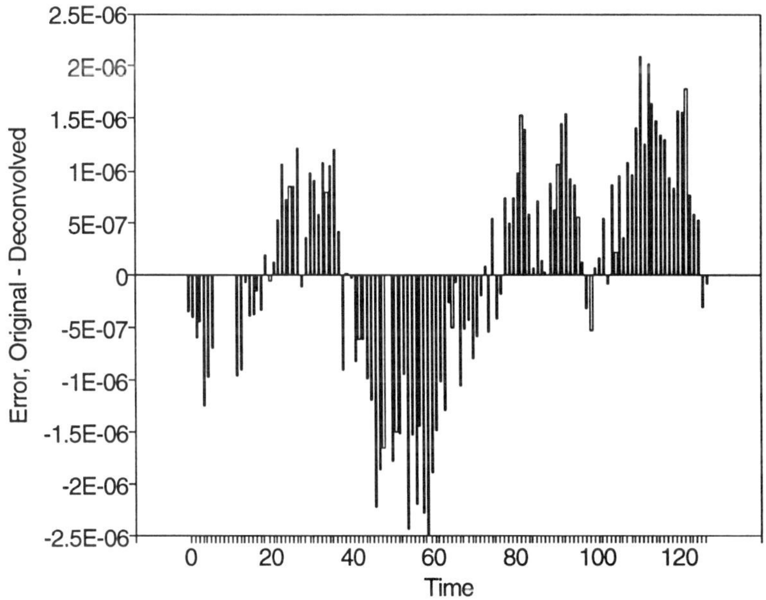

Figure 7-19. Errors in the deconvolution shown in Fig. 7-18.

7.18 Data Windows

We have already used data windows to improve spectrum resolution in some of the previous worksheets. Windowing is a form of convolutional pre-processing of data prior to the FFT operation. We have also used checkerboarding as a form of pre-processing but its purpose was esthetic, to center spectra so they would look "normal" like analytically calculated spectra.

The spectrum leakage that accompanies selecting a block of data can be regarded as a form of generalized diffraction (see Figs. 6-5 and 6-6). The abrupt start and stop of a data sequence is responsible for this diffraction, just as the sharp edge of a lens or mirror blurs an optical image due to diffraction. In fact, windowing in optics is known as "apodization," and refers to a neutral density gradient filter applied to a lens or mirror near the edge. Removing some of the energy near the boundary decreases diffraction effects and enhances image resolution.

Softening the start and stop of a data sequence in various ways will reduce this "diffraction" in frequency but we have enough experience to know that we will have to give up something to gain something. Actually, there are several trade-offs, including sidelobe level, equivalent noise bandwith, and coherent gain. These trade-offs are the reason that so many different windows have been invented.

Using no window is actually using a rectangular window, and this is the worst choice of all. Consider selecting a block of data from a sine wave. If the data block includes an integer number of half-periods then the data will rise slowly from zero at the start and fall slowly to zero at the end. It is unusual to be able to select such a block of data; most of the time we select an integer power of two data points at some sampling frequency and then process the data. Windowing the data block before the FFT can produce astounding improvements in spectrum resolution, as we will see.

The windowing process usually involves multiplying a data point by a window function data point. We know that multiplication in time is equivalent to convolution in frequency, so windowing is a form of convolution.

The DATWINDO worksheet, shown in Fig. 7-20, has a small library of window functions that you can use. You can add new windows of your choosing. The built-in windows are rectangular, von Hann, Hamming, Blackman-Harris, and Kaiser-Bessel. The von Hann and Hamming windows are particularly simple. To select a window, copy the formula from the window cell to Cells A10 . .DX10. For example, to select the Hamming window, copy Cell A30 to row 10. Figure 7-21 shows a two-frequency test signal with a rectangular window, and Fig. 7-22 shows the same signal seen through a Hamming window. Figure 7-23 compares the spectra obtained using these windowed data. Figure 7-24 shows Blackman-Harris results.

Properties of more than twenty common windows are catalogued in the famous paper "On the use of windows for harmonic analysis with the discrete Fourier transform," F. J. Harris, *Proceedings of the IEEE* **66,** pp. 51–83 (1978) and in *The Fast Fourier Transform and Its Applications*, E. O. Brigham (Prentice-Hall 1988).

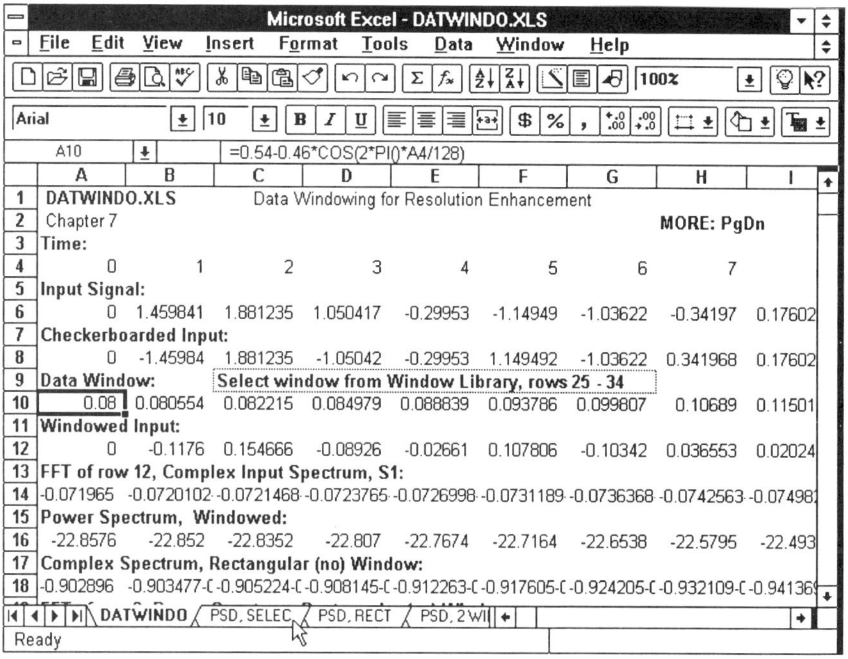

Figure 7-20. Home screen of the DATWINDO worksheet. Hamming window formula is shown in Cell A10. Note effect of window in row 12.

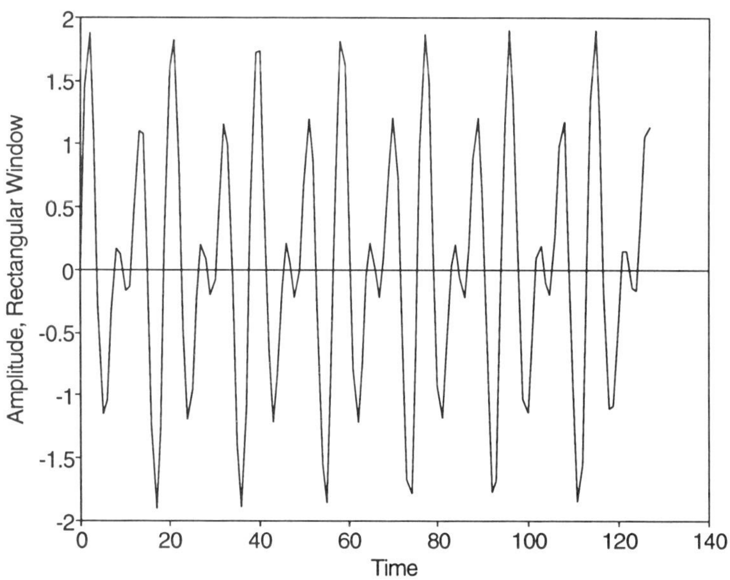

Figure 7-21. Two-frequency test signal seen through a rectangular window. Note abrupt end of 128-point data sequence.

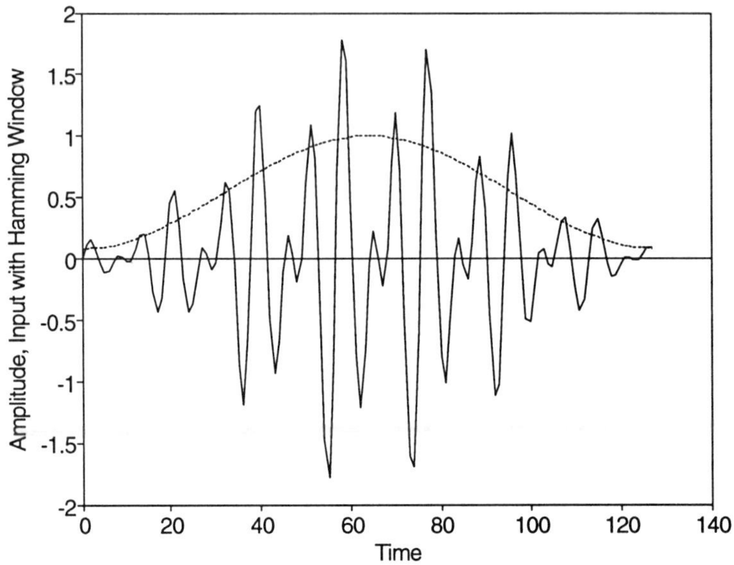

Figure 7-22. This is the same signal shown in Fig. 7-21, seen through a Hamming window *(solid curve)*. The window is shown as a dotted curve.

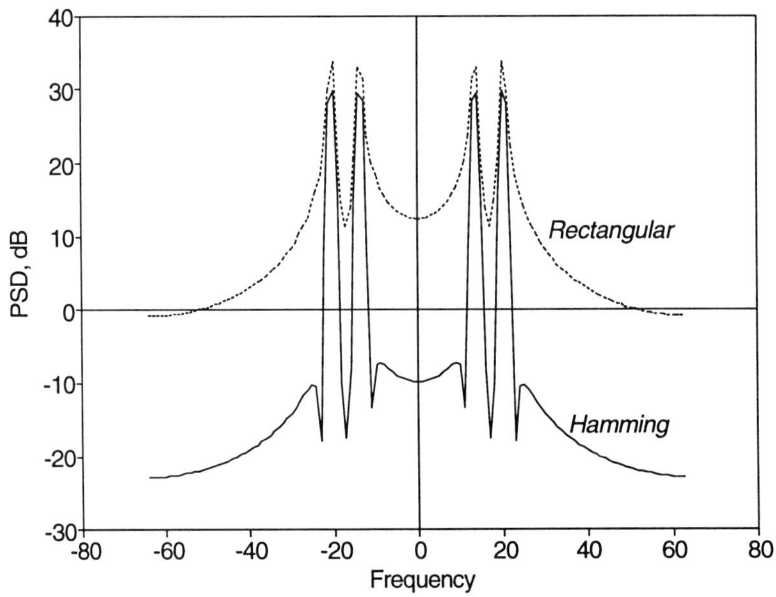

Figure 7-23. Comparison of rectangular window and Hamming window results. No vertical displacement has been imposed; these are the actual spectra. Note good spectral resolution and low leakage of the Hamming window.

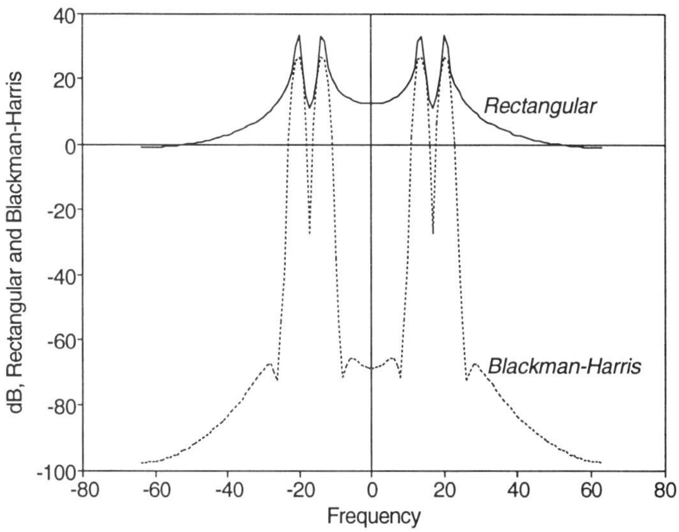

Figure 7-24. Spectra seen through the Blackman-Harris window and the rectangular window. This is the two-frequency test signal shown in Fig. 7-21. Compare with results of the Hamming window, Fig. 7-23.

Results of the Blackman-Harris window are impressive, with improvement of better than 90 dB on the wings of the spectral lines and excellent resolution of the two lines. The simpler Hamming window, Fig. 7-23, has good resolution but less improvement out on the wings. Notice the small price that must be paid with the Hamming and Blackman-Harris windows, in the slightly lower peaks of the spectra compared with the rectangular window. Figures 7-23 and 7-24 show why all spectra are really *estimates* of spectra.

As in life, what you see depends on what kind of window you look through. The choice of the "best" window depends on what you are looking for, and it is a good idea to look at spectra through several windows. The problem of spectral estimation involving random data with the FFT is a subject of current research interest.

7.19 Central Limit Theorem

The fundamental nature of the convolution process extends far beyond its manifold applications in system response and windowing. For example, it is obvious that in general you cannot simply add two probabilities because the total probability must always be ≤ 1.

An elementary result in statistics is that the sum of the probability density functions (Chapter 10) of two independent random variables is the convolution of the probability density functions. In other words, joining two probability distributions together requires their convolution. This simple statement and the central limit theorem have profound implications in both classical and quantum measurement processes.

The central limit theorem tells us that, for a large class of functions, the end result of an infinite number of self-convolutions is a Gaussian function. This remarkable, useful result is another unique property of the ubiquitous Gaussian function.

Even though an infinite number of convolutions is required mathematically, three or four convolutions are often enough to produce a respectable approximation to a true Gaussian function. Figure 7-25 shows five successive convolutions of a rectangular function. The rectangular function appears to "melt" due to convolutions.

The central limit theorem does not apply to one of our favorite, familiar functions, the "$\sin(x)/x$" function which, from Chapter 6, is the Fourier transform of a rectangular function. But, if multiple convolutions of this function do not produce a Gaussian, then what do they produce? You can get the result using a little thought, without computation. Recall that convolution in one domain is equivalent to multiplication in the other domain. Therefore, convolving two "$\sin(x)/x$" functions in one domain is the same as multiplying two rectangular functions in the other domain. Multiplying two rectangular functions produces another rectangular function, and this corresponds to another "$\sin(x)/x$" function; see Fig. 7-26.

Try this in your spreadsheet. Remember that, although these functions go on forever, they must be truncated in a computer, so don't expect to get perfect rectangular boxes and don't expect to get exact replicas of the "$\sin(x)/x$" function.

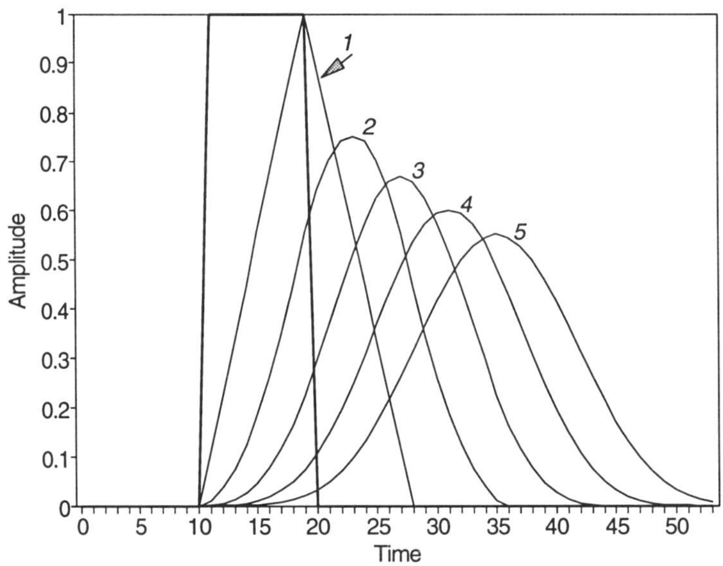

Figure 7-25. Example of the central limit theorem. Here, a rectangular function is subjected to multiple convolutions which approach a Gaussian form. The first convolution is a triangle. Subsequent convolutions appear more Gaussian.

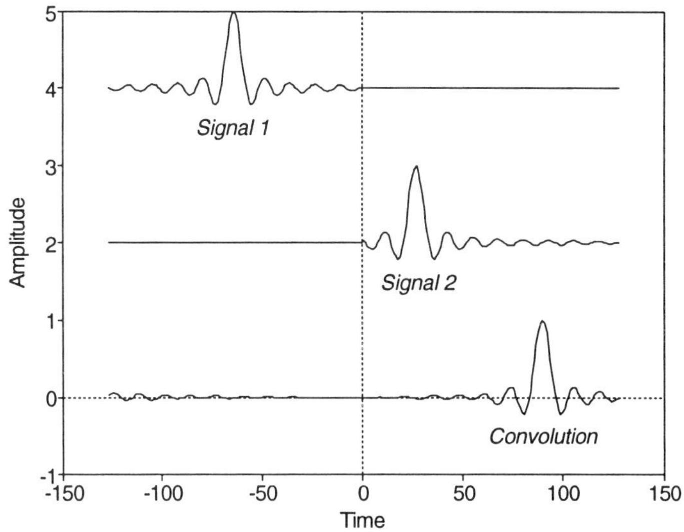

Fig. 7-26. Convolution of a sin(x)/x function with a similar function results in another sin(x)/x function. The results shown here used the FFT in *Excel*. Note the apparent lack of causality (non-zero convolved signal for $t < 0$). This apparently non-causal result is due to the circularity of the FFT. Actually, the convolved signal for $t < 0$ belongs with the truncated right side of the signal for $t > 0$. Signals 1 and 2 have been displaced vertically for graphical clarity.

It's worthwhile mentioning that another one of our favorites, the Dirac δ-function, does not obey the central limit theorem. For the convolution of two δ-functions we have

$$\delta_1(t - t_1) \otimes \delta_2(t - t_2) = \delta[t - (t_1 + t_2)]. \tag{7-52}$$

The δ-function or impulse function is certainly not a function in the usual sense. It was finally put on firm ground by the theory of distributions, that is, it is now considered a distribution, not a function. This is a subtle subject that challenged mathematicians for many years. For a brief introduction and summary, see *The Fast Fourier Transform and Its Applications*, E. O. Brigham, (Prentice Hall, Englewood Cliffs, NJ, 1988), pages 386–393.

In experiments, the noise in measured data is often due to contributions from many unknown sources. When these are convolved in a system (an oscillator, for example) it is not surprising that errors appear to have a Gaussian distribution.

Adding any number of random variables can be considered as adding two at a time. Add two, then add this sum to a third, and so on. The result of superimposing independent normal random variables leads to another normal random variable. Furthermore, this can be generalized to the case of dependent normal random variables, providing opportunities for wild speculation in the quantum domain.

What's Next?

Now you have a set of practical tools that you can use to analyze oscillators, and linear systems in general. The invention of the FFT has enabled implementation of extraordiarily powerful methods analysis in the frequency domain.

The two-frequency signal that we used as a test of window performance prepares us for spectrum splitting when we couple two oscillators in Chapter 9. Before we couple oscillators, however, we need to review methods of advanced dynamics, which give us powerful ways of analyzing more than one oscillator. Lagrangian and Hamiltonian methods will provide other perspectives, and they will push us to the classical limit.

Chapter 8

Methods of Advanced Dynamics

Everything should be made as simple as possible, but not simpler.
Albert Einstein

Using Newton's laws of motion is straightforward for a simple system. As the complexity of the system increases, the application of these laws become more difficult, because a considerable amount of cleverness is often involved in eliminating unknown reactions. Lagrange and Hamilton devised general methods of attacking complicated systems, although one must pay a higher mathematical price in the use of partial differential equations instead of ordinary differential equations. There is no way of deciding whether the Newtonian, Lagrangian, or Hamiltonian formulation is the most fundamental, because if you start with any one of them you can derive the other two. Hamilton's principle helps with the eternal question, "Of the infinite ways to go from A to B, which one do we actually observe?"

Lagrangian and Hamiltonian methods have six major advantages over the direct use of Newton's laws of motion:

1. There are no unknown internal reactions that require elimination.
2. There are a *minimum* number of variables.
3. The equations are *independent* of the type of coordinate system.
4. Deeper physical insights are often provided.
5. Problem solving becomes a routine procedure.
6. The methods are applicable to electrical, electromagnetic, acoustic, and mechanical systems, both classical and quantum.

8.1 Generalized Coordinates

For a system with n degrees of freedom it would be convenient to find a representation in terms of n coordinates, the minimum number necessary. Suppose we have Cartesian coordinates that are functions of some unspecified coordinates with n dimensions, as in Eq. (1-1).

$$x_i = x_i(q_1, q_2, q_3, \dots, q_n)$$

$$y_i = y_i(q_1, q_2, q_3, \dots, q_n)$$

$$z_i = z_i(q_1, q_2, q_3, \dots, q_n)\,. \tag{8-1}$$

It is not necessary that generalized coordinates have units of length. They may even be dimensionless, like angles. Let's find a way to transform back and forth between coordinate systems. An increment along the x-axis is given by

$$dx_i = \frac{\partial x_i}{\partial q_1} dq_1 + \frac{\partial x_i}{\partial q_2} dq_2 + \dots + \frac{\partial x_i}{\partial q_r} dq_r + \dots + \frac{\partial x_i}{\partial q_n} dq_n \tag{8-2}$$

and similarly for y and z.

Coordinate Transformation

As a familiar example of the general procedure outlined above, let's transform between Cartesian and plane polar coordinates,

$$x = x(r, \theta) = r\cos(\theta)$$

$$y = y(r, \theta) = r\sin(\theta). \tag{8-3}$$

Using Eq. (8-2) we have

$$dx = \frac{\partial x}{\partial r} dr + \frac{\partial x}{\partial \theta} d\theta = \cos(\theta)\, dr - r\sin(\theta)\, d\theta$$

$$dy = \frac{\partial y}{\partial r} dr + \frac{\partial y}{\partial \theta} d\theta = \sin(\theta)\, dr + r\cos(\theta)\, d\theta\,. \tag{8-4}$$

Equations (8-4) can be written compactly in matrix form,

$$\begin{pmatrix} dx \\ dy \end{pmatrix} = \begin{pmatrix} \cos\theta & -\sin\theta \\ \sin\theta & \cos\theta \end{pmatrix} \begin{pmatrix} dr \\ r\,d\theta \end{pmatrix}. \tag{8-5}$$

The inverse matrix is used to transform from Cartesian to plane polar. The two column matrices can be regarded as vectors. Note that this transformation is a special type called an *orthogonal transformation* which has unique and desirable properties.

- An orthogonal transformation preserves the magnitudes of vectors and the angles between vectors.
- You can always identify an orthogonal transformation because the determinant of the matrix has the value +1.

This is obviously the case for the transformation matrix in Eq. (8-5), because $\cos^2(\theta) + \sin^2(\theta) = +1$.

8.2 Generalized Momenta

Now let's turn our attention to momenta and their time rate of change. Time derivatives of coordinates can be written as

$$\frac{dx_i}{dt} = \dot{x}_i = \frac{\partial x_i}{\partial q_1}\dot{q}_1 + \frac{\partial x_i}{\partial q_2}\dot{q}_2 + \ldots + \frac{\partial x_i}{\partial q_r}\dot{q}_r + \ldots + \frac{\partial x_i}{\partial q_n}\dot{q}_n , \tag{8-6}$$

where we have used Newton's notation of a dot above the coordinate to indicate a time derivative. Similar expressions hold for y_i and z_i. It is clear that $\dot{x}_i$ is a linear function of the $\dot{q}$'s . The coefficients $\partial x_i / \partial q_r$ are functions of the coordinates, and are calculated from Eq. (8-1). Since $\dot{x}_i$ is a function of the q's and the $\dot{q}$'s we can differentiate Eq. (8-6) to obtain,

$$\frac{\partial \dot{x}_i}{\partial \dot{q}_r} = \frac{\partial x_i}{\partial q_r}, \qquad r = 1, 2, 3, \ldots , n . \tag{8-7}$$

We will need Eq. (8-7) shortly. The kinetic energy of a system is the total kinetic energies of all of the particles that make up the system,

$$K = \sum_{i=1}^{N} \frac{1}{2} m_i \left(\dot{x}_i^{\,2} + \dot{y}_i^{\,2} + \dot{z}_i^{\,2}\right) . \tag{8-8}$$

From Eq. (8-6) it is clear that the kinetic energy is a quadratic function of the generalized velocities $\dot{q}_i$ and the coefficients depend on the generalized coordinates q_i. Let's see how the kinetic energy depends on the generalized coordinates and velocities.

$$\frac{\partial K}{\partial q_r} = \sum_{i=1}^{N} m_i \left(\dot{x}_i \frac{\partial \dot{x}_i}{\partial q_r} + \dot{y}_i \frac{\partial \dot{y}_i}{\partial q_r} + \dot{z}_i \frac{\partial \dot{z}_i}{\partial q_r}\right) \tag{8-9}$$

$$\frac{\partial K}{\partial \dot{q}_r} = \sum_{i=1}^{N} m_i \left(\dot{x}_i \frac{\partial \dot{x}_i}{\partial \dot{q}_r} + \dot{y}_i \frac{\partial \dot{y}_i}{\partial \dot{q}_r} + \dot{z}_i \frac{\partial \dot{z}_i}{\partial \dot{q}_r}\right) \tag{8-10}$$

Using Eq. (8-7) we can rewrite (8-10) as

$$\frac{\partial K}{\partial \dot{q}_r} = \sum_{i=1}^{N} m_i \left(\dot{x}_i \frac{\partial x_i}{\partial q_r} + \dot{y}_i \frac{\partial y_i}{\partial q_r} + \dot{z}_i \frac{\partial z_i}{\partial q_r}\right). \tag{8-11}$$

Now let's define the generalized momentum p_i conjugate to the generalized coordinate q_i as

$$p_i = \frac{\partial K}{\partial \dot{q}_i}. \tag{8-12}$$

Generalized momenta do not necessarily have units of kg·m/s. This should not be surprising because generalized coordinates (angles, for example) may not have units of length. In Cartesian coordinates, Eq. (8-12) has the familiar form

$$p_x = \frac{\partial K}{\partial \dot{x}} = \frac{\partial}{\partial \dot{x}}\left(\frac{1}{2}m\dot{x}^2 + \frac{1}{2}m\dot{y}^2 + \frac{1}{2}m\dot{z}^2\right) = m\dot{x},$$

and similarly for p_y and p_z. In plane polar coordinates $K = K(q,\dot{q})$ so we have,

$$p_r = \frac{\partial K}{\partial \dot{r}} = \frac{\partial}{\partial \dot{r}}\left[\frac{1}{2}m\dot{r}^2 + \frac{1}{2}m(r\dot{\theta})^2\right] = m\dot{r}$$

and

$$p_\theta = \frac{\partial K}{\partial \dot{\theta}} = \frac{\partial}{\partial \dot{\theta}}\left[\frac{1}{2}m\dot{r}^2 + \frac{1}{2}m(r\dot{\theta})^2\right] = mr^2\dot{\theta}.$$

The radial momentum is p_r and the angular momentum is p_θ.

8.3 Lagrange's Equations

Joseph Louis Lagrange (Universities of Berlin and Paris) derived a Legendre transformation of Newton's laws into generalized coordinates and velocities that provided exactly the right number of independent equations needed to solve problems, that is, a number of equations that was equal to the number of degrees of freedom. (There was gossip that Lagrange was one of Marie Antoinette's favorite persons, but I won't repeat it here.) We are not going to derive this transformation; it is done in most classical mechanics texts. Also, we are only going to discuss conservative systems.

Lagrange defined a function L as the difference between kinetic energy and potential energy,

$$L(q_r,\dot{q}_r) = K(q_r,\dot{q}_r) - U(q_r) \tag{8-13}$$

and he found the required set of n second-order partial differential equations which describe classical systems in terms of this function,

$$\frac{d}{dt}\left(\frac{\partial L}{\partial \dot{q}_r}\right) - \frac{\partial L}{\partial q_r} = 0. \tag{8-14}$$

This is remarkable. Evidently L contains all of the information about the system. We can extract this information by operating on L with differential operators. In Equations (8-14) we have a set of n scalar equations that describes classical systems in a completely general way, involving a scalar function in terms of kinetic and potential energy. *The Newtonian concept of force does not appear.*

Harmonic Oscillator

Let's use Lagrange's equations to describe the simple harmonic oscillator. The kinetic energy is

$$K = \frac{1}{2} m\dot{q}^2 \tag{8-15}$$

and the potential energy is

$$U = \frac{1}{2} \kappa q^2 \tag{8-16}$$

so the Lagrangian can be written as,

$$L = K - U = \frac{1}{2} m\dot{q}^2 - \frac{1}{2} \kappa q^2. \tag{8-17}$$

Now we can operate on L with the prescribed differential operators to find,

$$\frac{\partial L}{\partial \dot{q}} = m\dot{q} \tag{8-18}$$

and

$$\frac{d}{dt}\left(\frac{\partial L}{\partial \dot{q}}\right) = m\ddot{q}. \tag{8-19}$$

We also need

$$\frac{\partial L}{\partial q} = -\kappa q. \tag{8-20}$$

If we insert these derivatives in Lagrange's equations we obtain

$$m\ddot{q} + \kappa q = 0. \tag{8-21}$$

This is our old simple harmonic oscillator equation. It looks like we have taken a long, hard route to get this. Simplc problems are easier using Newton's laws, but complicated problems, like a group of coupled oscillators, are easier using Lagrange's equations.

8.4 Hamilton's Equations

Hamilton sought a transformation of Lagrange's n second-order partial differential equations into $2n$ first-order partial differential equations. In generalized coordinates q and velocity $\dot{q}$ we have written the Lagrangian for conservative systems as

$$L(q,\dot{q}) = K(q,\dot{q}) - U(q) \tag{8-22}$$

where the momentum is

$$p_r = \frac{\partial K}{\partial \dot{q}_r} = m\dot{q}_r = \frac{\partial L}{\partial \dot{q}_r} . \tag{8-23}$$

Hamilton derived a transformation from generalized coordinates and generalized velocities to generalized coordinates and generalized momenta,

$$H(p_i,q_i) = \sum_r p_r\dot{q}_r - L(q_i,\dot{q}_i). \tag{8-24}$$

The Hamiltonian function $H(p_i,q_i)$ may appear to be a function of the generalized velocities but it is actually only a function of generalized momenta and coordinates, as you can show by differentiation with respect to the generalized velocities,

$$\frac{\partial H}{\partial \dot{q}_r} = p_r - \frac{\partial L}{\partial \dot{q}_r} = 0. \tag{8-25}$$

For a conservative system

$$H(p_i,q_i) = K(p_i,q_i) + U(q_i). \tag{8-26}$$

The classical Hamiltonian is a *function*, with a numerical value equal to the total energy E. Also, the Hamiltonian has phase-space "coordinates" of p and q so you might guess that it deserves a fundamental position in our description of the physical world. We can also express the Hamiltonian as,

$$H = 2K - L = 2K - (K - U) = K + U \Leftrightarrow E. \tag{8-27}$$

We can find the derivatives,

$$\frac{\partial H}{\partial q_r} = -\dot{p}_r \tag{8-28}$$

and

$$\frac{\partial H}{\partial p_r} = \dot{q}_r . \tag{8-29}$$

Equations (8-28) and (8-29) are Hamilton's equations. They are really quite remarkable. They were derived from Lagrange's equations, so they convey the same information but they provide a different way of thinking about dynamics. Equations (8-28) and (8-29) look like gradients in phase space, with H as a super potential.

Harmonic Oscillator

Let's see how Hamilton's equations describe a simple one-dimensional harmonic oscillator. The Hamiltonian is

$$H(p, q) = \frac{p^2}{2m} + \frac{\kappa q^2}{2} = \frac{p^2}{2m} + \frac{m(\omega_o q)^2}{2} \tag{8-30}$$

so, by Hamilton's equations,

$$\dot{q} = \frac{\partial H}{\partial p} = \frac{p}{m} \tag{8-31}$$

and

$$-\dot{p} = \frac{\partial H}{\partial q} = \kappa q = m\omega_o^2 q. \tag{8-32}$$

The equation for $\dot{q}$ tells us that $p = m\dot{q}$ which is not very exciting by itself but when it is combined with the equation for $\dot{p}$ we obtain something interesting,

$$\dot{p} = m\ddot{q} = -\kappa q = -m\omega_o^2 q. \tag{8-33}$$

This is our harmonic oscillator equation, in generalized coordinates. To show the greater power of this method, consider the torsion oscillator used in the Coulomb, Cavendish, Eötvos, Hull and Williams, and Dicke experiments. For this we have

$$H(p,q) = \frac{p_r^2}{2m} + \frac{p_\theta^2}{2I} + \frac{1}{2}\kappa_\theta \theta^2 \tag{8-34}$$

where $p_\theta = mr^2\dot{\theta}$ and $I = mr^2$. Let's assume that the bar is rigid so $r =$ constant and $p_r = 0$. Then,

$$\dot{q} = \dot{\theta} = \frac{\partial H}{\partial p_\theta} = \frac{p_\theta}{I} \tag{8-35}$$

and

$$-\dot{p}_\theta = \frac{\partial H}{\partial q} = \kappa_\theta \theta. \tag{8-36}$$

Combining these two we have our old harmonic oscillator again,

$$-\dot{p}_\theta = -I\ddot{\theta} = \kappa_\theta \theta \tag{8-37}$$

or

$$\ddot{\theta} = -\frac{\kappa_\theta}{I}\theta = -\omega_o^2\theta, \tag{8-38}$$

It seems that all roads lead to Rome, but some are more direct than others.

8.5 Poisson Brackets

Poisson brackets in classical mechanics are analogous to commutator brackets in quantum mechanics. For two (unspecified) dynamic quantities g and h that may depend on momenta, coordinates, and time, Poisson defined the bracket $[g, h]$ as

$$[g, h] \equiv \sum_i \left(\frac{\partial g}{\partial q_i}\frac{\partial h}{\partial p_i} - \frac{\partial g}{\partial p_i}\frac{\partial h}{\partial q_i} \right). \tag{8-39}$$

It is not difficult to prove that the total time derivative of a dynamic quantity g is given by

$$\frac{dg}{dt} = [g, H] + \frac{\partial g}{\partial t}, \tag{8-40}$$

where H is the Hamiltonian. The following Poisson brackets are of interest:

$$[q_i, H] = \dot{q}_i$$

$$[p_i, H] = \dot{p}_i$$

$$[p_i, p_j] = 0$$

$$[q_i, q_j] = 0$$

$$[q_i, p_j] = \delta_{ij}$$

where δ_{ij} is the Kronecker delta symbol, which is zero when the indices are different, and which is one when they are the same. (Leopold Kronecker was a businessman, pianist, gymnast, and a friend of the composer Felix Mendelssohn.) The first two brackets in the list above are Hamilton's equations, expressed in Poisson bracket notation.

By Eq. (8-40), if a dynamic quantity does not depend explicitly on time, and if its Poisson bracket with H is zero, then it is a constant of the motion. In quantum mechanics, if a quantity commutes with the Hamiltonian, then it is conserved.

In 1925 Dirac was looking for a connection between Heisenberg's formulation of quantum mechanics and the Hamilton-Jacobi equation (Section 8.7). Dirac wrote

"During a long walk on a Sunday it occurred to me that the commutator might be the analogue of the Poisson bracket, but I did not know very well then what a Poisson bracket was. I had just read a bit about it, and I did not have any book at home which gave Poisson brackets, and all the libraries were closed. So I had to wait impatiently until Monday morning when the libraries were open and check on what Poisson bracket really was. Then I found that they would fit, but I had one impatient night of waiting."

8.6 Hamilton's Principle

The number one position on everybody's Top Ten List of Abstruse Equations is occupied by the Euler-de Moivre theorem, $-e^{j\pi} = 1$, which implies that all of the real positive integers are composed of transcendental, imaginary, and negative numbers. In Section 6.8 we mentioned Feynman's high regard for this, "the most remarkable formula in mathematics." The full meaning of this is not now known.

A strong candidate for the number two position in The List is Hamilton's principle, which contains essentially all of classical mechanics in one equation, and is closely connected to optics and quantum theory. Sir William Rowan Hamilton was Irish; it has been said, perhaps by James Joyce, that all Irishmen are mystics. Hamilton was also a child prodigy whose genius extended into adulthood. At age 13 he was so fluent in 13 languages that he could write poetry in all of them. However, William Wordsworth (what a name for a poet!) said that Hamilton's work could hardly be called poetry. Queen Victoria appointed Wordsworth poet laureate in 1843, but earlier (1841) Hamilton presented the world with his poetic principle,

$$\delta S = \delta \int_{t=1}^{t=2} L(q_i, \dot{q}_i, t)\, dt = 0, \tag{8-41}$$

where δ is not an angle or the δ–function, but as used here it is the standard symbol for "variation" of the integral. Hamilton's principle says that the integral S is an extreme value, a maximum or a minimum. Often it is a minimum. When the integral is a minimum, this is called *the principle of least action*. Hamilton's principle is strange and wonderful, and perhaps a bit mystical and poetic.

It has been said that Eq. (8-41) is a vague concept based on an obscure concept. In a purely mathematical sense, this is not a valid statement because finding a function that minimizes or maximizes an integral is a basic exercise in the calculus of variations, just as finding a minimum or maximum of a function is a basic exercise in elementary calculus.

In a classical physical sense, Eq. (8-41) is somewhat puzzling because the end points are fixed and any variation of the trajectory would violate conservation of energy and, probably, momentum. In other words, there are no other classical paths, so how could the path be varied? Of all paths from A to B, which one does a particle take? It's as if some guiding intelligence tried all trajectories and chose the proper one that is actually observed. How does a lightning bolt know what is the path of least resistance? At Cape Kennedy, lightning struck a 100 m steel tower at its base.

Hamilton's principle looks more like optics and Fermat's principle than all of the mechanics that came before it. Recall that Fermat's principle states that the path that a light *ray* travels between two points is a minimum time; this is generally not the shortest distance when the index of refraction is not constant in the path. This is how a lens works, from a geometrical optics viewpoint.

If we pursue this line of inquiry eventually we come to the question of blocking off all unused paths. If they are not used, then are they needed? To put this in concrete terms, consider reflection of light from a mirror and reflection of a billiard ball from the side of a billiard table. In both cases, the law of reflection holds: The angle of incidence is equal to the angle of reflection, a consequence of conservation of momentum. In both cases, the observed path is a minimum-time path. For light, it might seem that we could make the mirror arbitrarily small, but if we make it quite small we find that the reflected light spreads out in many directions.

- Evidently the parts of the mirror are necessary where the angle of incidence is not equal to the angle of reflection, where it seems as if momentum is not conserved, in order for momentum to be conserved in the law of reflection.

Apply this reasoning to Young's double slit experiment and to driving to work in the morning. If there are two ways to go, and one is blocked off, your choice is restricted and you are forced to go the other way. If the variation of the paths in Hamilton's principle is restricted, then this has physically observable consequences.

The function S has units of Planck's constant, but Max Planck was not born until 1858. It's as if Eq. (8-41) were waiting for Planck to arrive in the twentieth century, and later for Feynman. However, Planck used thermodynamics to derive his distribution function, not Eq. (8-41).

The familiar *brachistochrone* problem of the Bernoulli brothers is an example, dating from 1696, in which it is required to find the shape of the curve that minimizes the time for a mass to fall from one point to another in a vertical plane. This is often done at science fairs with ball bearings and tracks or tubes; some people are surprised to find that in general the shortest time is not on a straight line joining the points. The Bernoullis examined this problem 92 years before Lagrange, so they didn't use the Lagrangian function; they were interested in the shape of the curve, not the dynamics of the motion.

Apparently Johann Bernoulli, who proposed the *brachistochrone* problem, hoped his brother Jakob would not be able to solve it. Jakob solved it in several months; it is said that Newton solved it the same night that he read it.

- In retrospect, Jakob Bernoulli's great contribution was not so much in solving the problem, but in recognizing that the *global* minimum property must also be a *local* minimum property.

In other words, the function that is the shortest time overall must also be the shortest time between all intermediate pairs of points. This facilitates analysis of the problem because it leads to a differential equation, and this was a path that Feynman took later with quantum mechanics.

Before Hamilton, in 1788 Lagrange elaborated on Euler's work and used the calculus of variations to find the differential equation that must be satisfied to make an integral like the one in Eq. (8-41) an extremum. The Euler-Lagrange equation is

$$\frac{\partial L}{\partial q} - \frac{d}{dt}\left(\frac{\partial L}{\partial \dot{q}}\right) = 0. \tag{8-42}$$

As an example, consider $L = 0.5mv^2 - 0.5\kappa x^2$ for the harmonic oscillator. Then Eq. (8-42) yields $-\kappa x - d/dt\,(mv) = 0$ which looks familiar. Clearly, this is another way of deriving Lagrange's equations. Hamilton saw the importance of $\delta S = 0$. It is possible to push this formalism to the limit of obtaining a wave-like interpretation, with surfaces of constant S phase, but we will not do this.

In Richard Feynman's Nobel Prize acceptance speech (1965) he described how he found the connection between the Lagrangian and quantum mechanics using Hamilton's principle. This led to his development of the "sum over histories" or path integral method and got him a round-trip ticket from Pasadena to Stockholm.

So, the variation in S is zero for observed paths, but what is the equation to find S itself? The answer is, it's the Hamilton-Jacobi equation.

8.7 Hamilton-Jacobi Equation

Like Hamilton, Carl Gustav Jacob Jacobi (Universities of Berlin and Königsberg) excelled in languages as well as mathematics. Jacobi was a friend of Dirichlet, and he helped Hermite (Chapter 12) gain the recognition he deserved. We are not going to derive the Hamilton-Jacobi equation; it is done in most books on classical mechanics (see Bibliography). The equation for S can be expressed as

$$\frac{\partial S}{\partial t} + \frac{1}{2m}\left(\frac{\partial S}{\partial x}\right)^2 + U(x) = 0. \tag{8-43}$$

The first term is $-E$ and the second term is the square of the momentum divided by $2m$, which is the kinetic energy. We can regard the Hamilton-Jacobi equation as another expression for the conservation of energy, stated in terms of the action function S. In other words, the Hamilton-Jacobi equation says $K + U = E$ which is not new information. Nevertheless, the solution of this equation for the S function, and the phase space action J, sheds new light on dynamics and optics.

Let's use the Hamilton-Jacobi equation to describe the classical harmonic oscillator. This will not be so easy, and it will be somewhat like using a sledge hammer to crack a peanut, but we will sketch the method. In coordinates q and velocity $\dot{q}$ we can write the Lagrangian as

$$L(q,\dot{q}) = K - U = \frac{1}{2}m\dot{q}^2 - \frac{1}{2}\kappa q^2 \tag{8-44}$$

where, as before, the momentum is

$$p = \frac{\partial L}{\partial \dot{q}} = m\dot{q}. \tag{8-45}$$

The Hamiltonian for this case is

$$H(p,q) = \sum p_r\dot{q}_r - L = p\dot{q} - \left(\frac{1}{2}m\dot{q}^2 - \frac{1}{2}\kappa q^2\right) = \frac{p^2}{2m} + \frac{1}{2}\kappa q^2. \tag{8-46}$$

Now we can write the Hamilton-Jacobi equation as

$$\frac{\partial S}{\partial t} + \frac{1}{2m}\left(\frac{\partial S}{\partial q}\right)^2 + \frac{1}{2}\kappa q^2 = 0. \tag{8-47}$$

We are more interested in J than S, although we could solve Eq. (8-47). One complete cycle of coordinate q goes from $-\sqrt{2E/\kappa}$ to $+\sqrt{2E/\kappa}$ and returns to the initial position $-\sqrt{2E/\kappa}$. The action J is twice the one-way trip,

$$J = \oint p\,dq = 2\int_{-\sqrt{2E/\kappa}}^{+\sqrt{2E/\kappa}} \sqrt{2m(E - 0.5\kappa q^2)}\,dq \tag{8-48}$$

or four times a quarter-cycle,

$$\oint p\,dq = 4\int_0^{+\sqrt{2E/\kappa}} \sqrt{2m(E - 0.5\kappa q^2)}\,dq = 2\pi E\sqrt{m/\kappa}. \tag{8-49}$$

Notice that J is the area of the phase space ellipse. *It can be quantized.* Finally,

$$E = H = \frac{J}{2\pi}\sqrt{\kappa/m} \tag{8-50}$$

and the frequency is

$$f = \frac{\partial H}{\partial J} = \frac{1}{2\pi}\sqrt{\kappa/m}. \tag{8-51}$$

All of this was known in the nineteenth century, but then it seemed somewhat of an elegant theoretical curiosity. Experimental information was lacking. In the twentieth century, the action J was to become the basis of the Wilson-Sommerfeld- Ishiwara method of quantizing periodic systems (Chapter 11). Later, Arnold Sommerfeld said that the Hamilton-Jacobi equation was "the royal road to quantization." In retrospect, this was clearly a bridge between classical theory and semi-classical quantum theory, but with the rise of the full quantum theory the Hamilton-Jacobi formulation was no longer on the leading edge of research.

As things turned out, it was the Hamiltonian function and Poisson brackets that proved to be of major usefulness in the discovery of the modern quantum theory. The Hamiltonian function translated into the Hamiltonian operator (Table 12-1), and Poisson brackets translated into the commutator brackets (Section 12.15) of quantum mechanics.

8.8 Symmetry of the Hamiltonian Function

We have repeatedly written the Hamiltonian function for the harmonic oscillator in generalized coordinates and momenta as

$$H(p,q) = \frac{p^2}{2m} + \frac{1}{2}\kappa q^2. \qquad (8\text{-})$$

It is certainly trivial to rewrite this as

$$H(p,q) = \frac{p^2}{2m} + \frac{q^2}{2\kappa^{-1}}, \qquad (8\text{-})$$

but now we can see the perfect *symmetry in phase space* which is a truly unique property of the harmonic oscillator. A duality exists between p and q in phase space. It is simple to transform the Hamiltonian into new "coordinates" and "momenta" in which the roles of q and p are interchanged. A similar symmetry (and its concomitant duality) exists between the electric and magnetic fields of classical electrodynamics.

Symmetry in configuration space (shift invariance in translation and rotation) is commonly understood to lead to conserved dynamic quantities (Noether's theorem). What we have here is something less familiar, more subtle, a symmetry in phase space and a rotation in phase space. It is now known that this symmetry leads, in a fundamental sense, to the equal spacing of the energy levels of the quantum harmonic oscillator (Chapters 11 and 12). I wish I could say (like Pierre Fermat) that I have a simple proof but there is not enough room for it in the margin of this page.

Phase space symmetry is directly related to the Wigner distribution. Eugene Wigner, who was Dirac's brother-in-law, received the Nobel Prize in 1963 for his contributions to the theory of the atomic nucleus and fundamental particles through the discovery and application of symmetry principles. It may seem odd that this abstract research has practical applications. The Wigner distribution was discovered again in the context of joint-time-frequency analysis (Section 6.12) where it is known as the Wigner-Ville distribution. This is now available in digital signal processing software for personal computers.

What's Next?

Advanced dynamics can be quite abstract; we had no use for graphics in this chapter. Except for Lagrange's equations, some of what we have discussed may seem to be little more than sophisticated mathematical manipulation and transformations to alternative descriptions, of doubtful practical application. These transformations could have become mere exercises in mathematical virtuosity, as "... useless formal transformations," in Fourier's words. (Fourier's remark was not about advanced dynamics, however, but about integral transforms.)

All we started with were Newton's laws, yet we ended up with a description of motion that makes mechanics look like optics. Or, does it make optics look like mechanics? The distinction becomes blurred. We now have a concept that implies that geometrical optics is to wave optics as classical mechanics is to quantum mechanics.

Eliminating the concept of force now has some experimental justification. In the Aharonov-Bohm effect, particles are influenced where *the field is zero* but its potential function is not zero. Some people now think that potentials are "elements of reality" and when we consider forces we are dealing with derivatives of these elements, not with the physical entities themselves. Later we will see that potentials, not their associated force fields, are involved in Schrödinger's equation. Still, force stays in our vocabulary but its meaning has been modified to indicate "interaction." Generalized forces can be included in Lagrangian and Hamiltonian formulations.

Hamilton's equations, Hamilton's principle, and the Hamilton-Jacobi equation provide muffled whispers that there could be more to come, that our knowledge may be incomplete. This is all the more remarkable because advanced dynamics was originally derived from Newton's laws, long considered to be the epitome of the description of a clockwork universe. Of course, it is now known that classical mechanics includes the possibility of deterministic chaos. This oxymoron can be demonstrated with a simple pendulum.

As the number of particles in a system increases, eventually the detailed description of individual motions gives way to the necessity of computing group properties. Lagrangian and Hamiltonian dynamics turn out to be extraordinarily powerful methods in statistical mechanics. In Chapter 10 we will look into statistical descriptions. Ironically, statistical considerations eventually lead us back to new ways of thinking about a single harmonic oscillator.

But, before we discuss large groups, in Chapter 9 we'll see what happens when we connect a few oscillators together. Sometimes it appears as if the whole is greater than the sum of the parts. Isn't that how biology works?

Chapter 9

Coupled Oscillators

There is a concatenation of all events in this best of all possible worlds. . .
Voltaire (François Marie Arouet) in *Candide*

When you look at the stars at night, a single photon from a single wiggling electron in a far-off galaxy stimulates another electron in your retina. Voltaire seems to have anticipated Mach's principle and quantum field theory; all events are connected. Perhaps Voltaire's statement about "this best of all possible worlds" anticipated the many-worlds interpretation of quantum mechanics.

Up to this point we have studied the properties of one classical harmonic oscillator. However, no oscillator is disconnected from the rest of the universe, despite Fig. 1-1. Now we are going to connect some oscillators together.

In Chapter 6 we saw that Laplace's equation suggests that space behaves *as if* it were filled with harmonic oscillators. Later, Planck's work suggested a similar situation. Huygens' construction and the classical electromagnetic wave equation (Laplace's equation in four dimensions) suggest that spacetime behaves *as if* it were filled with harmonic oscillators. Laplace's equation ($\nabla^2 U = 0$) implies that the value of a field at a point is the average value of the field around that point, like a spider web.

This is the essence of the field concept: All points of a field are connected to (and eventually influence) all other points. All charges and masses are connected to (and eventually influence) all other charges and masses in the universe. The vacuum permittivity ε_o and permeability μ_o and the gravitational force constant G are measures of the strengths of these connections. The speed of light $(\varepsilon_o \mu_o)^{-1/2}$ is a measure of delay in information transmission and causality in the fields.

Connecting systems transfers energy, momentum, and information between them. So, let's connect some oscillators and see what happens. For simplicity we will couple two oscillators with a linear restoring force. The strength of coupling has important effects on the spectrum of the oscillations, and we will see how natural it is for normal modes to arise and for energy levels to split.

9.1 Two Coupled Oscillators

Let's start with two undamped harmonic oscillators, in one dimension. This arrangement begs to be described by Lagrangian or Hamiltonian methods (Chapter 8) but we are going to use Newton's laws to postpone abstractions. Each mass is attracted to its equilibrium position by a restoring force with force constant κ. Let's couple the oscillators by a simple harmonic restoring force, with force constant k. (This is analogous to two capacitatively-coupled LC resonators). Now we can write the equations of motion for each oscillator as

$$m_1 \frac{d^2 x_1}{dt^2} = -\kappa x_1 + k(x_2 - x_1) \tag{9-1}$$

$$m_2 \frac{d^2 x_2}{dt^2} = -\kappa x_2 + k(x_1 - x_2). \tag{9-2}$$

When $k = 0$ the oscillators are uncoupled. Let's try phasor solutions,

$$x_1 = Ae^{j\omega t} \quad \text{and} \quad x_2 = Be^{j\omega t}. \tag{9-3}$$

When we substitute these into Eqs. (9-1) and (9-2) we obtain

$$(-m_1\omega^2 + \kappa)A = k(B - A)$$

$$(-m_2\omega^2 + \kappa)B = k(A - B). \tag{9-4}$$

Let's rearrange these equations for a convenient matrix representation,

$$(-m_1\omega^2 + \kappa + k)A - kB = 0$$

$$-kA + (-m_2\omega^2 + \kappa + k)B = 0. \tag{9-5}$$

Equation (9-5) can now be expressed in matrix form,

$$\begin{pmatrix} -m_1\omega^2 + \kappa + k & -k \\ -k & -m_2\omega^2 + \kappa + k \end{pmatrix} \begin{pmatrix} A \\ B \end{pmatrix} = 0. \tag{9-6}$$

The necessary and sufficient condition for these equations to have a solution is that the determinant of the matrix must be zero,

$$\begin{vmatrix} -m_1\omega^2 + \kappa + k & -k \\ -k & -m_2\omega^2 + \kappa + k \end{vmatrix} = 0. \tag{9-7}$$

When we expand the determinant we obtain a quadratic equation in $(\omega^2)^2$,

$$m_1 m_2(\omega^2)^2 - (m_1+m_2)(\kappa+k)\omega^2 + (\kappa+k)^2 - k^2 = 0. \tag{9-8}$$

Equations (9-7) and (9-8) are forms of the *secular equation*, a name that owes its origin to similar calculations in celestial mechanics. Equation (9-8) has two real, distinct roots,

$$\omega^2 = \frac{(m_1+m_2)(\kappa+k)}{2m_1m_2} \pm \left[\frac{(m_1-m_2)^2(\kappa+k)^2}{4m_1^2m_2^2} + \frac{k^2}{m_1m_2}\right]^{1/2} \tag{9-9}$$

Normal Mode Frequencies

For the special case when the masses are identical we have

$$\omega^2 = \frac{\kappa+k}{m} \pm \frac{k}{m} \tag{9-10}$$

so the lower frequency is

$$\omega_{low} = \left(\frac{\kappa}{m}\right)^{1/2} \tag{9-11}$$

and the higher frequency is

$$\omega_{high} = \left(\frac{\kappa+2k}{m}\right)^{1/2}. \tag{9-12}$$

These are the frequencies of the *normal modes*. To get a physical picture of the motions corresponding to these frequencies, consider the case of the negative sign in Eq. (9-10). For this case, the lower of the two frequencies, the coupling has no effect and the frequency is the same as two isolated identical oscillators, Eq. (9-11). Evidently the two masses must be moving in synchronism such that their displacements are identical at every point in time. In other words, the coupling spring is not stretched or compressed in this mode, so it could be removed. This is the *even* or *symmetric* mode in which the two masses move in phase.

Now consider the case of the upper sign, corresponding to the higher of the two frequencies, Eq. (9-12). The effect of the coupling constant is doubled, indicating that the masses must be moving in opposite directions. In other words, the coupling spring is stretched simultaneously and subsequently compressed simultaneously by both masses. This is the *odd* or *anti-symmetric* mode, in which there is a phase difference of 180°.

Figure 6-7 shows an experiment with a baby elephant. I advised against doing this experiment at home, but there is an experiment with coupled oscillators that you can do in your kitchen. Get some duct tape or mylar tape, some string, a broomstick, and two identical unopened cans of soup. Tape the string to the cans, making two pendula, and tape the other ends of the strings to the broomstick. Keep the lengths of the strings the same. Suspend the broomstick between the backs of two chairs, tape the broomstick securely to the chairs, and experiment with the system. There will probably be enough coupling in the broomstick (due to flexing) without the need of a separate coupling between the soup cans. If you want to increase the coupling, tape a slack piece of heavy string between the cans. You can vary the coupling by changing the slack in the string.

You can excite the symmetric mode by displacing the pendula by the same amount, in the same direction, and start from rest. The anti-symmetric mode is excited by displacing the pendula by the same amount, in opposite directions, starting from rest. The superposition of the two modes is excited by any other set of initial conditions. For example, hold one pendulum fixed at its equilibrium position and displace the other pendulum. Observe the exchange of energy between the two oscillators. You will find that it is difficult to set initial conditions so that only one normal mode is excited; most of the time you will get a superposition of modes.

Because of the masses of the pieces of duct tape and the possible differences in masses of the soup cans amd string lengths it is difficult to make the two pendula identical, but you can come close enough to observe all of the phenomena. Think about how you would apply a driving force of variable frequency.

9.2 Magnetically Coupled Resonators

Very accurate measurements can be made with an electrical analog of two coupled pendula. Coupled electromagnetic resonators have been investigated in minute detail for many years because of their importance in radio, TV, and microwave systems. Let's consider a special case, two *LCR* circuits coupled magnetically through the mutual inductance of the inductors. This is easy to build, easy to tune, easy to vary the mutual inductance by spatial separation, and easy to measure because you can see the waveforms on an oscilloscope. Let's assume that both resonators have identical values of L, C, and R, and are arranged in a parallel configuration.

We can excite the system with a variety of waveforms using a signal generator in one circuit, but we're only going to discuss sinusoidal excitation. Let's put the signal generator in circuit #1 and measure circuit #2 relative to #1 as we vary the frequency of excitation. The coupling is characterized by the mutual inductance M.

After a little use of Kirchoff's law around the loops, and a bit of algebra, we find

$$\frac{V_1}{V_2}\left(\frac{\omega_o}{\omega}\right)^2\frac{M}{L} = \left[1-\left(\frac{\omega_o}{\omega}\right)^2 - j\frac{R}{\omega L}\right]^2 - \left(\frac{M}{L}\right)^2 . \tag{9-13}$$

With more algebraic manipulation this can be put into a usable form in terms of the magnitude,

$$|V_1/V_2|^2\,(Wk)^2 \;=\; [(1-W)^2 - k^2 - 1/Q^2]^2 + [2/Q\,(1-W)]^2, \tag{9-14}$$

where $W = (\omega_o/\omega)^2$, $Q = \omega L/R$, and $k = M/L$. The voltages V_1 and V_2 are measured across each inductor with respect to ground potential.

The phase is given by

$$\tan(\theta) \;=\; \frac{2(1-W)/Q}{k^2 + 1/Q^2 - (1-W)^2}\;. \tag{9-15}$$

9.3 COUPLOSC, Worksheet for Coupled Oscillators

Figure 9-1 shows the Home screen of this worksheet. This computes the magnitude and phase of two identical systems for two values of mutual inductance M, so you can compare them. The box contains computed auxiliary quantities. For low values of coupling the resonance curve is sharp and there is only a hint of spectrum splitting.

As you increase the coupling, the spectrum splits into a high-frequency wing and a low-frequency wing, the splitting increases as the coupling increases. Increasing R blurs the separation and broadens the bandwidth, as you would expect when the Q is decreased.

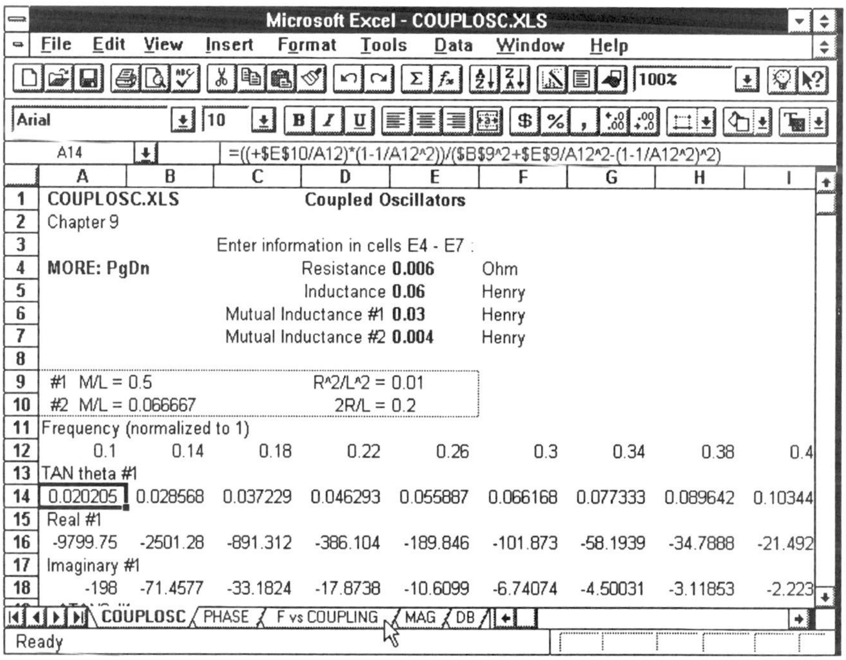

Figure 9-1. Home screen for two magnetically coupled oscillators. Press PgDn for more. Use your computer to make changes in parameters and view graphs.

The effects shown here are the classical analog of splitting which occurs as two atoms or molecules are brought together. These effects also give rise to the band theory of solids.

The magnitude of the frequency responses are shown in Fig. 9-2, for the parameters shown in Fig. 9-1. Observe how increased coupling, keeping R constant, broadens the bandwidth and splits the resonance between the high-frequency wing and the low-frequency wing.

The phase is shown in Fig. 9-3. For graphical clarity, only one system is shown. For higher Q the phase changes very abruptly, as with a single resonator. Figure 9-4 shows the frequency splitting as a function of the coupling. This is qualitatively similar to the atomic case. For multi-level quantum systems this leads to conduction bands and valence bands.

9.4 Time Evolution, Weak Coupling

Equations (9-1) and (9-2) are linear so the principle of superposition holds and any sum of solutions is also a solution. This enables us to write,

$$x_1(t) = A\cos(\omega_{low}t) + B\cos(\omega_{high}t)$$
$$x_2(t) = A\cos(\omega_{low}t) - B\cos(\omega_{high}t). \tag{9-16}$$

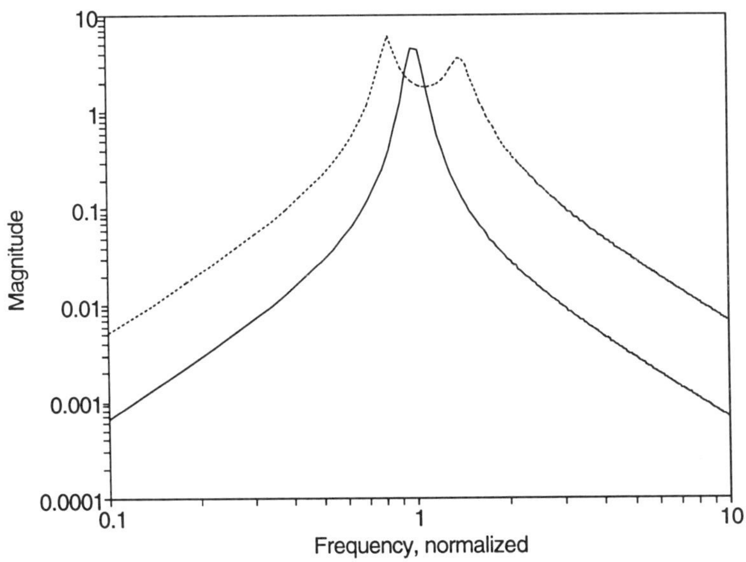

Figure 9-2. Splitting of spectrum as coupling is increased, for parameters shown in Fig. 9-1. *Dotted line:* M = 0.03 H. *Solid line:* M = 0.004 H

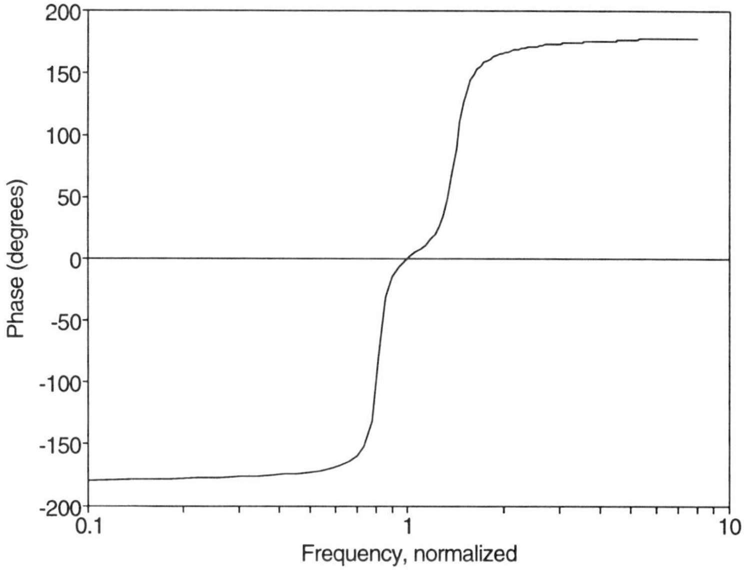

Figure 9-3. Phase response for two coupled oscillators, for $M = 0.03$ H. Other parameters are shown in Fig. 9-1.

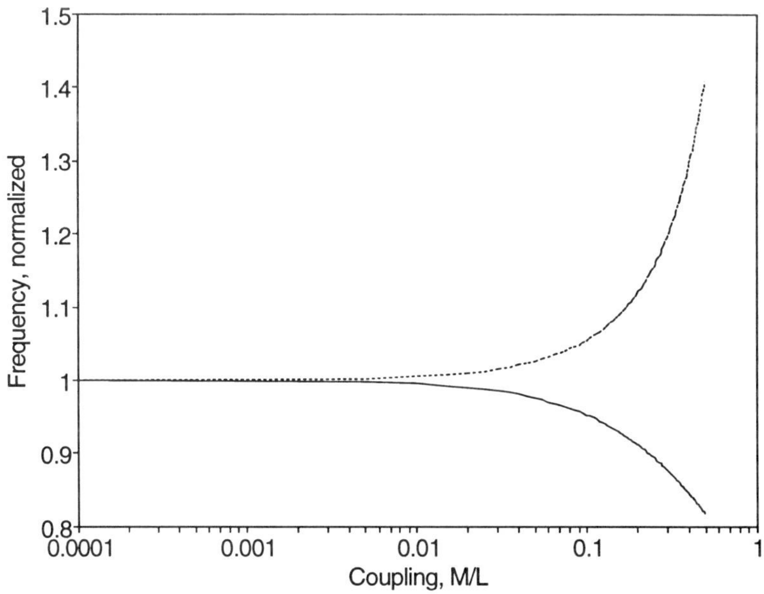

Figure 9-4. Spectrum splitting of two coupled oscillators as a function of coupling factor. This is similar to spectrum line splitting at the atomic and molecular level.

In Eq. (9-16), A and B are constants that depend on the initial conditions. If either A or B is zero, then the motion is a normal mode.

A simplification occurs when the coupling force is much weaker than the restoring forces of the oscillators and $A = B$. In this case the normal mode frequencies are nearly the same. For compact notation let's write,

$$\omega_{av} = \frac{1}{2}(\omega_{high} + \omega_{low}) \tag{9-17}$$

$$\Delta\omega = \frac{1}{2}(\omega_{high} - \omega_{low}). \tag{9-18}$$

So,

$$\omega_{low} = \omega_{av} - \Delta\omega \tag{9-20}$$

$$\omega_{high} = \omega_{av} + \Delta\omega. \tag{9-21}$$

In this approximation, ω_{av} is the average of the normal mode frequencies and $\Delta\omega$ is the deviation from the average. We are going to use this weak-coupling approximation and a trigonometric identity to change Eq. (9-16) into an interesting form that looks like amplitude modulation. Substitute Equations (9-20) and (9-21) into (9-16), and recall the trigonometric identity,

$$\cos(a \pm b) = \cos(a)\cos(b) \pm \sin(a)\sin(b).$$

After a little simplification the results are,

$$x_1(t) = [2A\cos(\Delta\omega\, t)]\cos(\omega_{av}t)$$

$$x_2(t) = [2A\sin(\Delta\omega\, t)]\sin(\omega_{av}t). \tag{9-22}$$

The motions described by Eqns. (9-22) are in the worksheet COUPLOS2. At time $t = 0$ the displacement $x_1 = 2A$ and $x_2 = 0$. As time goes by, energy flows from the first oscillator to the second oscillator and then the system returns to its original state when all of the energy returns to the first oscillator, as with the soup can experiment. The exchange time τ_{ex} for one complete cycle of energy transfer is given by

$$\tau_{ex} = \frac{\pi}{\Delta\omega} \tag{9-23}$$

and the corresponding frequency for energy exchange is

$$\omega_{ex} = \frac{2\pi}{\tau_{ex}} = 2\,\Delta\omega. \tag{9-24}$$

This cyclic exchange of energy also has a quantum analog.

9.5 COUPLOS2, Worksheet #2 for Coupled Oscillators

Figure 9-5 shows the Home screen for this worksheet, which enables you to experiment with two coupled oscillators with and without damping, in the time domain. This worksheet uses the FFT for the frequency domain. You can experiment with different data windows, as in Chapter 7, to observe the spectra. Adjust the time increment and the time start to zoom in and out of a particular part of the waveforms.

Figure 9-6 shows the waveforms for the parameters given in Fig. 9-5. Figure 9-7 shows the spectrum of Fig. 9-6, and Fig. 9-8 shows the polar plot of oscillator #2 as a function of oscillator #1 for the same parameters.

9.6 Normal Coordinates

Equations (9-1) and (9-2) can be uncoupled in a formal way, following Eq. (9-16), that leads to a general result involving *normal coordinates*. Let,

$$\Psi_1 = x_1 + x_2 \quad \text{and} \quad \Psi_2 = x_1 - x_2 \tag{9-25}$$

so,

$$x_1 = \frac{1}{2}(\Psi_1 + \Psi_2) \quad \text{and} \quad x_2 = \frac{1}{2}(\Psi_1 - \Psi_2). \tag{9-26}$$

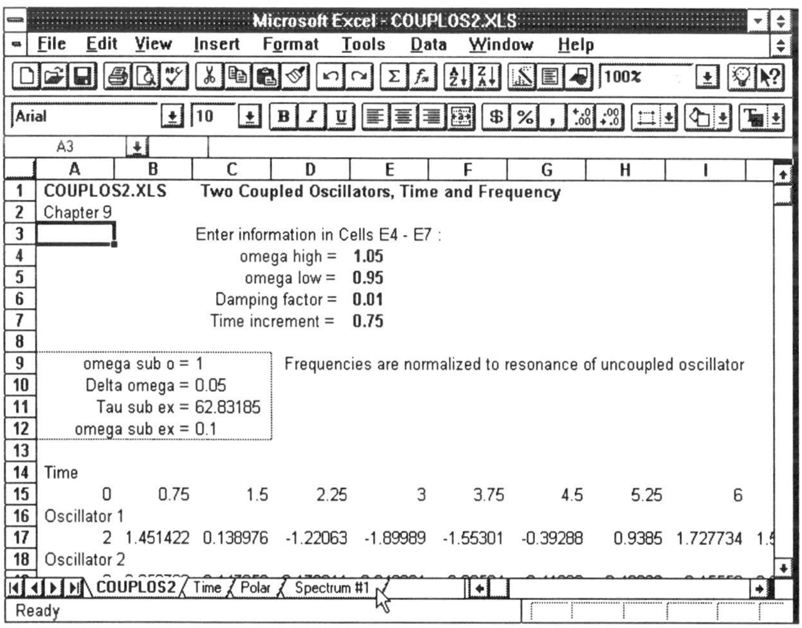

Figure 9-5. Home screen for COUPLOS2. FFT must be recalculated when parameters are modified.

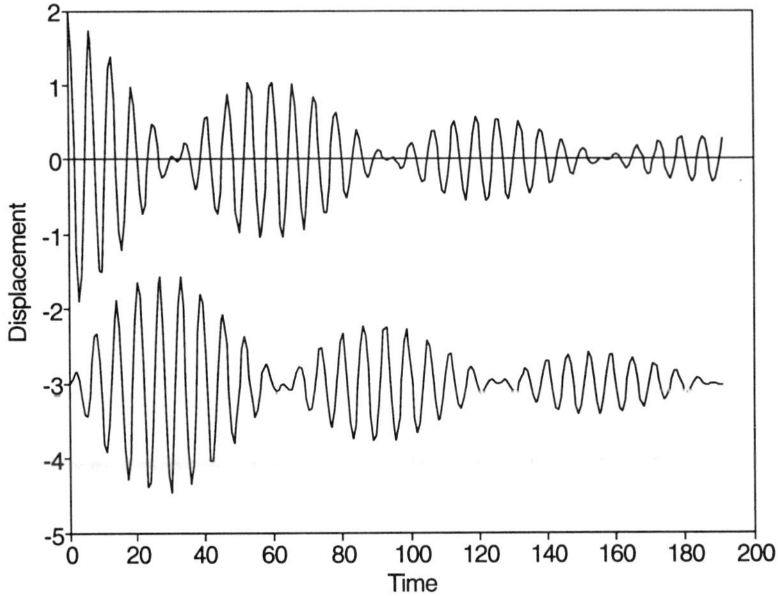

Figure 9-6. Displacements of two damped, coupled, identical oscillators for parameters shown in Fig. 9-5. Note periodic exchange of energy.

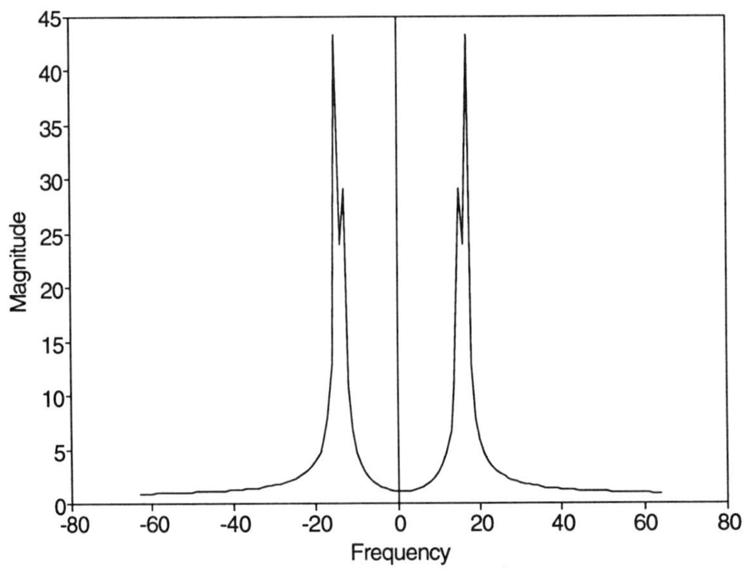

Figure 9-7. FFT spectrum magnitude of displacement waveform of oscillator #1, shown in Fig. 9-6. A rectangular window was used in the 128-point FFT.

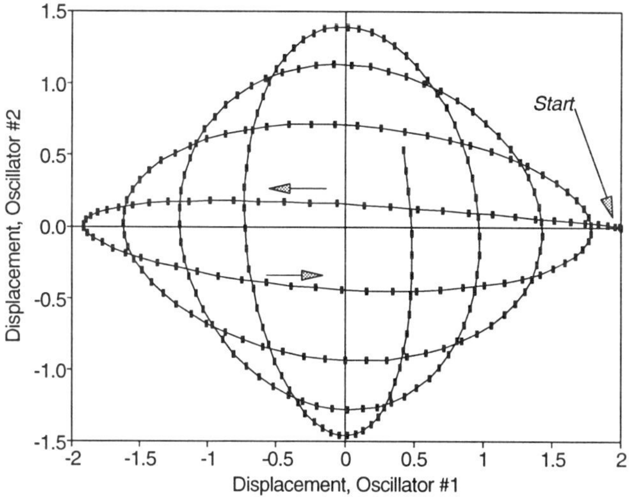

Figure 9-8. Polar plot of oscillator #2 vs. oscillator #1, for waveforms shown in Fig. 9-6. Note shifting phase difference. Time interval between markers is constant.

Next, substitute Eqs. (9-26) into Eqs. (9-1) and (9-2) to obtain,

$$\frac{m}{2}(\ddot{\Psi}_1 + \ddot{\Psi}_2) = -\frac{\kappa}{2}(\Psi_1 + \Psi_2) - k\Psi_2$$

$$\frac{m}{2}(\ddot{\Psi}_1 - \ddot{\Psi}_2) = -\frac{\kappa}{2}(\Psi_1 - \Psi_2) + k\Psi_2\,. \tag{9-27}$$

When we add these two equations, and then subtract them, we find,

$$m\ddot{\Psi}_1 = -\kappa\Psi_1$$

$$m\ddot{\Psi}_2 = -(\kappa + 2k)\Psi_2\,, \tag{9-28}$$

which are separated into the new normal coordinates. Notice that each normal coordinate for this case is a superposition of two states, and a normal coordinate does not correspond to a spatial coordinate in any ordinary sense. The transformation to normal coordinates is a mathematical trick that decouples and simplifies the equations of motion, and this trick is applicable to more complicated systems.

In Eq. (9-28), Ψ_1 has the frequency of an uncoupled oscillator; evidently it corresponds to the symmetric mode. The equation for Ψ_2 corresponds to the anti-symmetric mode. For the coupled *LCR* resonators, the symmetric mode has the currents going in the same direction and the anti-symmetric mode has the currents going in opposite directions.

9.7 Oscillating Galaxies

One, two, three, many. This Polynesian number system could be our guide for extending this chapter to larger physical systems. We mentioned in Section 5.4 that results of computer models involving approximately 4×10^5 particles interacting through the gravitational force suggest that galaxies may undergo harmonic oscillations. Real galaxies have tens of thousands of times more stars than this model, but the model has enough to show the main features.

Two fundamental modes have been identified in the model. The first fundamental mode involves expansion and contraction of the entire galaxy, in which matter farther from the center is faster than matter close to the center. In a rotating spiral galaxy this radial motion affects velocities of hydrogen atoms moving in the galaxy, and this can be measured by radio telescopes.

The second fundamental mode involves matter closer to the center, moving in a direction opposite to that of matter farther away. In this mode, matter piles up in the center to produce a bulge or takes matter away to create a hole. Estimated periods of both modes of oscillation are in the range of 10^9 years, so observation on a single galaxy may be difficult. However, bulges and holes in known galaxies could be the result of observing an ensemble of galaxies in various stages of oscillation.

These two galactic fundamental modes are analogous to the even and odd modes of the two coupled oscillators we discussed.

9.8 Synchronized Oscillators

The Dutch physicist Christian Huygens was ill during the winter of 1665 and stayed in his room for several days. He had made two pendulum clocks of different design and they were mounted on a common support in his room. In a letter to his father, Huygens wrote that he observed that the pendula were always 180° out of phase and that the clocks kept perfect time with each other. He disturbed the pendula so that they were no longer in opposite phase, but within 30 minutes they returned to apparently perfect synchronism with the characteristic 180° phase difference.

Huygens realized that the clocks were influencing each other through the common support or possibly through air currents set up by the pendula. To test this hypothesis he moved one clock to a wall on the opposite side of the room from the other clock. The clocks lost synchronism. One clock gained 5 seconds/day compared to the other clock, the pendula did not oscillate at the same frequency, and the phase drifted. When the clocks were returned to the common support, the synchronization reappeared.

It is now known that oscillator synchronization occurs in such diverse phenomena as electronic circuits, superconducting Josephson junctions, and neuron networks in the brain, the heart, groups of crickets, frogs, and fireflies.There is also evidence of synchronized oscillation in Belousov-Zhabotinsky chemical reactions. Studies show that synchronization of non-identical oscillators involves increasing the frequencies of slow oscillators and decreasing the frequency of fast oscillators. The end result is that the ensemble of non-identical oscillators is phase-locked at some group rate determined by the dynamics of the system.

Yoshiki Kuramoto (University of Tokyo) developed a simple, solvable model in which synchronization is somewhat analogous to a phase transition, like the freezing of water into ice. An ensemble of oscillators, initially unsynchronized with random phases and a spread of frequencies, spontaneously freezes in frequency when the frequency spread is less than a critical value.

The time evolution of synchronization is interesting. Suppose two oscillators near the center frequency of the final state become locked through accidental phase alignment. If this coherent oscillation is above the noise level it pulls other frequency-adjacent oscillators into synchronization. This freezing feeds on itself until the entire ensemble is synchronized. A possibly unphysical assumption in Kuramoto's simple model is that each oscillator has the same coupling to every other oscillator in the ensemble, not just the nearest neighbors or with some coupling constant that decreases with distance. In our solar system, planet Jupiter is the major influence that appears to pull smaller objects into orbital synchronization. Can you conceive of synchronized galaxies?

Synchronization of oscillators is an area of current research interest. One effect that interferes with synchronization is in special relativity, where the twin paradox predicts that two *identical* synchronized oscillators become unsynchronized when one is displaced and subsequently returned to its initial position. In the famous "flying clocks" experiments, atomic clocks became unsynchronized after being flown completely around planet Earth in opposite directions. The time difference was consistent with relativity. Synchronization of oscillators in deep-space missions is made even more difficult by general relativity and the Doppler shift.

Crabs are known to have a temperature cycle that is synchronized to the local diurnal cycle. In the famous "flying crabs" experiments, crabs on the east coast of the United States were equipped with temperature monitors and sealed in a light-tight, electromagnetically-shielded, temperature-controlled box. The crabs were flown to the west coast of the United States and their temperature was continuously monitored during the journey and after their arrival. The phase of their temprature cycle was observed to shift gradually into synchronism with the new local day and night cycle as the crabs experienced their new environment. It is not known how the crabs sensed their new longitude, but shielding from very low frequency magnetic fields is a difficult task. Several life forms, including marine bacteria, are known to contain particles of magnetite as a magnetic sensor. It is believed that magnetic sensing assists bacteria in finding food by navigation toward the ocean floor, but it is not now known whether crabs have this type of sensor. In any case, the synchronizing signal for the flying crabs must be very small.

Synchronize Your PC to an Atomic Clock

If you have a modem, a program named *Atomic Clock™ 2.0* for Windows® enables you to synchronize your PC's clock to any one of 20 atomic clocks around the world.

What's Next?

Even though we can now perform numerical computations on systems with hundreds of thousands of components, we still need insights into the properties of a multitude of coupled oscillators that only analytical methods can provide. Figure 9-9 shows a screen from a software package that can be used for studying coupled oscillators and many other classical systems. Numerical methods and analytical methods reinforce each other in forming an intuitive feeling for complicated systems.

Consideration of an ensemble of harmonic oscillators led Max Planck, reluctantly, to the quantization of energy, $E = hf$. Planck wrote, ". . . I can characterize the whole procedure as an act of despair, since, by nature I am peaceable and opposed to doubtful adventures. . . ."

In the next chapter we will look at some implications of the statistical point of view where we lose track of individual particles and measure group properties.

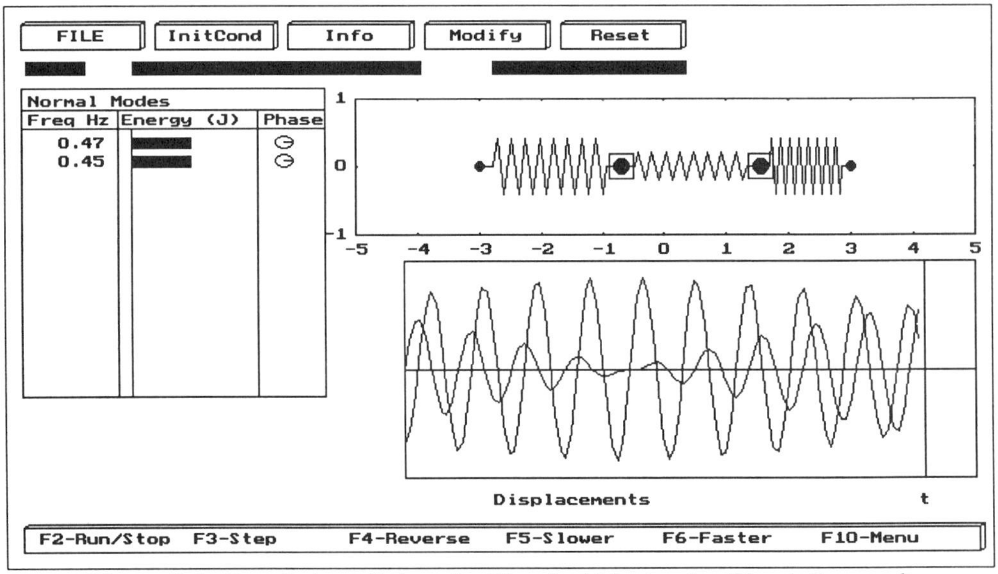

Figure 9-9. Coupled oscillators in *Classical Mechanics Simulations* software. (See Bibliography). Note shifting phases in waveforms at lower right. The waveforms and the physical model are animated in this program. In the configuration shown here, the end masses have been chosen to be infinite, the two central masses are identical, and the two end springs are identical.

Chapter 10

Statistical Perspective

In this life nothing is certain but death and taxes.
Benjamin Franklin

Classical oscillators often consist of very large numbers of particles. Because we lack information about the coordinates and momenta of individual particles and/or it is impossible, impractical, or even undesirable to follow individual particles, we use statistical methods (recall Sections 1.1 and 1.6). Statistical ideas also appeared in Section 2.6 as the equipartition of energy, where we showed that the time-average kinetic energy of an oscillator is equal to the time-average potential energy. We used the concept of r.m.s. in discussing phase space and in comparing power associated with different functions of time, and power spectral density in Chapter 6. The subject of this chapter could (and does) fill several books.

The quantum domain is inherently statistical in nature. This was pointed out by Professor Max Born (grandfather of singer-actress Olivia Newton-John), who received the Nobel Prize in 1954 for his statistical interpretation of the quantum wavefunction. Quantum oscillators can now be made with a single electron; in such systems, uncertainty relations prevent us from knowing associated coordinates and momenta as precisely as we desire. Other uncertainty relations in energy and time, angular momentum and its associated angle, or any pair of canonically conjugate dynamic variables (Chapter 8), impose similar restrictions on these quantities.

10.1 Distributions and Moments

Let's start with mass distributions, because they are easy to visualize. It has proven fruitful to treat a large group of particles as a single particle by describing the group statistically. We have already used the idea of average mass density,

$$\rho_{av} = \frac{M}{V}, \tag{10-1}$$

where M is the total mass and V is the volume occupied by the mass. If the mass is not uniformly (homogeneously) distributed then we need the density at a point,

$$\rho(x, y, z) = \frac{dM}{dV} . \tag{10-2}$$

The first moment of the mass distribution, normalized to the total mass, is called the *center of mass*. The center of mass is useful in describing the translational motion of a group. In this way we can treat the motion of the center of mass as a single particle of mass M, although we lose information about the motion of the members of the group.

The x-coordinate of the center of mass of a collection of discrete masses is defined in terms of a spatial average as

$$x_{av} = \frac{\sum_{i=1}^{N} x_i m_i}{\sum_{i=1}^{N} m_i} = \frac{1}{M} \sum_{i=1}^{N} x_i m_i . \tag{10-3}$$

Similar expressions hold for y_{av} and z_{av}. Here, the total mass M has the appearance of a normalizing factor.

If we go to the useful (but unrealistic) limit of a continuous mass distribution, then the sums become integrals. For example, the x coordinate of the center of mass becomes

$$x_{av} = \frac{\int x \, dm}{\int dm} = \frac{1}{M} \int x\rho \, dV, \tag{10-4}$$

where we have used $dm = \rho dV$ to transform the integrals over mass to integrals over volume. This is a useful procedure in calculating moments.

Moments to Remember

The quantities Mx_{av}, My_{av}, and Mz_{av} are *first moments* of the mass distribution, just as $|\boldsymbol{r} \times \boldsymbol{p}| = |mvr \sin\varphi|$ is the first moment of momentum, which is angular momentum.

For rotational motion, as in a torsion pendulum oscillator, the *second moment* of the mass distribution, called the *moment of inertia*, is the important quantity. For combined translation and rotation a great simplification occurs because:

- It is possible to separate the general motion of a group of particles into translational motion of the center of mass and rotational motion about the center of mass.

- Another important simplification occurs if the collection of particles can be treated as a rigid body. In a rigid body, the distance $\mathbf{r}_{ij}$ is constant between any mass i and any other mass j.

The inertia tensor has nine components but only six are independent. Furthermore, no matter how unsymmetrical the mass distribution is in space, you can always find a reference frame in which the off-diagonal components vanish, leaving three components. These special axes are called the *principal axes*. Diagonalization of the inertia tensor (in matrix form) is called the *principal axis transformation*. We are not going to pursue this further, but mention it because of an analogous diagonalization operation later in quantum mechanics. Table 10-1 summarizes three typical moments of mass distributions.

Table 10-1. Some Moments of Mass Distributions

Moment Number	*Type*	*Name*	*Definition*
0	Scalar	Total mass	$M = \sum x_i^0 m_i$
1	Vector	Center of mass	$x_{av}M = \sum x_i^1 m_i$
2	Tensor	Moment of inertia	$I_{yy} = \sum (x_i^2 + z_i^2) m_i$

The first moment can be positive, negative, or zero, but the second-moment diagonal components (I_{xx}, I_{yy}, I_{zz}) cannot be less than zero; they are intrinsically positive with minimum values when referred to center-of-mass coordinates. We can define other distributions and their moments for velocity, momentum, energy, probability, and so on. These turn out to be as useful as their spatial analogs.

To visualize a distribution function in a concrete way let's imagine that we have a collection of bricks. Bricks are composed of many particles, but suppose we can't measure anything smaller than a single brick. We can arrange or distribute the bricks in many ways, like a wall, an outhouse, or an office building.

If we start with a disorderly pile of bricks then we could describe the system by a random distribution function. If we do work on the pile of bricks and use intelligent information then we can transform some of the local disorder to local order. The orderly system is less probable than the disorderly system. System entropy is greater in the disorderly system and work is required to reduce local entropy. Even though our work decreased local entropy, some work was lost due to friction and heat during the building process, so the total entropy of the universe increased due to our wall-building exercise.

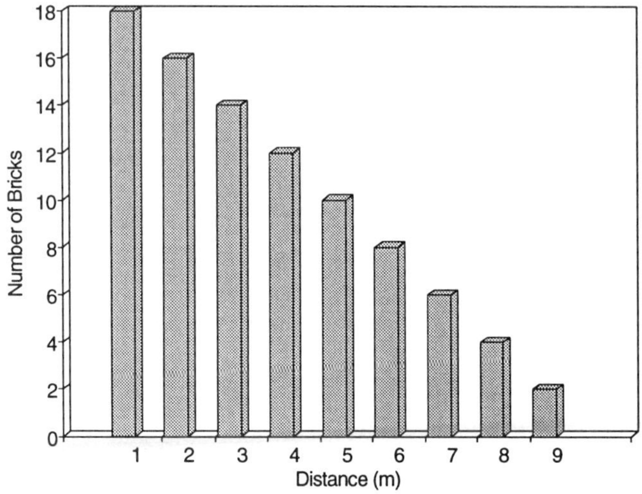

Figure. 10-1. A non-uniform distribution of bricks.

Let's suppose that we stack bricks on top of each other, as in Fig. 10-1. This is not the best way to build a wall, but let's line the bricks up vertically. In the natural course of geological time the wall will deteriorate, the bricks will be dispersed and returned to a more probable state, and all of our work and energy will be transformed into nonrecoverable forms, just as oscillators run down. *Sic transit gloria.* During a coffee break in the building process the wall might look like Fig. 10-1.

The number of bricks in the first stack is 18; in general the number of bricks in a stack is n_i. The total number of bricks in the wall in this state is

$$N = \sum n_i \, . \tag{10-5}$$

We can define a linear density of bricks as the mass of bricks Δm_i per location Δx_i,

$$\rho_i = \frac{\Delta m_i}{\Delta x_i} \, . \tag{10-6}$$

The probability of finding a brick is greatest where there are more bricks. It is eight times more probable that you will find a brick in stack 2 than in stack 9, inferring that there is a close connection between distribution functions and probability. Willie Sutton, the infamous bank-robber, was familiar with spatial probability distributions. When asked "Why do you rob banks?" he replied, "Because that's where the money is."

Let each brick be one kg and the center of each brick be 1 m from the centers of adjacent bricks. What is the first moment of the mass distribution in Fig. 10-1?

$$x_{av}M = \sum x_i^1 m_i \tag{10-7}$$

$$= 18\times1 + 16\times2 + 14\times3 + 12\times4 + 10\times5 + 8\times6 + 6\times7 + 4\times8 + 2\times9 = 330$$

so the center of mass is located at

$$x_{av} = \frac{330 \text{ m·kg}}{90 \text{ kg}} = 3\tfrac{2}{3} \text{ m}.$$

Similarly, we can compute the moment of inertia with respect to a vertical axis through the origin, a horizontal axis at ground level, or any other axis we choose.

The stacks of bricks in Fig. 10-1 may have artistic merit, but most people want a wall that looks like Fig. 10-2. Now the bricks are arranged in a uniform distribution over an interval, and the center of mass is now located at $x = x_{av} = 4.5$ m.

10.2 Accuracy and Precision

Accuracy and precision are associated concepts but they are fundamentally different. Accuracy refers to the error or difference between a measurement of a quantity and the true value of the quantity. Precision refers to the spread of a group of measurements about the average value of the group.

Consider archery. High accuracy infers that the average position of the holes in the target is near the bull's eye. High precision infers that the spread in the positions of the holes is small. You can imagine a situation in which the accuracy is high (the average position of the holes is at the center of the bull's eye), but the precision is low (not a single arrow hit the bull's eye). This is like the center of mass; for some mass distributions, like a diatomic molecule, there could be no mass at x_{av}.

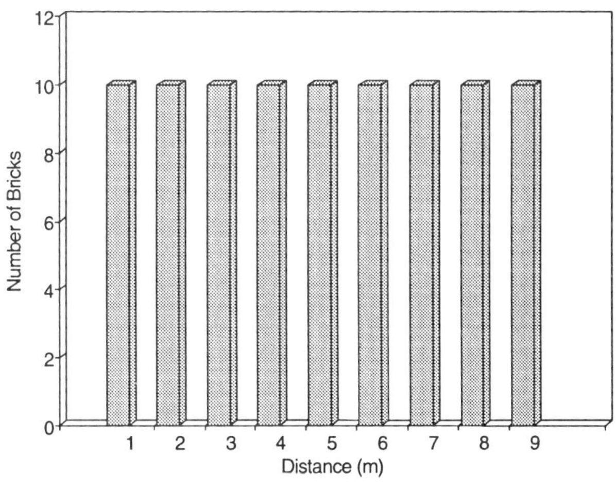

Figure 10-2. A distribution of bricks that is uniform over an interval.

Measurements can have high accuracy and low precision; they can also have low accuracy and high precision. Naturally we hope to achieve high accuracy and high precision.

We now believe that some physical quantities cannot be measured as precisely as we choose. There are certain limitations imposed by Nature; we have already seen this in the classical uncertainty principle for time and frequency in Chapter 6. Others, of quantum origin, will be explored in Chapter 12.

10.3 Parable of the Bunny Rabbits

Nature does not inherently limit accuracy or precision in counting. Usually people have ten fingers and ten toes, for a total of twenty. You would not say you have 20.3 digits ± 1.7. Nevertheless, errors often exis,t even in counting experiments.

Ensemble preparation

Let's prepare an ensemble of identical systems consisting of fenced carrot patches. Now suppose you release 500 indistinguishable, classical white bunny rabbits into each carrot patch and they diffuse, as bunnies do. You count the bunnies one by one as you let them in so you know there are 500 in each carrot patch.

Measurements

When the bunnies are hopping around happily inside the carrot patches you start counting them again. Bunnies must be counted swiftly because their number tends to increase. A difficulty arises in counting because the bunnies are identical and they are hopping around rapidly.

Typical results of counting the bunnies many times are shown in Fig. 10-3. A striking feature here is that you would not expect identical results in measuring an ensemble of systems that were prepared identically.

The state of each system in the ensemble could be described by the coordinates, momentum, orientation, angular momentum, and energy of each bunny. Measuring the state of a system would be more difficult, and would require more data, than merely counting occupants. The time evolution of the ensemble is controlled by Newton's laws; these are classical rabbits. *Degenerate states* have the same energy but different configurations.

Presumably, if the bunnies did not move and we were skilled at counting, we could obtain the accurate value of 500 with infinite precision. However, if we made an infinte number of measurements with hopping bunnies (and their number did not change) we might expect an approximate Gaussian or normal distribution. For a continuous distribution this has the form,

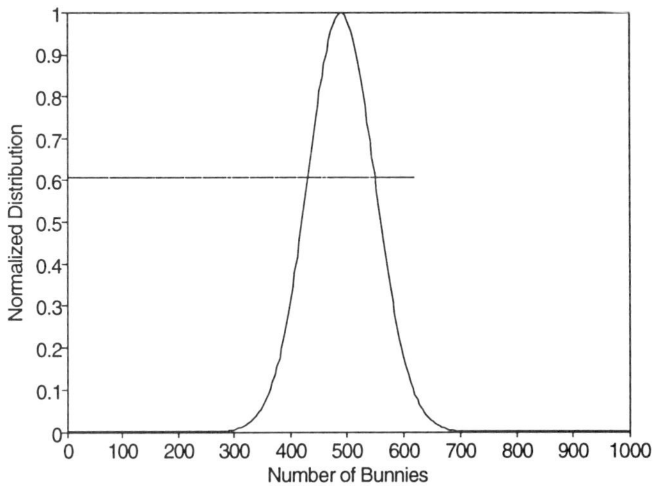

Figure 10-3. Results of counting bunnies many times. Here it is known that there are 500 bunnies, but the results of many measurements indicate that the average value is 490. The standard deviation is 59.

$$f(x) = \frac{1}{\sqrt{2\pi}\,\sigma} \exp[-(x - x_{av})^2/2\sigma^2] \tag{10-1}$$

where x is the quantity being measured and σ is the standard deviation,

$$\sigma = \sqrt{(x^2)_{av} - (x_{av})^2}\,. \tag{10-2}$$

The standard deviation is a common measure of precision; its spreadsheet function is STD. There is also a *sample standard deviation* (STDS) in spreadsheets. When N is large they are practically the same; consult Help for information on these functions. In a normal distribution approximately 68 percent of measurements lie within $\pm\sigma$. When x_{av} is zero (origin of coordinates is centered on the average value) then the standard deviation is the r.m.s. value. This provides some justification for using r.m.s. values as measures of uncertainty, as we did in Section 1.9.

10.4 Displacement Probability Density

Let's find the probability that we can observe a harmonic oscillator as it oscillates through its allowed range of displacement. You might think of this as an experiment using a movie of a mechanical harmonic oscillator. Each frame of the movie is a sample of data that shows the position of the oscillator during the short time interval Δt that the film was exposed. If we make a histogram of the number of times the oscillator is in some small space interval between x and $x + \Delta x$, we would expect to find it most often where it was moving slowly, and we would expect to find it fewer times where it was moving rapidly. Obviously, it spends less time at places where it's moving fast! The velocity is maximum at the equilibrium point and is zero at the ends, as shown in Figs. 2-4 and 2-5.

Another way you might think of this as an experiment is to put a sine wave into the vertical channel on an oscilloscope and stop the horizontal sweep. The vertical trace will be brightest at the top and bottom (where the beam spends more time). The light is dimmest at the zero crossings. If the light intensity from the phosphor were a linear function of the number of electrons hitting it, then a light meter moved up and down the trace would show readings proportional to the probability of finding a particular voltage of the sine wave. You might ask, "What is the probability that I will measure a voltage between 1.1V and 1.15V?"

To find the probablility analytically, let's rewrite the equation for the elliptical orbit in phase space, Eq. (2-23). We have earlier identified this as an expression for the total energy of the undamped harmonic oscillator,

$$0.5m\omega^2x^2 + 0.5mv^2 = 0.5m\omega^2X^2 = E.$$

We can use this to find the velocity $v(x, E)$,

$$v(x,E) = \omega\sqrt{X^2 - x^2} = \omega\left(\frac{2E}{m\omega^2} - x^2\right)^{1/2}. \tag{10-2}$$

The probability of finding the oscillator in a range between x and $x + dx$ is $P(x)dx$, which is proportional to the time spent in this range. Since $dx = v\,dt$ we can write $dt = dx/v$. Using this, we can express $P(x)dx$ as

$$P(x)dx = C\,dt = C\frac{dx}{v(x,E)} \tag{10-3}$$

where C is a constant of proportionality. Using Eq. (10-2) we can write,

$$P(x)\,dx = \frac{C\,dx}{\omega\sqrt{X^2 - x^2}}. \tag{10-4}$$

The probability of finding the oscillator in its total range from $-X$ to $+X$ is 1,

$$\int_{-X}^{+X} P(x)\,dx = 1. \tag{10-5}$$

This normalization of the probability, Eq. (10-5), enables us to find the constant of proportionality C,

$$\int_{-X}^{+X} P(x)\,dx = \int_{-X}^{+X} \frac{C\,dx}{\omega\sqrt{X^2 - x^2}} = 1. \tag{10-6}$$

From a table of integrals we find,

$$\int \frac{dx}{\sqrt{a + bx + cx^2}} = -\frac{1}{\sqrt{c}} \arcsin\left(\frac{-2cx - b}{\sqrt{b^2 - 4ac}}\right). \tag{10-7}$$

Comparing Eq. (10-7) with (10-6) we find,

$$\int \frac{C\,dx}{\omega\sqrt{X^2 - x^2}} = \frac{C}{\omega} \arcsin\left(\frac{2x}{2X}\right). \tag{10-8}$$

When this is evaluated from $-X$ to $+X$ we obtain,

$$\frac{C}{\omega}[\arcsin(1) - \arcsin(-1)] = \frac{C}{\omega}\left(\frac{\pi}{2} - -\frac{\pi}{2}\right) = 1. \tag{10-9}$$

From this the constant is determined as

$$C = \frac{\omega}{\pi}. \tag{10-10}$$

So, we finally have the displacement probability,

$$P(x)\,dx = \frac{dx}{\pi\sqrt{X^2 - x^2}} = \frac{dx}{\pi\sqrt{2E/(m\omega^2) - x^2}}. \tag{10-11}$$

Integrate Eq. (10-11) between two points to calculate the probability of finding the oscillator within that interval. For example, as in Figs. 10-4 and 10-5, let the amplitude-squared be $X^2 = 2E/(m\omega^2) = 1$. Let's compute the probability of finding the oscillator between $x = -0.1$ and $x = +0.1$. Equation (10-11) becomes,

$$\int_{-0.1}^{+0.1} P(x)\,dx = \frac{1}{\pi}\int_{-0.1}^{+0.1} \frac{dx}{\sqrt{1^2 - x^2}} = \frac{1}{\pi}\arcsin\left(\frac{x}{1}\right)\Bigg|_{-0.1}^{+0.1}$$

which is evaluated as

$$\int_{-0.1}^{+0.1} P(x)\,dx = \frac{1}{\pi}[\arcsin(0.1) - \arcsin(-0.1)] = 0.06376856.$$

So, for these conditions the probability of finding the oscillator in this range is about 64 chances out of 1000. Notice that the minimum probability density, for a given Δx, is at the equilibrium point $x = 0$ where $v(x,E)$ is maximum. The minimum value is

$$Minimum\ P(x) = \frac{1}{\pi X} = \frac{\omega}{\pi\sqrt{(2E/m)}}. \tag{10-12}$$

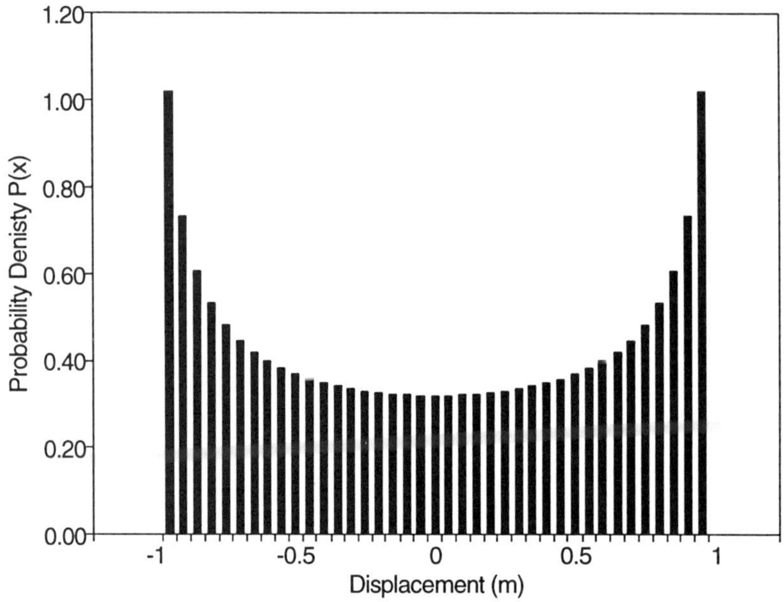

Figure 10-4. Discrete displacement probability density for the classical simple harmonic oscillator. A large number of measurements are taken at a finite number of points over a small time interval at each point, as in a movie.

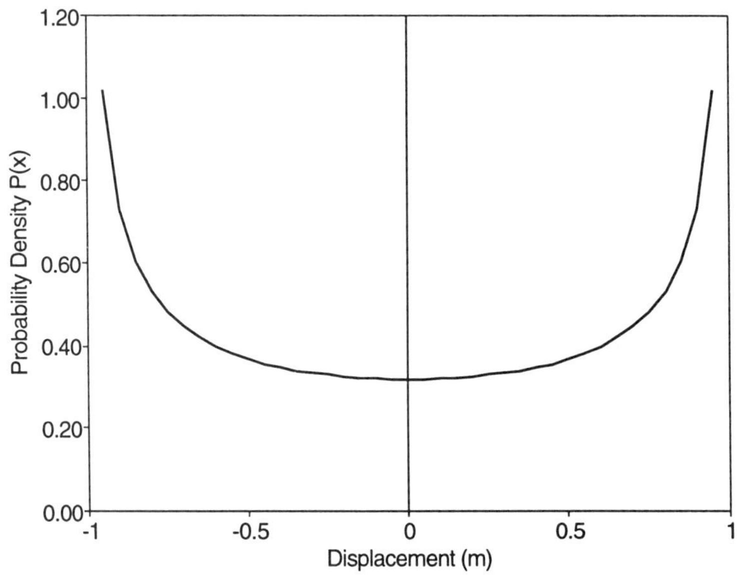

Figure 10-5. Continuous probability density is associated with an infinite number of measurements at every position occupied by the oscillator.

PROBABIL, Classical Oscillator Probability Worksheet

The worksheet for this Section is named PROBABIL. After starting your spreadsheet in the normal way, open this file and you should see the upper left-hand corner (Home position) shown in Fig. 10-6.

Two of the graphs connected to this worksheet are named *Discrete P(x)* and *P(x)*. The probability density is in error at the limits of the displacement because of overflow when x approaches X in Eq. (10-11). You can approach very close to the limits of displacement by modifying the increment in x in Row 5.

10.5 Uncertainty Due to Thermal Motion

We saw in Section 2.6 that, in a classical system, energy is distributed in such a way that every degree of freedom has a time-average energy $0.5kT$, where k is Boltzmann's constant and T is the absolute temperature. This is equipartition of energy at thermodynamic equilibrium.

For a simple harmonic oscillator at temperature T this implies that the time-average kinetic energy is

$$(0.5\, mv^2)_{av} = 0.5\, kT. \tag{10-13}$$

So, the r.m.s. velocity at temperature T is

$$v_{rms} = (kT/m)^{1/2}. \tag{10-14}$$

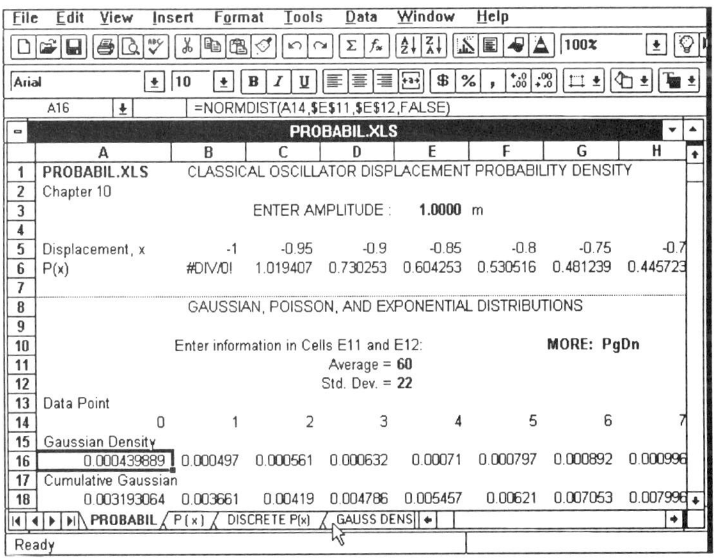

Figure10-6. Home screen of the worksheet PROBABIL. Enter the oscillator amplitude in Cell E3. Three common probability distributions are also available.

Equation (10-14) shows that even in the classical domain there is a finite, non-removable r.m.s. velocity at any temperature above absolute zero. This is a restless universe. Similarly we find that there is a finite non-removable r.m.s. displacement above absolute zero,

$$(0.5\ \kappa x^2)_{av} = 0.5\ kT. \tag{10-15}$$

So,

$$x_{rms} = \left(\frac{kT}{\kappa}\right)^{1/2} = \left(\frac{kT}{m\omega^2}\right)^{1/2}. \tag{10-16}$$

Figure 10-7 shows thermal effects in a classical oscillator (see the THERMAL and THERMAL2 worksheets). The torsion oscillator is one our most sensitive measuring devices (see Section 5.15). For a torsion oscillator with torsion constant κ_θ and moment of inertia I_{yy}, the results of Equations (10-14) and (10-16) translate into

$$\omega_{rms} = \left(\frac{kT}{I_{yy}}\right)^{1/2}$$

and

$$\theta_{rms} = \left(\frac{kT}{\kappa_\theta}\right)^{1/2} = \left(\frac{kT}{I_{yy}\omega^2}\right)^{1/2}.$$

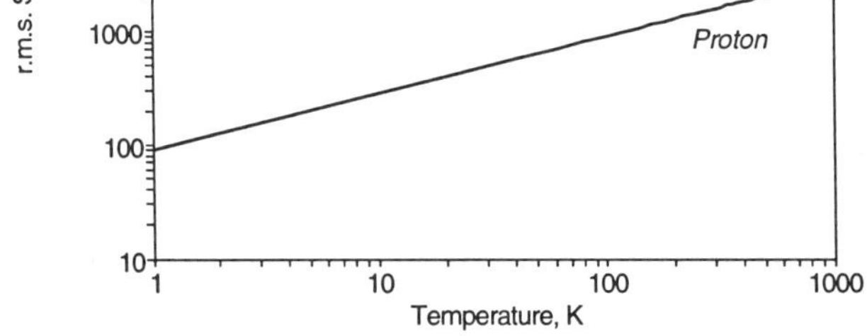

Figure 10-7. Thermal motion of classical oscillators. Classical results are inaccurate at low temperatures. Oscillators must be corrected for relativity at high temperatures.

This shows the importance of keeping oscillators cool if you want to make precise measurements. Proposed measurements of the gravitational force constant G with a torsion oscillator will be performed at a temperature of approximately 4K.

Uncertainty Principle, Part 3

As in Section 1.9, let's take r.m.s. values as the uncertainties in velocity,

$$\Delta v = \frac{\sqrt{kT}}{\sqrt{m}} \tag{10-17}$$

and displacement,

$$\Delta x = \frac{\sqrt{kT}}{\sqrt{\kappa}}. \tag{10-18}$$

We can rewrite Eq. (10-17) as

$$\Delta v = \sqrt{m}\,\frac{\sqrt{kT}}{m} \tag{10-19}$$

so the uncertainty in momentum can be written as

$$\Delta p = \Delta mv = \sqrt{m}\,\sqrt{kT}. \tag{10-20}$$

Multiplying Eq. (10-20) by Eq. (10-18) gives

$$\Delta p_x \,\Delta x = (\sqrt{m}\,\sqrt{kT}\,)\left(\frac{\sqrt{kT}}{\sqrt{\kappa}}\right) = \frac{kT}{\omega_o}. \tag{10-21}$$

Equation (10-21) is a very important result. Since kT is the total energy we can rewrite this as

$$\Delta p_x \,\Delta x = \frac{E}{\omega_o}. \tag{10-22}$$

Do you recall the adiabatic invariant E/ω from Section 2.10? Equation (10-22) tells us that the *minimum area* in phase space is an adiabatic invariant, a constant with units of J·s.

- Phase space is fuzzy *even in the classical domain*. According to Eq. (10-22) squeezing a particle in displacement (localization) expands the uncertainty in momentum. Sharpening our information about momentum causes us to lose information about location.

We must regard Eq. (10-22) as the best that we can do, because the r.m.s. values are the *minimum* uncertainties. Minimum uncertainties can be achieved only if the variables are Gaussian-distributed and all other errors have been eliminated. In realistic measurements we need to relax Eq. (10-22) to the condition,

$$\Delta p_x \, \Delta x \ \geq \ \frac{E}{\omega_o} \ . \tag{10-23}$$

Equation (10-23) is classical and we have seen how it arises naturally due to thermal motion as well as in a completely deterministic oscillator. However, it is a disguised quantum expression because $E = hf_o$. This leads to $E/\omega_o = hf_o/(2\pi f_o) = h/(2\pi)$, and Eq. (10-23) becomes,

$$\Delta p_x \, \Delta x \ \geq \ \frac{h}{2\pi} \ . \tag{10-23}$$

We saw this expression in Section 6.13, and we will see this again.

10.6 More Distribution Functions

We have been using distribution functions since Chapter 1, but we didn't identify them as such. Actually, we can regard $x(t)$, $p(t)$, and so on as distributions in time. Now let's look at some distributions in velocity. As usual, we define a system stationary state in terms of the smallest number of variables needed to give a complete description of motion. The number of particles dn_v in a particular velocity range dv is proportional to the probability F of finding those particles in the infinitesimal velocity element $dv_x dv_y dv_z$ in "velocity space"

$$dn_v \ = \ F \, dv_x dv_y dv_z. \tag{10-24}$$

The distribution function depends only on energy E of the state. The number of states dS_E in dE is proportional to the density of states $g(E)$,

$$dS_E \ = \ g(E) \, dE, \qquad (10\text{-})$$

and the number of particles in dE is

$$dn_E \ = \ Fg(E) \, dE \ = \ f_E \, dE, \qquad (10\text{-})$$

where $Fg(E) = f_E$ is the energy distribution function. In classical systems,

$$g(v) = 4\pi v^2 \text{ and } g(E) = (2^{5/2}\pi/m^{3/2}) \, E^{1/2}. \qquad (10\text{-})$$

For weakly interacting particles it is usual to assume (as a first approximation) that $U(r) = 0$ so the total energy is entirely kinetic. Many different states can have the same energy (this is called *degeneracy*). For some unspecified quantity $W(E)$ we can write $dn_W \ = \ Fg(W) \, dW$.

The probability F is of critical importance. For classical systems this is the Maxwell-Boltzmann probability F_{MB}. For quantum systems this is either the Fermi-Dirac probability F_{FD} or the Bose-Einstein probability F_{BE}. Table 10-2 lists expressions for these three probability distribution functions.

Table 10-2. Classical and Quantum Probability Distribution Functions

Name	*F*
Maxwell-Boltzmann (classical)	$F_{MB} = N\left(\frac{m}{2\pi kT}\right)^{3/2} \exp[-E/kT]$
Bose-Einstein (quantum, spin zero and integer spin, bosons)	$F_{BE} = \frac{1}{\exp[\alpha]\exp[E_i/kT] - 1}$
Fermi-Dirac (quantum, spin one-half and half-integer spin, fermions)	$F_{FD} = \frac{1}{\exp[\alpha]\exp[E_i/kT] + 1}$

The classical distribution function gives the probability of finding the number of particles with energy between E and $E + dE$. The quantum distribution functions give the probability of finding the number of particles that occupy the state i with energy E_i. For bosons with numbers that are not conserved (like photons), $\alpha = 0$. For fermions, α is a function of T. For $kT >> E_i$ both quantum distribution functions approximate the classical distribution function. This is another way that classical mechanics appears as an approximate upper limit of quantum mechanics.

When using distribution functions to compute a physical quantity it is important to note that the most probable value, the average value, and the r.m.s. value are usually different. For example, for Maxwell-Boltzmann statistics the most probable speed, $\sqrt{2kT/m}$, is $\sqrt{2/3}\, v_{rms}$ and $v_{rms} = \sqrt{3\pi/8}\, v_{av}$.

Table 10-3 shows some typical expressions involving a normalized velocity distribution function.

10.7 Natural Sources of Randomness

Let's calculate the r.m.s. thermal voltage across an inductor in a high-Q LCR resonator, where R is the resistance of the inductor. The energy stored in the inductor is

$$E = \int_0^I V_L\, i\, dt = \int_0^I L \frac{di}{dt}\, i\, dt = \int_0^I L\, i\, di = 0.5\, LI^2, \tag{10-24}$$

which is a familiar result. By equipartition of energy,

$$E_{av} = 0.5\, L(I^2)_{av} = 0.5kT. \tag{10-25}$$

Table 10-3. Computations Based on Velocity Distribution Function

Name	*Continuous*	*Discrete*
Particles in *dv*	$dn = f(v)\, dv$	$n_i = f_i \Delta v_i$
Particles in the range, *v(max) – v(min)* (zero moment)	$N = \int_{v(min)}^{v(max)} v^0 f(v)\, dv$	$N = \sum v_i^0 f_i \Delta v_i$
Average *v* (first moment)	$v_{av} = \frac{1}{N} \int_{v(min)}^{v(max)} v^1 f(v)\, dv$	$v_{av} = \frac{1}{N} \sum v_i^1 f_i \Delta v_i$
Mean square *v* (second moment)	$(v^2)_{av} = \frac{1}{N} \int_{v(min)}^{v(max)} v^2 f(v)\, dv$	$(v^2)_{av} = \frac{1}{N} \sum v_i^2 f_i \Delta v_i$

If the Q is large, then the energy spectrum is compactly centered at the resonant frequency $\omega_o = (LC)^{-1}$ and the impedance of the inductor is practically equal to the inductive reactance. The inductor Q is $\omega_o L/R$ so the emf across the inductor is

$$V_L = ZI = (\omega_o L)I = QRI. \tag{10-26}$$

Therefore the mean square emf is

$$(V_L^2)_{av} = (\omega_o L)(I^2)_{av} = (QR)^2(I^2)_{av}, \tag{10-27}$$

or

$$(V_L^2)_{av} = \omega_o^2 LkT. \tag{10-28}$$

Finally, we have

$$(V_L)_{rms} = \omega_o \sqrt{LkT} = \sqrt{\omega_o QRkT} = \left(\frac{kT}{C}\right)^{1/2}. \tag{10-29}$$

The average noise power is

$$\frac{(V_L^2)_{av}}{R} = \omega_o QkT. \tag{10-30}$$

Notice the similarity, and especially the difference, between the noise in a high-Q system and the noise in a pure resistance (Johnson noise). The average open-circuit thermal noise power in a pure resistance is $4kTB$, where B is the bandwidth of the measuring system in Hz. Some sources of electrical noise are summarized in Table 10-4.

Table 10-4. Natural Sources of Electrical Noise

Name	*Controlling Equation*	*Description*
Thermal noise (also called Johnson noise, white noise, or Nyquist noise)	$V_{rms} = \sqrt{4kTBR}$ Open-circuit rms voltage across R.	The origin of thermal noise in *R* is the random motion of atoms and electrons (Brownian motion)
Shot noise (also called Schottky noise)	$I^2_{noise} = 2qI_{dc}B$ Mean square noise current. q is electron charge in coulombs, I_{dc} is the dc current in amperes, and B is the bandwidth of the measuring system in Hz.	Shot noise, first studied theoretically by W. Schottky, is due to the discrete nature of the charge carriers. It is independent of frequency, but depends on the flow of current.
$1/f$ noise ("one over f noise," power inversely proportional to frequency)	$(V^2)_{av} = A\dfrac{I^2}{f}$ The constant *A* is empirical, and depends on the type of material, its physical shape, and sometimes its history.	1/*f* noise is most important in semiconductors and carbon resistors, where it may be the dominant noise. This type of noise is not well understood.
Blackbody radiation	Planck distribution (see Chapter 11)	Electromagnetic thermal radiation.

High-Q resonators find a multitude of applications in modern technology, especially at microwave frequencies. For example, superconducting microwave cavities are used in particle accelerators. High-Q resonators ring constantly due to thermal excitation so it is a good idea to keep them as cool as possible, even if they are not superconducting.

10.8 Poisson, Exponential, and Gaussian Distributions

These three distributions are available in the *Excel* spreadsheet, and their use is demonstrated in the worksheet PROBABIL (see Fig. 10-6). The exponential distribution is the classical Maxwell-Boltzmann distribution, except for the normalizing factor.

Poisson Distribution

The Poisson distribution often describes *discrete*, rare physical events that occur in an interval of time or space. Radioactive decays and electron currents seem to fit Poisson statistics, but careful measurements of electrons (fermions) show that, due to the exclusion principle, they are likely to be found farther apart than classical particles. Photons (bosons) are likely to be found closer together than classical particles, as demonstrated in Brown-Twiss experiments. A practical measure of "rare" events is that, in at least 50 trials, the number of trials times the probability of an event occuring is less than about 5. The characteristics of the Poisson distribution are:

- The probability of a single event is directly proportional to the size of the interval.
- The probability of two or more events in an interval is very small if the interval is sufficiently small.
- Events are independent in intervals that do not overlap.

In Microsoft *Excel*, the Poisson distribution is available with the POISSON function. It's syntax is POISSON(*x, mean, cumulative*), where *cumulative* is a logical value that can be set as TRUE or FALSE. For cumulative = FALSE, the Poisson probability density is given in the *Excel* on-screen Help notation as

$$POISSON = \frac{e^{-\lambda}\lambda^{x}}{x!} \tag{10-31}$$

where λ is the expected mean number of counts.(The notation here is used in*Excel's* on-screen Help.) For cumulative = TRUE, the cumulative Poisson distribution is

$$CUMPOISSON = \sum_{k=0}^{x} \frac{e^{-\lambda}\lambda^{x}}{k!} \; . \tag{10-32}$$

The difference between the Gaussian and Poisson distributions is shown in Fig. 10-8. For large N the discrete Poisson distribution can be approximated by the continuous Gaussian distribution. The standard deviation of the Poisson distribution is $\sqrt{\lambda}$.

Exponential Distribution

This distribibuition is similar to the classical Maxwell-Boltzmann distribution (Table 10-2). The probability density is given by

$$DENSITY = \lambda e^{-\lambda x}, \tag{10-33}$$

and the cumulative probability is given by

$$CUMULATIVE \;\; F(x;\lambda) = 1 - e^{-\lambda x}. \tag{10-34}$$

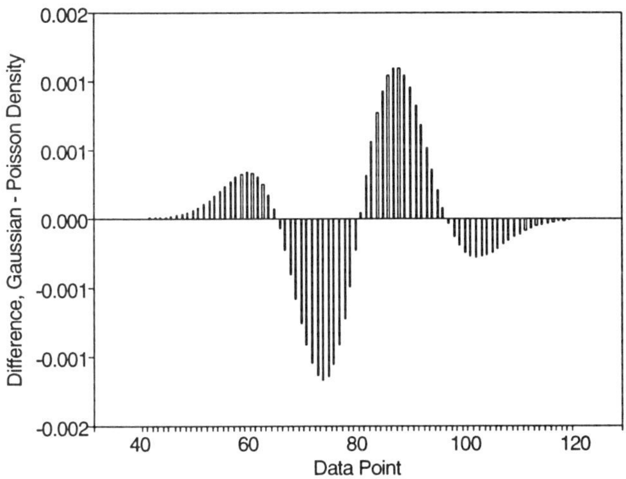

Figure 10-8. Gaussian density minus Poisson density. Mean = 81, standard deviation = 9. Discrete Poisson approximates continuous Gaussian for large *N*.

Gaussian (Normal) Distribution

Excel has two functions related to the Gaussian distribution. NORMDIST returns the normal cumulative distribution for specified mean and standard deviation; NORMSDIST returns the standard normal cumulative distribution function with a mean of zero and a standard deviation of one.

The syntax for NORMDIST is NORMDIST(*x, mean, standard dev, cumulative*), where x is the value for which you want the distribution and *mean* is the arithmetic mean of the distribution. *Cumulative* is a logical value, which returns the cumulative distribution function when set to TRUE and the ordinary distribution function when set to FALSE.

The NORMDIST density function is defined as usual as

$$NORMAL = \frac{1}{\sqrt{2\pi}\,\sigma} \exp[-(x - x_{av})^2/(2\sigma^2)]. \tag{10-35}$$

The NORMSDIST syntax is NORMSDIST(x), where x is the value for which you want the distribution. The standard normal density function is

$$f(x,0,1) = \frac{1}{\sqrt{2\pi}} \exp[-x^2/2]. \tag{10-36}$$

The Poisson, exponential, and normal distributions are set up in the worksheet PROBABIL, and other distributions are available in *Excel*. Consult the on-screen Help for additional information.

10.9 Random Excitation

The Poisson distribution naturally leads to a discussion of random excitation. The Tacoma Narrows suspension bridge collapse on November 7, 1940, is a good (or bad) example of a classical resonant system excited by random events. Let's consider a general system excited by random impulses, a topic addressed by Campbell's theorems. N. R. Campbell showed that the average response of a linear system to a sequence of single events is given by

$$V_{av} = r_{av} \int_0^{\infty} f(t)\, dt \tag{10-37}$$

where r_{av} is the average rate of events (events/time) and $f(t)$ is the response to a single event. For example, an event could be the arrival of an electron or a photon at an oscillator. Let's assume an event to be a delta function, so the impulse response $h(t)$ of a system is of critical importance. If we know a system's impulse response then we can determine the response for almost any other input.

The second of Campbell's theorems tells us that the corresponding mean square response is given by

$$\Delta^2 = [(V - V_{av})^2]_{av} = r_{av} \int_0^{\infty} [h(t)]^2\, dt. \tag{10-38}$$

From our experience with linear systems in Chapter 7 and Parseval's theorem in Chapter 6 we can express Eq. (10-38) in the frequency domain as

$$\Delta^2 = \frac{2}{\pi} r_{av} A^2 \int_0^{\infty} [|H(\omega)|^2]\, d\omega, \tag{10-39}$$

where $H(\omega)$ is the system function, the Fourier transform of the impulse response. The factor of 2 appears because we are only considering positive frequencies and the spectrum is symmetric.

We are not going to derive Campbell's theorems, but we will sketch the path and discuss some results of practical importance. For simplicity, we will start with a δ-function impulse of amplitude $A\delta(t)$ arriving at $t = 0$. As in Chapter 7 we can write the system's response to this single impulse as $Ah(t)$, where $h(t)$ is the system's unit impulse response.

Suppose a group of impulses with uniform amplitude A excites the system, and the impulses arrive at random times with an average rate of r_{av} impulses/s. During an interval dt the average number of impulses is $r_{av}dt$, so the average response at time delay τ is $Ah(t - \tau)\, r_{av}d\tau$. The average response due to all previous impulses is

$$V_{av} = \int_{-\infty}^{\tau} Ah(t - \tau) r_{av}\, d\tau. \tag{10-40}$$

(This can be generalized to a non-uniform distribution of amplitudes.) We can modify this slightly by letting $t' = t - \tau$, so

$$V_{av} = A r_{av} \int_0^{\infty} h(t')\, dt'. \tag{10-41}$$

For example, consider an RC integrator (low-pass filter) with R and C in parallel. The mean voltage across the integrator due to shot noise is

$$V_{av} = A r_{av} \int_0^{\infty} C^{-1} \exp[-t'/RC]\, dt' = A r_{av} R. \tag{10-42}$$

It's also of interest to calculate the r.m.s. value of fluctuations about the mean. To do this we must make some assumption about the distribution. Let's assume Poisson statistics so the r.m.s. fluctuation in the number of impulses is $(r_{av} dt)^{1/2}$. Due to fluctuations, the average number of impulses in time interval dt is $r_{av} dt \pm (r_{av} dt)^{1/2}$. Here is another case where the r.m.s. value is a useful measure of the uncertainty of the average value. The mean square response is

$$\Delta^2 = A^2 r_{av} \int_0^{\infty} [h(t)]^2\, dt. \tag{10-43}$$

For an integrator consisting of R and C in parallel, Eq. (10-43) becomes

$$\Delta^2 = \frac{A^2 r_{av} R}{2C} = \frac{IqR}{2C}. \tag{10-44}$$

Next, let's consider shot noise voltage across a resonant circuit consisting of a capacitor C in parallel with an inductor L of effective resistance R. A current I consists of charge inputs $r_{av} = I/q$, where q is the discrete charge. Therefore,

$$A^2 r_{av} = Iq. \tag{10-45}$$

At $t = 0$ a charge q_o enters the resonator so $V_o = q_o/C$. You can translate the results from Chapter 4, Table 4-3, to find the voltage across the resonator,

$$V(t) = \frac{q_o}{C} \frac{e^{-Kt} \cos(\omega t - \theta)}{\cos(\theta)}, \tag{10-46}$$

where, as usual, $K = R/2L$, $\omega = (\omega_o^2 - K^2)^{1/2}$, $\omega_o^2 = 1/LC$, and $R/(2\omega L) = \arctan(\theta)$. The impulse response is then,

$$h(t) = \frac{V(t)}{q_o} = \frac{1}{C} \frac{e^{-Kt} \cos(\omega t - \theta)}{\cos(\theta)}. \tag{10-47}$$

Recall from Eq. (5-10),

$$\omega = \omega_o[1 - 1/(4Q^2)]^{1/2} \tag{10-48}$$

$$\sin(\theta) = \frac{1}{2Q}. \tag{10-49}$$

So, for high Q we have $\omega \approx \omega_o$ and $\theta \approx 0$. Therefore the impulse response becomes

$$h(t) = \frac{V(t)}{q_o} = \frac{1}{C}e^{-Kt}\cos(\omega_o t) \tag{10-50}$$

and

$$\int_0^\infty h(t)^2\,dt = \frac{1}{4KC^2} = \frac{L}{2RC^2}. \tag{10-51}$$

So,

$$\Delta^2 = \frac{IqL}{2RC^2}. \tag{10-52}$$

Shottky first derived this relation in 1922, and it was confirmed experimentally by Hull and Williams in 1925. Equation (10-52) gives another method of measuring the charge of an electron. The higher the Q, the better the results.

The National Institute of Standards and Technology is developing (1996) an atomic time and frequency standard somewhat analogous to our calculation of Campbell's theorem applied to a resonator. At NIST an atomic "fountain" shoots laser-cooled cesium atoms up into a microwave resonant cavity. Subsequently they fall back under the influence of gravity. It is anticipated that the low velocity of the falling atoms will reduce the uncertainty in frequency by a factor of five or ten compared with current standards. Does this remind you of an atomic reflex klystron?

10.10 Entropy

Entire books have been written about this subject, which started around 1850 with R. J. E. Clausius. It was later polished by Ludwig Boltzmann and others. Since its initial conception, entropy has baffled students as the Second Law of Thermodynamics. We are only going to touch on its probabilistic aspects as a measure of the disorder of a system and its relation to measurement and information.

There are many anecdotes associated with entropy. It is said that the First Law of Thermodynamics is "You can't win; the best you can do is break even," and the Second Law is "You can't even break even." It is also said that entropy is discussed by philosophers, understood by physicists, and used by engineers. When Claude Shannon (at Bell Laboratories, now Lucent Technologies) was developing information theory, he derived an expression like Eq. (10-53) and asked John von Neumann what he should call it. John von Neumann replied that Shannon should call it entropy, because nobody would know what he was talking about. So, they couldn't argue with him. Earlier, Boltzmann showed that the entropy function can be expressed as

$$H(P) = -C \sum_j P_j \ln(P_j) \tag{10-53}$$

where P_j is the probability that the system is in state j and C is a scale factor that depends on the system. For example, in a binary system C = 1/ln(2). Be careful about the notation in Eq. (10-53). This is standard notation, but don't confuse H with the Hamiltonian function or operator. In a physical system the units of entropy are joules per kelvin or eV/K. For a binary system $P_2 = 1 - P_1$ and Eq. (10-53) becomes

$$H(P) = -\frac{1}{\ln(2)} [P \ln(P) + (1 - P) \ln(1 - P)]. \tag{10-54}$$

Equation (10-54) is graphed in Fig. 10-9. A completely disorderly system is the most probable and is characterized by maximum entropy; oscillators run down. A resistor and a damped oscillator are entropy factories that transform orderly energy into disorderly (thermal) energy. Consider measurements of an oscillator. Today most measurements are done digitally so that information on the state and evolution of an oscillator can be represented by a sequence of zeros and ones. A random sequence would indicate that the oscillator was at thermodynamic equilibrium, executing random thermal motion, as in Section 10.5.

To make these ideas quantitative in a simple way, consider a two-state digital system or flipping a coin, heads or tails. If the two states are equally probable ($P_j = 0.5$) then the system entropy is maximum and is given by

$$H(p) = -\frac{1}{\ln(2)} [0.5 \ln(0.5) + 0.5 \ln(0.5)] = 1. \tag{10-55}$$

If the binary digital system is stuck on 0 or 1, or if the coin has two heads or two tails, then the entropy is zero, because ln(1) = 0. This leads to another fundamental idea:

- If there is no uncertainty, then the entropy is zero and we do not have to make a measurement, because we know what the result will be. No uncertainty, no information. No pain, no gain.

There is no need to flip a two-headed coin or to observe a stopped clock. Information is associated with uncertainty and entropy. Before we observe an oscillator, its displacement, momentum, frequency, phase, and energy are completely uncertain. The act of observation, coupling an observing system to an observable system, removes some uncertainty (but not all) and generates some information. When entropy is low, a small amount of information removes a relatively large uncertainty. When entropy is high, a larger amount of information is required, as in Fig. 10-9.

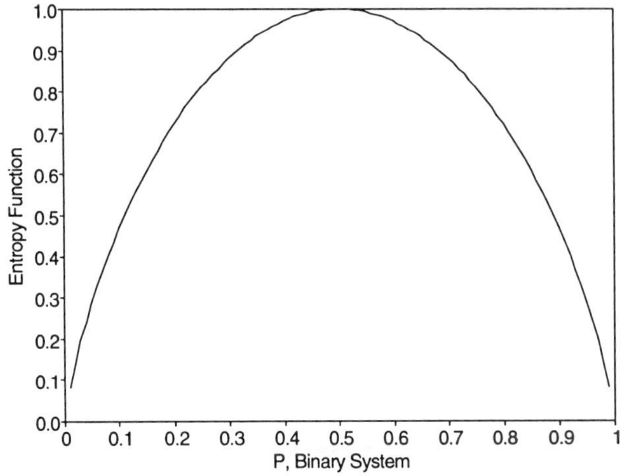

Figure 10-9. Entropy function for a binary system, eq. (10-54). At maximum entropy the probabilities of the two states are equal, $P_a = 1 - P_b = 0.5$.

But all of this is statistical, and almost everywhere we look we see systems with low entropy, far from thermodynamic equilibrium. Examples are abundant: biology, the sun in the morning, and the stars at night. Furthermore, events of low probability are possible and they are observable. For example, in a long sequence of plain English text the letter *e* occurs with the highest probability, about 0.13 (count the letters in this chapter). Nevertheless, in 1939 Ernest Vincent Wright published *Gadsby*, a 267-page novel written in plain English without using the letter *e*. The German poet Gottlob Burmann (1737–1805) wrote 130 poems, about 20,000 words, without using the letter *r*. In second harmonic generation using a laser, the probability of finding a photon with twice the frequency of the fundamental may be only 1 in 10^9, but if you have 10^{20} photons in a low-power laser, then the second harmonic will have about 10^{11} photons.

Entropy has been called The Arrow of Time, because it increases in the direction of increase of time, in the present epoch. There are some subtle questions that we will mention without further comment: Why do oscillators run down, why do living things age, why does entropy increase, and why is the sky dark at night (Olber's paradox)? The single answer seems to be that the universe is expanding; more states in Eq. (10-53) are available in the future than in the present or in the past. If the universe stops expanding and starts to collapse, then all of this would change.

10.11 Virial Theorem

Clausius is also responsible for this theorem in the classical case; the quantum derivation follows along the same lines. Let's define a function O as

$$O \equiv \sum_i \mathbf{p}_i \cdot \mathbf{r}_i , \tag{10-56}$$

and this has a time derivative given by

$$\frac{dO}{dt} = \sum_i (\mathbf{p}_i \cdot \dot{\mathbf{r}}_i + \dot{\mathbf{p}}_i \cdot \mathbf{r}_i). \tag{10-57}$$

Now we can argue that the time-average value of dO/dt can be made as small as we want (perhaps zero). It is easy to show that this is zero for a periodic system, but this approaches zero even for a non-periodic system for sufficiently long times, because O is a bounded function. As a result,

$$\left(\sum_i \mathbf{p}_i \cdot \dot{\mathbf{r}}_i\right)_{av} = -\left(\sum_i \dot{\mathbf{p}}_i \cdot \mathbf{r}_i\right)_{av} , \tag{10-58}$$

but from Chapter 8 the left-hand side is twice the total average kinetic energy $2K_{av}$ and the time rate of change of momentum on the right-hand side is a force,

$$2\left(\sum_i K_i\right)_{av} = -\left(\sum_i \mathbf{F}_i \cdot \mathbf{r}_i\right)_{av} \tag{10-59}$$

or

$$K_{av} = -\frac{1}{2}\left(\sum_i \mathbf{F}_i \cdot \mathbf{r}_i\right)_{av} . \tag{10-60}$$

Clausius named the right-hand side of Eq. (10-60) the *virial*. In words, the virial theorem says that the time-average kinetic energy of a system is equal to its virial. For a one-dimensional harmonic oscillator the virial can be written as

$$K_{av} = -\frac{1}{2}(-\kappa x \cdot x)_{av} = \frac{1}{2}(\kappa x^2)_{av} = U_{av}. , \tag{10-61}$$

which agrees with Eqs. (2-30) and (2-31), obtained directly from Newton's laws. The virial theorem applies to more than a harmonic oscillator; consider a potential energy function of the general form $U(x) = Cx^n$,where C is a constant. Then we have

$$K_{av} = -\frac{1}{2}(-nCx^{n-1} \cdot x)_{av} = \frac{n}{2}(Cx^n)_{av} = \frac{n}{2} U_{av} . \tag{10-62}$$

10.12 Liouville's Theorem

We are not going to derive this theorem, due to Joseph Liouville, but we will mention its implications for the statistical description of harmonic oscillators. In words, Liouville's theorem states that the density of points in phase space is constant in time. This implies that the points in phase space move like an incompressible fluid.

For harmonic oscillators the points in phase space move around ellipses. For an ensemble of identical oscillators with different initial conditions we can represent the motion in phase space as the motion of points associated with members of the ensemble. Furthermore, we can specify the statistical properties of the ensemble at time t by the density of points in phase space.

Liouville's theorem assures us that the density in the neighborhood of any point will be constant. This leads to the necessary and sufficient condition for statistical equilibrium. This is quite different from static equilibrium.

10.13 Information and Noise

Our discussion of entropy implies that, for digital data, a measurement boils down to deciding whether we have 0 or 1, and prediction involves a statement of whether the next digit will be be 0 or 1. Long-range prediction is more complicated, because it requires forecasting what a longer sequence of zeros and ones will be.

Total information (in bits) is defined as

$$\text{Information (bits)} = CB\Delta t\,, \tag{10-63}$$

where C is a constant depending on the system, B is the bandwidth in Hz, and Δt is the time interval during which the information was transmitted or received. Notice the presence of the bandwidth-time product in Eq. (10-63), reminiscent of the $\Delta\omega\Delta t$ uncertainty relation. For the same amount of information, you can use a small bandwidth and a long time or a large bandwidth and a short time. Radiotelegraph operators knew this from experience. In the early era of space exploration, bandwidths were so small that data had to be recorded aboard spacecraft and transmitted back to Earth stations over a long period of time. Televised images of astronauts in the first moon landing appeared jerky because sufficient information could not be transmitted in real time to achieve continuous movement.

An equivalent definition of information for a single event is

$$I(x_i) = \log_b[P(x_i)]^{-1}, \tag{10-64}$$

where the information is associated with the known occurrence of event x_i and $P(x_i)$ is the probability of the event; the logarithm base is 2 for binary digits or 10 for hartleys (for R. V. L. Hartley of Hartley oscillator fame). Entropy, Eq. (10-53), associated with information, is measured in bits per symbol for a binary system.

Shannon showed that the maximum information rate is

$$R\,(\text{bits/s}) = CB\log_b(1 + S/N), \tag{10-65}$$

where R is called the channel capacity, C is a constant depending on the system, B is the bandwidth in Hz, the base of the logarithm depends on the type of coding, and S/N is the signal-to-noise power ratio. For $b = 2$ the information rate is in bits/s as shown. For $S/N \ll 1$ the channel capacity goes to zero because $\log_b(1) = 0$.

Equation (10-65) shows that, for low S/N, you can increase the information rate most efficiently by increasing the S/N, which usually means increasing S. For high S/N you can increase the information rate most efficiently by increasing the bandwidth. It must be emphasized that Eq. (10-65) is the best you can hope for. It is an idealized upper limit that has never been achieved.

In the movie *Wall Street*, Gordon Gecko (played by Michael Douglas) said "Information is the most valuable commodity I know of."

10.14 Stochastic Resonance

This is a subject of current research, both theoretical and experimental. Stochastic resonance is not completely understood, but this has not prevented applications in extracting signals from noise and image enhancement. This type of resonance involves *adding random noise* to improve signal detection, which is certainly contrary to intuition, considering the effort that people expend to reduce noise. Until recently, commonly accepted wisdom was to increase the signal-to-noise ratio (SNR or S/N) in order to increase accuracy and precision. At present, stochastic resonance is only concerned with signal detection, the existence of an unknown signal, not with the rate at which information can be transferred, as demonstrated in Eq. (10-65).

Stochastic resonance appears to depend on three conditions:

1. A weak, coherent signal in a system with nonlinearity.
2. Random background noise.
3. The existence of a threshold, barrier, or decision level.

Experiments indicate that there is an *optimum* SNR for detection in observations governed by these three conditions. If the SNR is too low or too high, then detection is less probable. Quantum tunneling experiments with superconducting quantum interference devices (SQUIDs) appear to confirm existence of stochastic resonance effects on the atomic scale.

Biological systems also appear to make use of stochastic resonances. It has been known for many years that digitized images appear to be enhanced when some noise is added, even though this adds no information and actually decreases the SNR. Recent experiments with human subjects indicate that subthreshold images (images that are not detected) can be detected by adding optimum noise.

10.15 Phase Space Correlation and Covariance

The correlation *coefficient* discussed here should be distinguished from the correlation *function* discussed in Chapter 7. The correlation coefficent and the covariance are available in *Excel* as CORREL and COVAR. The correlation coefficient is a measure of a possible causal relationship between two data arrays, and covariance is the average of the products of deviations from the means of data point pairs.

The syntax for CORREL is CORREL(*array 1, array 2*), where the arrays have the same number of data points. If the arrays have different numbers of data points, the #N/A error value shows. If either array is empty, or if the standard deviation is zero, the #DIV/0! error value shows.

The syntax for COVAR is COVAR(*array 1, array 2*), where the arrays have the same number of data points. If the arrays have different numbers of data points, the #N/A error value shows. If either array is empty, the #DIV/0! error value shows. COVAR does not involve the standard deviations. Covariance is defined as

$$Cov(X,Y) = \frac{1}{n}\sum_{k=1}^{n}(x_k - m_x)(y_k - m_y), \tag{10-66}$$

where m_x and m_y are the mean values of the arrays. The correlation coefficient $\rho_{x,y}$ is defined as

$$\rho_{x,y} = \frac{Cov(X,Y)}{\sigma_x\,\sigma_y}, \tag{10-67}$$

where σ_x and σ_y are the standard deviations of the arrays.

Consider a horse race. Suppose all of the horses are lined up at the starting gate ($x = 0, t = 0$) and they start with different momenta. At the start there is no correlation between displacement and momentum, because all horses have the same displacement, regardless of momentum. As time goes by, the fundamental things apply and horses with larger momenta will acquire larger displacements. Horses with smaller momenta will acquire smaller displacements. In this horse race, displacement and momentum become increasingly correlated as time goes by. This is somewhat like the expanding universe and the cosmic red shift; the faster stars and galaxies (the ones with the greater Doppler shift) are more distant. This argument can be extended to the spreading of a free, localized quantum wavepacket.

An oscillator behaves differently. The periodicity of an oscillator produces periodic correlation and anti-correlation of displacement and momentum. You can use the worksheet PHASECOR to experiment with correlation in an oscillator. The Home screen of this worksheet is shown in Fig. 10-10, and a typical set of results is shown in Fig. 10-11. Observe how the phase relation between displacement and momentum results in a square wave for the correlation coefficient.

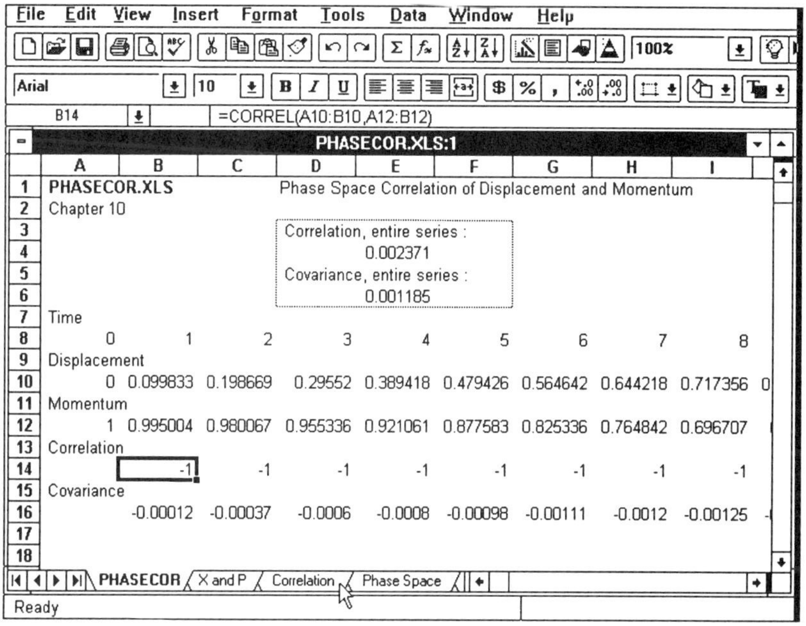

Figure 10-10. Home screen of PHASECOR, classical phase-space correlation of an undamped, free harmonic oscillator. Note the syntax of CORREL.

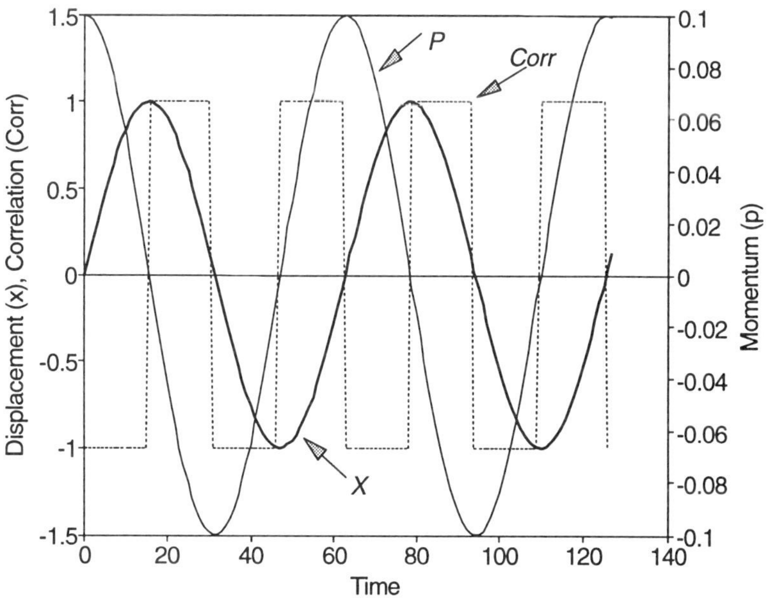

Figure 10-11. Displacement, momentum, and correlation coefficient between x and p, as a function of time. Correlation flips between +1 and –1.

10.16 Predictability

Consider one form of the solution to Eq. (1-4), the undamped oscillator,

$$x(t) = A\cos(2\pi f_o t + \theta). \tag{10-68}$$

Superficially this appears to be completely deterministic, completely predictable, but this is not true. Forget about phase space and thermal motion for a while. What other uncertainties underlie $x(t)$? Some of the uncertainties involve our ability to measure fundamental units of mass, space, time, and electric charge.

At the end of the twentieth century one atomic mass unit was defined as $1.6605402 \times 10^{-27}$ kg. There is an uncertainty of $\pm$ 10 in the last two decimal places. This definition is based on carbon-12, which, by international agreement, has been assigned a mass of 12 atomic mass units.

During the same epoch the length standard of 1 m was defined as the length of the path traveled by light in vacuum during the time interval of 1/299,792,458 of a second. (The reason for this is that the speed of light in vacuum is the standard by which many other things are measured, and it was defined as exactly 299,792,458 m/s.)

The time standard at the end of the twentieth century was defined as one second, which is 9,192,631,770 vibrations of microwave radiation emitted by an atom of cesium-133. (This radiation is in the X-band, about 3 cm wavelength, and is in a part of the electromagnetic spectrum commonly used by traffic-control radar and weather radar.)

(a) How accurately can we measure the amplitude A? The dimension of A is length, with a unit of meters, so our knowledge of $x(t)$ depends on how accurately we can measure length, or an analogous amplitude such as voltage.

(b) How accurately do we know the cosine or sine function? These functions can be expressed as

$$\cos(\theta) = \frac{(e^{j\theta} + e^{-j\theta})}{2} \quad \text{and} \quad \sin(\theta) = \frac{(e^{j\theta} - e^{-j\theta})}{2j}, \tag{10-69}$$

so our knowldege of these basic functions depends on how well we know the transcendental number e. An alternative expression for the cosine is the series,

$$\cos(\theta) = 1 - \frac{\theta^2}{2!} + \frac{\theta^4}{4!} - \frac{\theta^6}{6!} + \ldots, \tag{10-70}$$

but this is related to the transcendental number e through the series,

$$e^{\theta} = 1 + \theta + \frac{\theta^2}{2!} + \frac{\theta^3}{3!} + \frac{\theta^4}{4!} + \ldots. \tag{10-71}$$

Euler and de Moirve first discovered Eq. (10-71) by adding and subtracting the series for the sine and cosine. The number e can be computed to as many decimal places as wc choose, but we can never know it exactly because e in decimal form is an infinite, never-repeating sequence. Consequently, the accuracy of the sine and cosine functions is limited by our knowledge of e.

(c) We know the integer 2 exactly, because it can be obtained by counting. This is the thing we know best about the oscillator.

(d) The transcendental number π offers the same difficulties as the transcendental number e. In August, 1995, Yasumasa Kanada and his colleagues (University of Tokyo) computed π to 3.22×10^9 digits. Their work required about 36 hours on a supercomputer. Before that, π was known to 2.26×10^9digits, a record established in 1991 by David and Gregory Chudnovsky (Columbia University), but this is better than we know e. Of course, π and e are related by "the most remarkable formula in mathematics,"

$$e^{j\pi} + 1 = 0.$$

Max Born recognized this problem. In his Nobel Prize acceptance speech (1954) he said, "Is there any sense — and I mean any physical sense, not metaphysical sense — in which one can speak of absolute data? Is one justified in saying that the coordinate $x = \pi$ cm where $\pi = 3.1415\ldots$ is the familiar transcendental number that determines the ratio of the circumference of a circle to its diameter? As a mathematical tool the concept of a real number represented by a nonterminating decimal fraction is exceptionally important and fruitful. As the measure of a physical quantity it is nonsense."

(e) The product $f_0 t$ cannot be known as accurately as we may choose. This is a fundamental limitation in both classical and quantum domains. Recall the classical uncertainty relation from Chapter 6, $\Delta f\, \Delta t \geq 1/4\pi$, which is the *best* you can do if measurements in frequency and time are optimum (Gaussian-distributed). Most of the time you can't do this well. Even if you know π extremely well, there are still major uncertainties in f and t when "simultaneous" measurements are made. The frequency of the oscillator is related to mass or its analog. Since we can never know the mass exactly, the natural frequency is uncertain even without the frequency-time uncertainty relation.

(f) $\Delta p_x\, \Delta x \geq h/4\pi$ and $\Delta E\, \Delta t \geq h/4\pi$. These quantum uncertainties, introduced briefly in Section 6.13, will appear again in Chapter 12.

(g) The frequency f_0 depends on how accurately we can count in a time interval, but there is some uncertainty in knowing the time interval. The time interval Δt cannot be measured or known with infinite precision and/or accuracy. Furthermore, we cannot determine $t = 0$ exactly, so we do not even know what the "initial conditions" are, exactly.

(h) The phase angle θ requires measurement of displacement (or voltage) and time. For example, θ is related to x_0/A at $t = 0$, so we are limited by uncertainties in x_0, A, and t.

(i) Special relativity tells us that, under certain conditions, two observers cannot agree on measurements of simultaneity, length, time, and frequency, even if both observers are in inertial systems. This is a fundamental property of the Lorentz transform. In addition, there is the inevitable time delay due to the finite speed of light. The *twin paradox* tells us that two identical oscillators, initially synchronized at one point in spacetime, will be unsynchronized if one oscillator is moved and then returned to its initial position. The oscillator that moves will be retarded (younger) in time. We will not pursue this further, but synchronization of clocks over long distances can be a problem in space telecommunications and navigation. General relativity tells us that clocks slow down in strong gravitational fields. The Global Positioning System has corrections that take general relativity into account. This is probably the only practical application of general relativity at present.

(j) Nevertheless, the situation is more serious than the lack of exact knowledge of numerical values. Apparently Newton recognized that there was a fundamental chaotic quality beneath the seemingly deterministic surface of classical mechanics. Although the two-body central force problem can be solved exactly (in the sense that the solution is mathematically unique, not that the parameters and initial conditions are known exactly), three-or-more-body problems do not have unique solutions.

Ironically, this unpredictabilty led Wolfgang Pauli to predict the existence of the neutrino in beta decay, to account for the spread in energy of the decay products. The energy of the electron would have been well defined for a two-body decay. Pauli could have received the Nobel Prize for the neutrino, but he received this award in 1945 for discovering the exclusion principle of quantum mechanics. (The neutrino in beta decay is now called the anti-neutrino, for reasons that do not concern us at this point.) In 1988 Leon M. Lederman, Melvin Schwartz, and Jack Steinberger shared the Nobel Prize for their experimental discovery of the muon neutrino. In 1996 Lederman was awarded Honorary Membership in the Institute of Electrical and Electronics Engineers.

I don't want to spoil your day, but the situation is even more serious. We cannot predict from orbital measurements and calculations when a Texas-size meteor will strike planet Earth. It is now known that our own solar system is fundamentally chaotic [G. J. Sussman and J. Wisdom, *Science* **241**, 433 (1992); *Science* **257**, 56 (1992)]. The stability of the solar system is ephemeral if you take the long view. Since the days of Poincaré and Lyupanov, the subject of deterministic chaos has been a popular pastime for mathematicians, engineers, physicists, chemists, and economists.

(k) When we make electronic oscillators we are confronted with four additional types of randomness, thermal noise, shot noise, $1/f$ noise, and blackbody radiation (including cosmic noise). These are fundamental types of noise, but in addition there are induced electromotive forces due to changing magnetic fields (Earth's field and the solar wind) and non-random electromotive forces due to human activity (power grid and electromagnetic radiation used in communication).

With all of these uncertainties, it's a wonder that we can make a clock tell time at all. Oscillators are only "predictable" for short times, if you take the long view. In 1996 the best atomic clock on this planet was the NIST-7 at the National Institute of Standards and Technology (NIST). This oscillator operates with an uncertainty of less than one part in 10^{14}, equivalent to gaining or losing less than a second in over 3×10^6 years. It is anticipated that the uncertainty can be improved by a factor of 10^4. NIST has increased the accuracy of oscillators by a factor of 10^7 since the first NH_3 oscillator was placed in operation in 1948.

Because atoms perturb each other, the fewer the number of atoms in an atomic clock, the better the accuracy. At the Boulder, Colorado, laboratory of NIST a group has developed a linear ion trap to hold a few Hg atoms stripped down to one electron. The ions, stored in an electromagnetic trap, are irradiated with a laser tuned so that energy is extracted from the ions. In other words, the laser acts as a refrigerator, reducing the ion temperature close to absolute zero. In this way, uncertainty about the resonant frequency is reduced and, as a side effect, the observation time is increased. It is expected that the uncertainty of this system will be a few parts in 10^{18}.

Is all of this equivalent to asking how many angels can dance on the head of a pin? Not at all. Accuracy at these levels is required by users of secure communications systems, tests of fundamental physics, and deep-space navigation. Somewhat less accuracy is required for Earth-based electronic navigation systems.

10.17 Time Evolution and Predictions

Although they often differ quantitatively and qualitatively in their predictions, the logical structure of classical statistical mechanics and quantum mechanics are somewhat similar in terms of their three-stage prediction processes. This was pointed out by Willis E. Lamb, Jr. (Nobel Prize 1955, for measurement of the hydrogen spectrum fine structure). Table 10-2 shows a comparison of the two paths.

If the quantum path is not clear at this point, it may make more sense after Chapter 12. In the case of a harmonic oscillator, there are no interference effects in the classical case and probabilities simply add. For a quantum oscillator, the probability repeats itself with the period of the oscillator, instead of continuing to spread like a free particle. The lack of spreading in the classical case can be attributed to the fact that every oscillator in the ensemble has the same period, with the result that each associated point in phase space returns to its initial position at the same time.

Table 10-2. Classical Statistical Mechanics and Quantum Prediction Processes

Classical	*Quantum*
A. *Preparation.* Based on the history of the system, its representative ensemble at $t = 0$ is determined.	A. *Preparation.* Based on the history of the system, its state vector $\psi(0)$ at $t = 0$ is determined.
B. *Time development.* Based on the representative ensemble at $t = 0$, the representative ensemble at any later time is uniquely determined by Newton's laws.	B. *Time development.* Based on $\psi(0)$ at $t = 0$, the state vector at any later time is uniquely determined by the time-dependent Schrödinger equation.
C. *Probabilistic predictions.* The classical statistical predictions of measurements at time t can be obtained from the representative ensemble.	C. *Probabilistic predictions.* The quantum prediction of the result of measuring an observable at time t is obtained by expanding $\psi(t)$ in the eigenvectors of the operator corresponding to that observable.

10.18 Expectation Values: *Caveat Emptor*

Paul Ehrenfest's theorem relates classical and quantum time evolution of observable quantities, in terms of the time evolution of expectation values of the observables. "Expectation value" is the name often used for the mean value in statistics and quantum theory. For example, the center of mass (Section 10.1) is the expectation value of the mass distribution.

Don't expect to measure the expectation value in a single trial. This apparent contradiction turns out to have a simple meaning if you consider a mass distribution. There may be no mass at the center of mass. If an artillery shell explodes in flight, the center of mass will hit the target, even though no parts of the shell hit the target. If planet Earth explodes, its center of mass will continue serenely in its orbit around the sun, because internal forces cannot affect the motion of the center of mass. The *most probable value* is the result you are likely to get in a single measurement; this is associated with the maximum value of the distribution function.

This relates directly to some recent experiments in quantum tunneling that appeared to measure particles traveling faster than light. Apparently the distribution function became skewed in such a way that the most probable value indicated speeds greater than light. For a symmetric distribution function, the expectation value and the most probable value are the same, but if the distribution becomes skewed then the peak (most probable) could arrive before the center of mass (first moment). This is under current investigation; no one really knows what happens during tunneling.

10.19 Expert Opinions on Predictability and Causality

Even matters of chance seem most marvellous if there is an appearance of design as it were in them; as for instance the statue of Mitys at Argos killed the author of Mitys' death by falling down on him when a looker-on at a public spectacle; for incidents like that we think to be not without a meaning.
There is a great difference between a thing happening propter hoc *and* post hoc.

Aristotle
in *Nichomachean Ethics*, Book Two, translated by W. D. Ross

Bill Mead died in 1941 while filming the movie They Died With Their Boots On. *Riding beside Erroll Flynn, Mead's horse stumbled. Mead threw his sword forward to avoid falling on it. The sword stuck in the ground, hilt down. Mead was impaled and died with his boots on.*

Only fools, charlatans, and liars predict earthquakes.

Charles Richter
seismologist, nudist, and inventor of the Richter Scale

. . . while the past of a time series is accessible for examination, its future is not.

Norbert Wiener
in *Extrapolation, Interpolation, and Smoothing of Time Series*
John Wiley & Sons, New York, 1949

There are two kinds of economists, those who do not know the future and those who do not know that they do not know the future.

John Kenneth Galbraith
economist and former U.S. ambassador to India

The division between what is random in Nature and what is purposeful is extremely difficult to determine.

Tom Robbins
in *Jitterbug Perfume*, Bantam Books, New York, 1984

If you're going to predict, predict often.

Robert Dressler
physicist and financial wizard

Why do you want to predict? You would think that there is a predictive instinct which must be satisfied. . . . Many people don't fully and deeply realize that there is something missing. They are so used to doing statistical calculations, and saying that only statistics matters, that they do not notice that there is an actual, individual fact which is not accounted for.

David J. Bohm
in *A Question of Physics: Conversations in Physics and Biology*
Routledge and Kegan Paul, London, 1979

What's Next?

The (apparently) intrinsically statistical quantum domain is next, but our thought experiments with the idealized, noiseless, harmonic oscillator and the classical bunny rabbits have shown us that probability distributions arise without invoking quantum conditions. Using a *single* oscillator, completely deterministic, noiseless, and governed by Newton's laws, we found that there was a probability distribution that described measurements made on the displacement, and an uncertainty relation connecting momentum and displacement. An *ensemble* of classical, discrete, indistinguishable rabbits, hopping randomly but governed by Newton's laws, has similar probabilities associated with measurements on the ensemble. If we add thermal randomness (Brownian motion) to a single oscillator, or measure the properties of an ensemble of such oscillators, prepared identically, we would expect our measurements on the single oscillator and the ensemble to be probabilistic and there would be *minimum uncertainties* associated with the displacements and momenta.

There are seeds of uncertainty and probability contained in the seemingly innocent and simple expressions $E = hf$ and $p = h/\lambda$. In addition, there is a connection between special types of distribution functions and the spins of fundamental particles. We will see that the results shown in Figs. 10-4 and 10-5 approximate the average value of the quantum oscillator probability density, for large quantum numbers. This is another way that classical theory appears as an approximate limit to quantum theory. For small quantum numbers the classical probability density does not agree with the quantum probability density and experiments.

Blaise Pascal (1623–1662) noted that, although random events cannot be predicted, statistical patterns emerge. (This is the secret of the insurance business and quantum mechanics.) Most of what we have done so far can be regarded as a very good approximation under some conditions, but completely incorrect under other conditions. Harmonic oscillators played a central role in the discovery processes that we will discuss in Chapter 11.

Chapter 11

Quantum Perspective

I look upon quantum mechanics with admiration and suspicion.
Albert Einstein

The successes of James Clerk Maxwell's classical electromagnetic theory and the experimental verification by Heinrich Rudolph Hertz (formerly a student of Hermann von Helmholtz at the University of Berlin) were high points of the age of classical physics. It is ironic that the first measurements of electromagnetic oscillations and waves were also the beginning of the end of classical physics. During these experiments, Hertz recorded the first observation of the photoelectric effect, a pure quantum phenomenon. Measurements of the photoelectric effect were inconsistent with Maxwell's equations.

11.1 Classical Crisis

In addition to the photoelectric effect there were several other nagging difficulties.

- It was known that Maxwell's equations, which worked so well, were fundamentally inconsistent with Newton's laws, which worked so well.
- Spectra of light emitted from gases in the laboratory and observed in light from distant stars appeared to be *discrete* spectra with definite frequencies. From what was known about Fourier series and the Fourier transform, this indicated that there might be *periodic motion* deep within matter, perhaps some kind of harmonic oscillator.
- Pieter Zeeman's measurements of spectral line splitting in magnetic fields and Hendrik Lorentz's calculations suggested, quite remarkably, that there might be electric charges in periodic motion inside matter. Zeeman and Lorentz were awarded Nobel Prizes in 1902.
- Experiments with blackbody radiation showed continuous spectra with unexpected results at high frequencies (short wavelengths).

Blackbody Radiation

About the time that Hertz discovered electromagnetic waves, Josef Stefan was studying the radiation from ovens. A small hole in a closed oven is a good approximation to a blackbody, an object that is a perfect absorber and therefore a perfect emitter. Stefan found experimentally that the power density (W/m^2) emitted by the surface of a blackbody is given by

$$R = \sigma T^4 \tag{11-1}$$

where $\sigma = 5.6703 \times 10^{-8}$ W·m^{-2}·K^{-4} is an empirical constant. Ludwig Boltzmann derived this equation from theoretical considerations, and it became known as the Stefan-Boltzmann law. Typical results are shown in Fig. 11-1, from 1K to 10^8 K. This T^4 is a remarkably strong dependence on temperature. Doubling the absolute temperature increases the power per square meter by a factor of 16. This law says that the power radiated per unit area depends only on the absolute temperature and has nothing to do with the composition of the body. Any material radiating as a blackbody behaves like this; there is something universal here. Nothing about a harmonic oscillator depends on its composition. Could there be a connection?

It is a common observation that as things become warm they begin to glow, first in the far red. As their temperature rises the color becomes red, orange, then yellow, and very hot things appear to move toward the blue. However, the Stefan-Boltzmann law says nothing about the spectral distribution of the radiation.

About that time Wilhelm Wien was performing experiments on the spectral distribution of energy from blackbodies. He found something very unexpected, that the spectral distribution has a peak, and the peak changes with temperature.

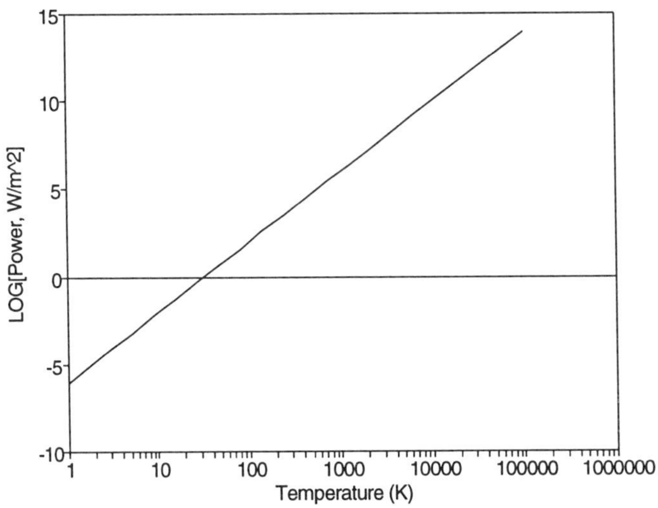

Figure 11-1. Stefan-Boltzmann law. Radiated power increases as T^4. The sun radiates approximately as a blackbody at 6000 K.

As the temperature increases the peak moves toward shorter wavelengths and the wavelength at which this maximum occurs is a function *only* of the absolute temperature,

$$\lambda_{max} = \frac{2.898 \times 10^{-3}\ \text{m·K}}{T}. \tag{11-2}$$

The numerical constant in Eq. (11-2) was determined experimentally. This appeared to be very strange and completely inconsistent with classical electromagnetic theory, which predicted *no maximum*, just increasing, power per unit area as the wavelength decreased. The result of Eq. (11-2) is shown in Fig. 11-2.

Despite remarkable successes it was known that there were some problems with classical electromagnetic theory and this was one of them. The continuously increasing power as the wavelength decreased was known as the *Rayleigh-Jeans law* (Lord Rayleigh and Sir James Jeans); it was also known as the *ultraviolet catastrophe*, because the total power was predicted to become infinite as the wavelength went to zero. It also turned out to be catastrophic for classical theory.

As we saw in Chapter 10, the average energy of a classical one-dimensional harmonic oscillator is kT. If $u(\lambda)\, d\lambda$ is the fraction of the energy per unit volume in the cavity, in the wavelength interval $d\lambda$, then the Rayleigh-Jeans law predicts that,

$$u(\lambda,T)\, d\lambda = n(\lambda)kT\, d\lambda = \frac{8\pi}{\lambda^4} kT\, d\lambda, \tag{11-3}$$

where $n(\lambda) = 8\pi\lambda^{-4}$ is the number of modes of oscillation per unit volume.

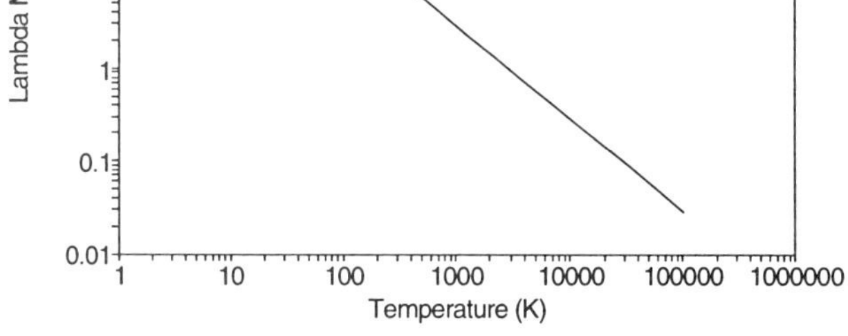

Figure 11-2. Wien displacement law. This experimental relationship shows the wavelength at which maximum blackbody radiation occurs as temperature increases. Wien received the Nobel Prize in 1911 for this work.

Equation (11-3) was found to agree well with experiments at low frequencies (like radio waves), but failed completely at high frequencies (like infrared). There is no prediction of a maximum; it blows up if you integrate it over all wavelengths. This leads to the silly result that a single candle would radiate an infinite amount of energy. Clearly such a paradox is an indication of fundamental error. Experimental results are shown in Figs. 11-3 and 11-4, together with the classical Rayleigh-Jeans prediction.

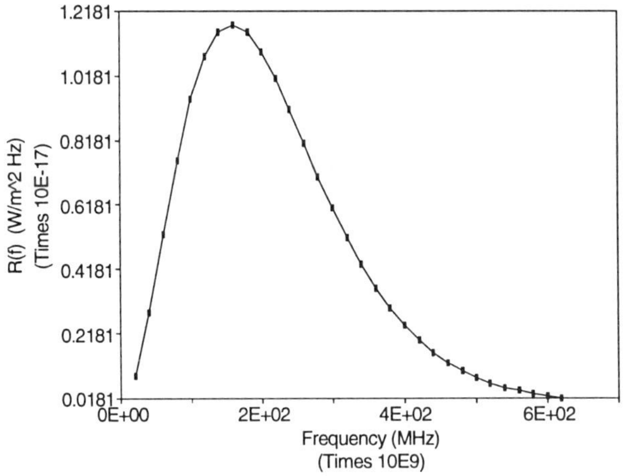

Figure 11-3. Cosmic background radiation which appears to be left over from the Big Bang is consistent with a blackbody at about 2.7 K. The COBE (Cosmic Orbiting Background Explorer) satellite recorded data similar to this Figure.

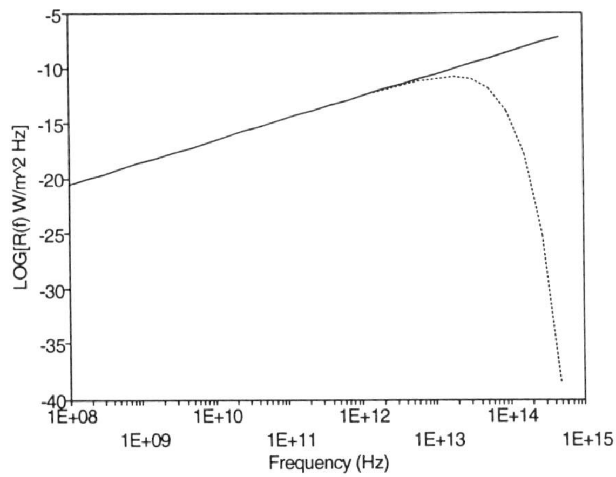

Figure 11-4. Rayleigh-Jeans prediction *(solid line)* and experimental results *(dotted line)*. Agreement is excellent in the radio, microwave, and far infrared at 300 K but there is catastrophic failure at higher frequencies.

11.2 A Collection of Oscillators

Disturbed by this gross discrepancy between theory and experiment, Max Planck (University of Berlin) decided reluctantly that the classical energy distribution function had to be modified. All he had to work with were good experimental results; there was no theory to guide him. Adiabatic invariance was unknown. He later described his efforts as the most intense seven days of work in his life.

We will sketch the path. To calculate the average energy for a group of simple harmonic oscillators, start with the Maxwell-Boltzmann distribution for the probability of finding an oscillator in a certain energy state,

$$P(E)dE = \frac{1}{kT} e^{-E/kT} dE, \tag{11-4}$$

where the energy E is considered to be a continuous variable, and dE is the energy range in which we are interested. Notice how kT is analogous to a "time constant" for the distribution function in the sense that $P(E)dE$ falls to $1/e$ of its maximum value when $E = kT$. The normalization constant is also kT.

Then, as usual with distribution functions, we calculate the first moment of the classical distribution (the average energy) as

$$E_{av} = \int_0^\infty E\, P(E)\, dE = \int_0^\infty E \frac{1}{kT} e^{-E/kT}\, dE = kT. \tag{11-5}$$

This is what we expected for classical one-dimensional oscillators at thermodynamic equilibrium. This classical distribution and average energy are shown in Fig. 11-5. To emphasize the exponential character of the energy distribution, Fig. 11-6 shows the data in Fig. 11-5 on a semi-log plot.

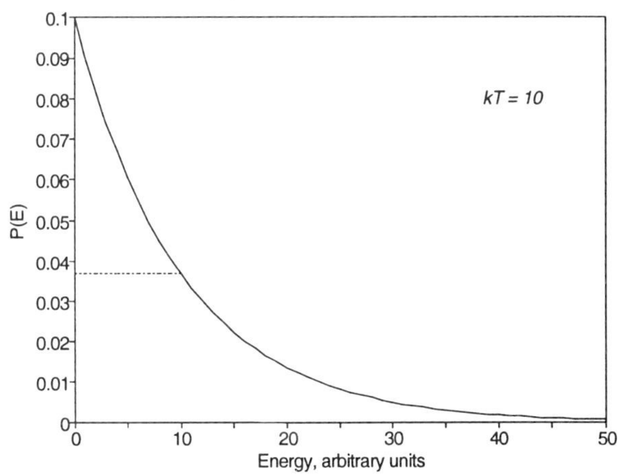

Figure 11-5. The classical continuous energy distribution allows any value of energy. The average energy kT is where $P(E)$ falls to $1/e$ *(dotted line).*

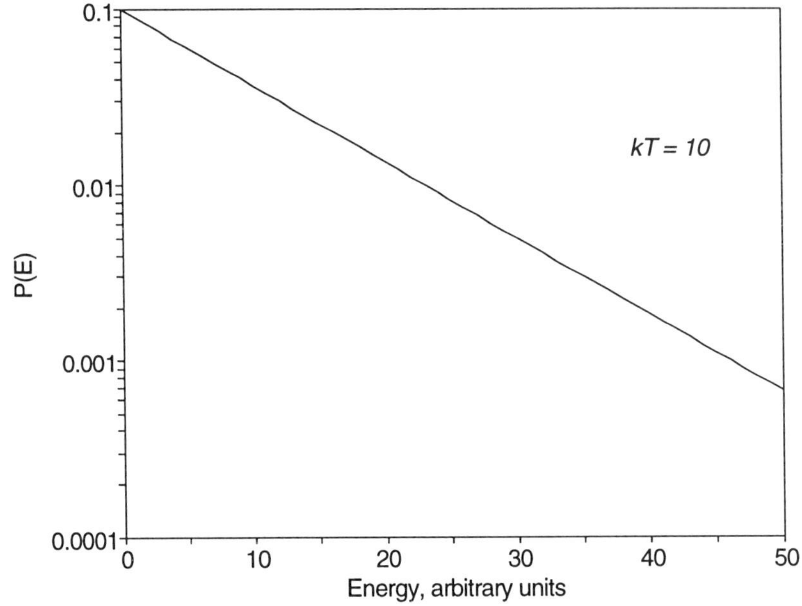

Figure 11-6. The classical, continuous energy distribution function is a pure exponential and all energy states are allowed.

Planck Distribution

Planck discovered that he could reproduce the empirical distribution function if he made two assumptions (with no justification, except that the end justifies the means). The first assumption was that the energy was a discrete variable, not a continuous variable. The second assumption was that the energy difference between adjacent states was a constant, $\Delta E = hf$, where f is the frequency in Hz and h is a constant to be determined by experiment. A discrete (quantized) energy can be described as

$$E_n = nhf, \quad n = 0, 1, 2, 3, \ldots . \tag{11-6}$$

At the time Planck had no reason to exclude $n = 0$. We will return later to the important question of zero-point energy. The discrete energy distribution becomes

$$P(E_n) = C \exp(-E_n/kT) = C \exp(-nhf/kT) \ . \tag{11-7}$$

Now the normalization constant C is determined by a sum, not an integral,

$$\sum_{n=0}^{\infty} P(E_n) = C \sum_{n=0}^{\infty} \exp\,[-nhf/kT] = 1. \tag{11-8}$$

The discrete energy distribution function is shown in Fig. 11-7.

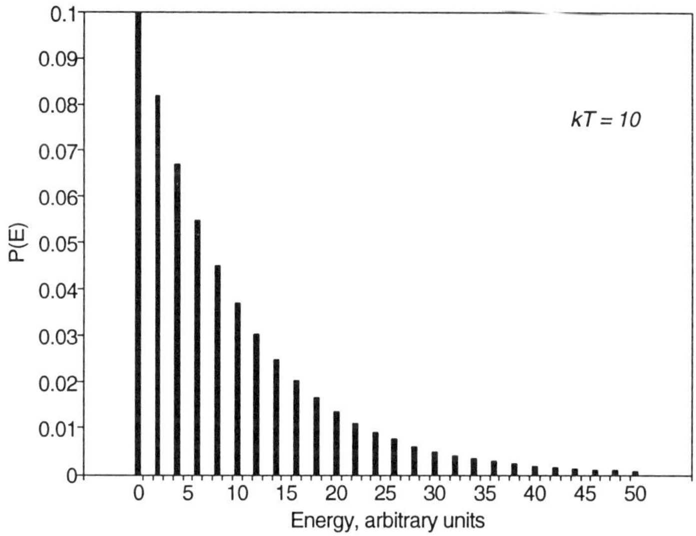

Figure 11-7. A discrete energy distribution function. Only certain energy states are allowed. For harmonic oscillators they are evenly spaced.

We need to find C based on this discrete distribution using the sum instead of the integral,

$$C = \frac{1}{\sum_{n=0}^{\infty} \exp\,[-nhf/kT]} \tag{11-9}$$

The evaluation of this sum is a bit tedious, but the result is well known. We also need to find the average value of the energy E_{av} based on the discrete distribution, using the sum instead of the integral,

$$E_{av} = \sum_{n=0}^{\infty} E_n P_n(E_n) = C \sum_{n=0}^{\infty} E_n \exp\,[-nhf/kT]\,. \tag{11-10}$$

After some more algebra the result is

$$E_{av} = \frac{hf}{\exp\,[hf/kT] - 1}\,. \tag{11-11}$$

In terms of wavelength $\lambda = c/f$,

$$E_{av} = \frac{hc/\lambda}{\exp\,[hc/\lambda kT] - 1}\,. \tag{11-12}$$

Finally, to get $u(\lambda)$ or $u(f)$ we need to know the number of modes of oscillation as a function of λ or f. This is given by

$$n(\lambda) = 8\pi\lambda^{-4}. \tag{11-13}$$

So,

$$u(\lambda)d\lambda = n(\lambda)E_{av}d\lambda = \frac{(8\pi\lambda^{-4})(hc/\lambda)}{\exp\,[hc/\lambda kT] - 1}\,d\lambda = \frac{8\pi hc\lambda^{-5}}{\exp\,[hc/\lambda kT] - 1}\,d\lambda. \tag{11-14}$$

Since $f = c/\lambda$, we have $df = -\,c/\lambda^2 d\lambda$ and $df/d\lambda = -\,c/\lambda^2$. This provides the conversion $u(\lambda) = u(f)\;c/\lambda^2$. In terms of frequency,

$$u(f)df = n(f)E_{av}df = \frac{8\pi hf^3c^3}{\exp\,[hf/kT] - 1}\,df. \tag{11-15}$$

The remarkable part is, ***this agrees with experiment!*** Planck's amazing assumption that energy is discrete instead of continuous appeared to have solved the puzzle of the ultraviolet catastrophe, and the result of this assumption contains the Stefan-Boltzmann law and the Wien displacement law as well. To see the Stefan-Boltzmann law, just integrate $u(\lambda)$ over all wavelengths. To see the Wien displacement law, just differentiate $u(\lambda)$ with respect to λ and set the derivative equal to zero to find λ_{max} . Even the strange empirical constants come out of this! It's almost enough to make you start to believe that quantized energy has some validity.

Planck's distribution agrees with classical theory in the limit of high temperature or low frequency (in general, $hf/kT \ll 1$). Look back at Fig. 11-4 and you can clearly see the region in which the quantum prediction agrees with the classical prediction, and where the classical prediction fails dramatically. To see this analytically, expand the exponential in the denominator of the quantum distribution function,

$$u(f)df = \frac{8\pi hc^{-3}f^3}{\exp\,[hf/kT] - 1}\,df \rightarrow \frac{8\pi hc^{-3}f^3}{1 + [hf/kT] + \ldots - 1}\,df \tag{11-16}$$

$$u(f)df \rightarrow \frac{8\pi hc^{-3}f^3}{hf/kT}\,df = 8\pi c^{-3}f^2kT\,df = (8\pi\lambda^{-4})kT = n(\lambda)kT, \tag{11-17}$$

and we're back to our old familiar kT and Rayleigh-Jeans. What a trick!

Planck was not convinced that the quantization of energy was much more than a mathematical trick which happened to describe blackbody radiation, but he received a Nobel prize for this in 1918. Einstein saw that it was more than a trick, and successfully applied it to such diverse phenomena as the photoelectric effect (for which he received the Nobel prize in 1921) and specific heats of solids. Nevertheless, Einstein always had reservations about the completeness and interpretation of the quantum description of Nature.

So, experiment indicates that harmonic oscillators are quantized, and the classical domain is a good approximation when $hf/kT \ll 1$. In view of the experimental evidence, how could they *not* be quantized? We have been calculating the energy density in the cavity, per unit wavelength or frequency interval. We usually measure the output power (W/m^2 per unit wavelength or frequency interval). This measured quantity is $c/4$ times the energy density $u(f)$ or $u(\lambda)$.

The Planck and Rayleigh-Jeans distributions are often plotted in terms of the *spectral radiant excitance*. This is the currently accepted name for this radiometric quantity, and the units are $W/(m^2Hz)$, which is power per unit area, per unit frequency interval. To find the total power per unit area in an interval requires integration of the distribution over that interval. Beware if you pick up older books; you will find other strange names for radiometric and photometric quantities, because the terminology seems to change every ten years or so.

11.3 PLANCK, Worksheet for Blackbody Radiation

The worksheet associated with this section is named PLANCK. After you start your spreadsheet, open this file and you should see the Home screen shown below. If you do not see this press the Home key.

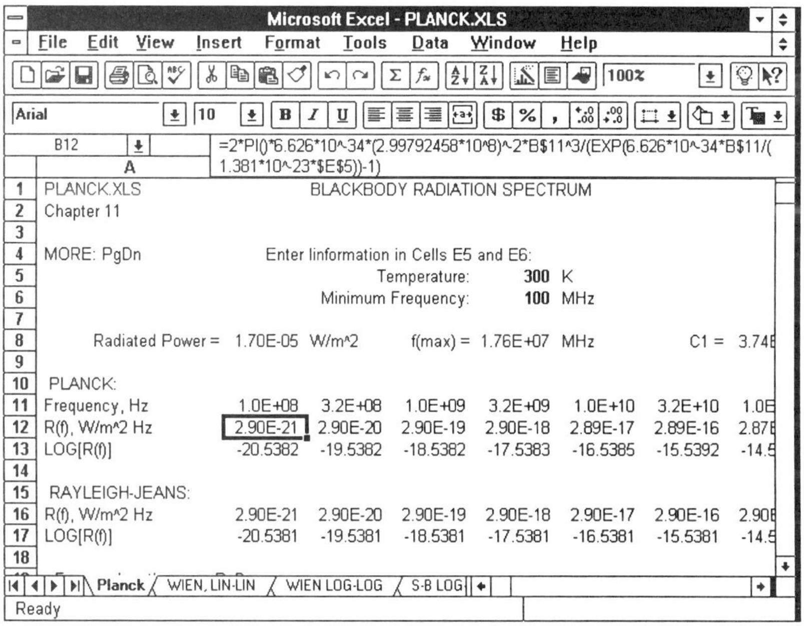

Figure 11-8. Home screen of the PLANCK worksheet. You can change the temperature by entering information in Cell E5. Very high frequencies and very short wavelengths may produce overflow and underflow error warning. Highlighted cell shows formula that computes value in Cell B12.

Overflow and underflow may occur due to the great range of values. If asterisks show in a cell the value in the cell is still valid, but if an error message is displayed the cell value is in error and calculations which depend on this cell are in error.

The named graphs connected to this worksheet are:
COMPARISON 1, Planck and Rayleigh-Jeans as functions of wavelength.
PLANCK, FREQ., Planck as function of frequency.
PLANCK, LAMBDA, Planck as function of wavelength.
PLANCK ZOOM F, Planck, zoom to narrow interval in frequency, around peak.
RAYLEIGH-JEANS1, Rayleigh-Jeans as function of wavelength.
RAYLEIGH-JEANS2, Rayleigh-Jeans as function of frequency.
S-B LIN-LIN, Stefan-Boltzmann law, linear on both axes.
S-B LOG-LOG, Stefan-Boltzmann law, logarithmic on both axes.
WIEN LIN-LIN, Wien displacement law, plotted linear on both axes.
WIEN LOG-LOG, Wien displacement law, plotted log on both axes.

11.4 Harmonic Oscillators and Radiation

Let's see why classical electromagnetic radiation looks like harmonic oscillators. We will use a normalization that looks strange at present, but it will be useful. For simplicity, consider a standing wave in a resonant cavity, volume V, with no sources. Let the electric field, with frequency ω, be polarized in the y direction. We can write the electric and magnetic fields in the standing wave as

$$\mathrm{E_y}(x,t) = \left(\frac{2N\omega^2}{\varepsilon_o V}\right)^{1/2} E(t)\sin(\omega x/c) \tag{11-18}$$

$$\mathrm{B_z}(x,t) = \left(\frac{2\mu_o}{NV}\right)^{1/2} B(t)\cos(\omega x/c), \tag{11-19}$$

where ε_o and μ_o are the vacuum permittivity and permeability, and N is a normalizing factor. Maxwell's equations for the vacuum provide two relations between $E(t)$ and $B(t)$,

$$\frac{dE(t)}{dt} = \frac{1}{N}B(t) \tag{11-20}$$

$$\frac{dB(t)}{dt} = -N\omega^2 E(t). \tag{11-21}$$

The total energy in the field is

$$U = 0.5\,N\omega^2 E^2 + 0.5 N^{-1} B^2. \tag{11-22}$$

The comparison between the electromagnetic field and a harmonic oscillator is shown in Table 11-1, The oscillator is written in Hamiltonian format, where $H(p,x)$ is the Hamiltonian, numerically equal to the total energy (see Chapter 8).

Table 11-1. Classical Harmonic Oscillator and Electromagnetic Field

	Harmonic Oscillator	*Electromagnetic Field*
Time evolution of x and E	$\frac{dx}{dt} = \frac{1}{m}p$	$\frac{dE}{dt} = \frac{1}{N}B.$
Time evolution of p and B	$\frac{dp}{dt} = -m\omega^2 x$	$\frac{dB}{dt} = -N\omega^2 E.$
Energy	$H(p,x) = \frac{1}{2m}p^2 + \frac{1}{2}m\omega^2 x^2$	$U = \frac{1}{2N}B^2 + \frac{1}{2}N\omega^2 E^2$

Now you can see the reason for the peculiar normalization used in Eqs. (11-18) and (11-19). Later, when we quantize the harmonic oscillator, it will be clear that we can use a similar method to quantize the electromagnetic field. Table 11-1 gives a sort of justification, in retrospect, for Planck's remarkable stroke of genius.

Feynman summed things up nicely: "That is one of the marvelous miracles of quantum mechanics. If one begins by considering a kind of state or condition for Bose particles which do not interact with each other (we have assumed that the photons do not interact with each other), and then considers that into this state there can be put either zero, or one, or two, . . . up to any number n of particles, one finds that this system behaves for all quantum mechanical purposes exactly like a harmonic oscillator. By such an oscillator we mean a dynamic system like a weight on a spring or a standing wave in a resonant cavity. And that is why it is possible to represent the electromagnetic field by photon particles."

Feynman continues, "From one point of view, we can analyze the electromagnetic field in a box or cavity in terms of a lot of harmonic oscillators, treating each mode of oscillation according to quantum mechanics as a harmonic oscillator. From a different point of view, we can analyze the same physics in terms of identical Bose particles. And the results of both ways of working are always in exact agreement. There is no way to make up your mind whether the electromagnetic field is really to be described as a quantized harmonic oscillator or by giving how many photons there are in each condition. The two views turn out to be mathematically identical. So in the future we can speak either about the number of photons in a particular state in a box or the number of the energy level associated with a particular mode of oscillation of the electromagnetic field. They are two ways of saying the same thing. The same is true of photons in free space. They are equivalent to oscillations of a cavity whose walls have receded to infinity."

11.5 Wilson-Sommerfeld-Ishiwara Quantization Rule

Back in Chapter 1 we saw that the area of phase space for a one-dimensional classical harmonic oscillator has units of Planck's constant h. This suggested to Professors W. Wilson, A. Sommerfeld, and J. Ishiwara (all working independently) that integrating around any closed path in phase space, like Eq. (8-49) with the Hamilton-Jacobi equation, should result in an integer number times h. In other words, periodic motion has a quantized or discrete phase space. The area in phase space can only change by discrete jumps, in units of h.

Simple Harmonic Oscillator

Let's examine the simple harmonic oscillator using this quantization rule. According to Wilson-Sommerfeld-Ishiwara,

$$\oint p_x \, dx = (n + n_o)h, \tag{11-23}$$

where n is an integer and the integral is around any *closed path* in phase space, and n_o is a constant whose value must be determined. Without more information, it is possible that n_o could be zero. In any event, the value of n_o will not affect our subsequent calculation of transitions between energy levels, because $n_o hf$ would be added to each level and would cancel upon taking the energy difference between levels. This is not a trivial problem, because the value of n_o sets the zero-point energy of a system. In 1912 Max Planck found $n_o = 0.5$. The zero-point energy of the electromagnetic field has important theoretical and experimental consequences; it can be measured.

For the harmonic oscillator,

$$x = X\cos(\omega t) \tag{11-24}$$

$$dx = -\omega X \sin(\omega t)\, dt. \tag{11-25}$$

So,

$$p_x = m\frac{dx}{dt} = -m\omega X \sin(\omega t). \tag{11-26}$$

When we put this in the Wilson-Sommerfeld-Ishiwara rule we get,

$$\oint p_x \, dx = \oint [-m\omega X \sin(\omega t)]\,[-\omega X \sin(\omega t)]\, dt = nh. \tag{11-27}$$

We recognize $m(\omega X)^2$ as twice the total energy, or $2E$. So, we can write,

$$2E\oint \sin^2(\omega t)\, dt = nh. \tag{11-28}$$

Let $\theta = \omega t$, $d\theta = \omega\, dt$ and Eq. (11-28) becomes

$$\frac{2E}{\omega}\int_0^{2\pi}\sin^2(\theta)\,d\theta = nh. \tag{11-29}$$

Do you remember E/ω, the adiabatic invariant from Section 2.10? Integrating Eq. (11-29) we find,

$$\frac{2E}{\omega}\pi = nh. \tag{11-30}$$

Let's rewrite this and put a label n on E to remind ourselves that each particular energy level corresponds to a particular value of n,

$$E_n = nh\frac{\omega}{2\pi} = nhf. \tag{11-31}$$

This is equivalent to the Planck-Einstein relation.

Transitions between energy levels involve emission or absorption of quanta by the oscillator. Emission of energy, going from level n to level $n-1$ looks like this:

$$\Delta E = E_{n-1} - E_n = (n-1)hf - nhf = -hf \tag{11-32}$$

and absorption of energy, going from level n to level $n+1$, looks like this:

$$\Delta E = E_{n+1} - E_n = (n+1)hf - nhf = +hf. \tag{11-33}$$

The energy levels of the quantum harmonic oscillator are evenly spaced and the frequency difference between each level is the same. These are the *only* energy levels of the oscillator; there are no levels between these. These results are shown graphically in Fig. 11-9.

- Oscillator energy increases in a quantum jump $+hf$. Oscillator energy decreases in a quantum jump $-hf$.
- Unlike the classical oscillator, the amplitude of the quantum oscillator changes in quantized steps, because the energy changes in quantized steps.

A Particle in a One-Dimensional Box

Deeper insight into the unique nature of the harmonic oscillator can be obtained if we look at another system, a particle confined to a definite region of the x-axis (a particle in a one-dimensional box, Section 2.10). In the box from $x = 0$ to $x = L$ there is no force on the particle, but it cannot go beyond these limits. Applying the Wilson-Sommerfeld-Ishiwara rule (still with $n_o = 0$) we find,

$$\oint p_x\,dx = nh. \tag{11-34}$$

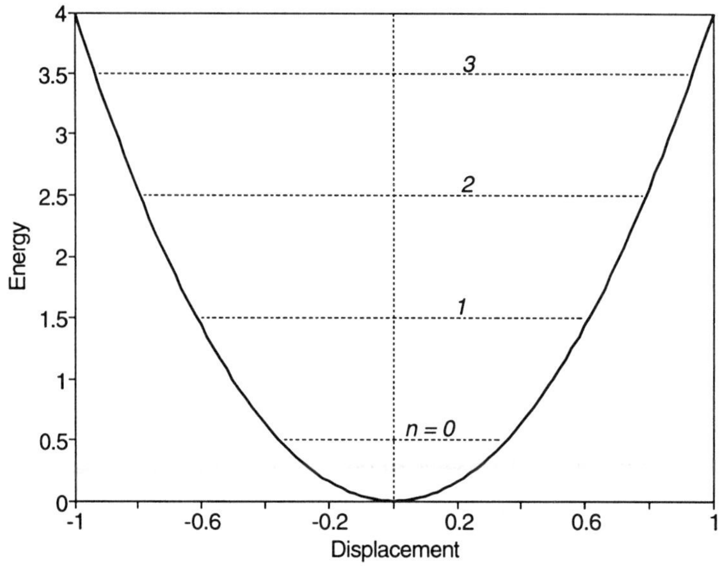

Figure 11-9. Low energy levels of the harmonic oscillator. In this figure, energy is measured in units of *hf* and n_o has been set to the known value of 0.5 (the ground state). The energy difference between adjacent levels is *hf*.

Because there is no force between 0 and L, this integral becomes

$$\oint p_x\,dx = \int_0^L mv\,dx + \int_L^0 -mv\,dx = nh \tag{11-35}$$

or,

$$mv\int_0^L dx - mv\int_L^0 dx = mvL - (-mvL) = 2mvL = nh. \tag{11-36}$$

so,

$$mv_x = p_x = n\frac{h}{2L}\,. \tag{11-37}$$

- Confining a particle in x results in quantization of p_x.
- Squeezing a particle increases its momentum, because $p \propto 1/L$. We noticed this effect when we first examined classical phase space.

Moreover, Eq. (11-37) means that the energy must also be quantized as

$$E = \frac{p_x^2}{2m} = \frac{[n\,(h/2L)]^2}{2m}\,. \tag{11-38}$$

Since E is different for each n we should label E with n,

$$E_n = \frac{n^2 h^2}{8mL^2} \tag{11-39}$$

or,

$$E_n = n^2 E_1 , \tag{11-40}$$

where the first energy level is

$$E_1 = \frac{h^2}{8mL^2} . \tag{11-41}$$

This is an interesting result, and there are two important points worth emphasizing:

- Energy levels are *not* evenly spaced as they are in the harmonic oscillator. For a particle in a box, they are proportional to n^2 so they are closer at lower energy levels than at higher levels.
- Energy levels are inversely proportional to L^2, which means that confining the particle to a smaller space has a very strong effect on increasing the energy of each level.
- Squeezing a particle increases its energy. This is an important result, *not* restricted to our "particle-in-a-box" calculation.

Using the atomic force microscope, atoms have been arranged in a "quantum corral." (See Bibliography, *Selected Papers from Other Journals*.) An electron in this corral clearly demonstrates the energy increase with confinement. According to Eq. (10-33) what would be the minimum energy of an electron confined to a space of the size of the nucleus, about 10^{-15} m? (This is one reason why electrons are not in nuclei.) Actually, the energy is so high that you are in error by a factor of about 100 if you don't do the calculation using relativity.

11.6 Summary of Energy Level Spacing

Different force laws give rise to different spacing of energy levels. Using an inverse square force (the Coulomb force) and assuming the quantization of angular momentum, Niels Bohr was able to derive the main feature of the hydrogen spectrum, that is, the energy levels in terms of their spectroscopic appearance. We will derive the energy levels of hydrogen shortly, but now let's summarize the influence of the force laws on the spacing of energy levels in Table 11-2.

Table 11-2. Energy Level Spacing for Different Force Laws		
System	*Force*	*Energy Level Spacing*
Particle in a box	$0, -L < x < +L$	n^2
Harmonic oscillator	$-Cx$	n^1
Hydrogen atom	$-Cx^{-2}$	n^{-2}

Relative to the harmonic oscillator, the particle in the box has its energy levels more closely spaced near the ground state, and the hydrogen atom has its energy levels more widely spaced near the ground state.

The harmonic oscillator is unique in having *evenly spaced* energy levels, like the rungs on a ladder. This is exclusively due to the linear restoring force $F = -\kappa x$ and its associated potential energy function $U(x) = 0.5\kappa x^2 = 0.5m\omega^2 x^2$. Only a potential energy like this expands exactly right so that, as the amplitude goes up with energy, the levels remain evenly spaced. Equal spacing of the harmonic oscillator energy levels has fundamentally profound results. In quantum electrodynamics and other integer-spin quantum fields, by means of field mode oscillators, equal spacing produces the particle (boson) picture.

11.7 Quantization of Angular Momentum

The Wilson-Sommerfeld-Ishiwara quantization rule can be expressed in generalized momenta and generalized coordinates as we did in Chapter 8,

$$\oint p_i \, dq_i = nh, \tag{11-42}$$

where as before the integral is around a closed path in phase space. We can let $p = p_\varphi$, the magnitude of the angular momentum, and $q = \varphi$, the angle associated with the angular momentum. For central forces p_φ is a constant of the motion so,

$$\oint p_\varphi \, d\varphi = p_\varphi \oint d\varphi = p_\varphi \, 2\pi = nh. \tag{11-43}$$

This gives the Bohr condition for angular momentum quantization,

$$p_\varphi = mvr = n\frac{h}{2\pi} = n\hbar, \tag{11-44}$$

where $\hbar$ (pronounced "h bar") is $h/2\pi$. Recall that Niels Bohr (Nobel Prize 1922) had to *assume* that angular momentum was quantized as in Eq. (11-44) in order to derive the experimentally known spectrum of the hydrogen atom, which was given by

$$\frac{1}{\lambda} = R\left(\frac{1}{n_f^2} - \frac{1}{n_i^2}\right), \tag{11-45}$$

where $R = 1.0973732 \times 10^7\ \mathrm{m}^{-1}$ was an experimentally determined constant and the n quantities are *integers*. This astonishing formula involving integers was discovered in 1885 by Johann J. Balmer, a school teacher in Switzerland, using experimental data of Anders Ängstrom. The empirical constant was very accurately measurable, and equally mysterious. You can't miss the historical analogy with Kepler's discovery of his three empirical laws from the planetary data of Tycho Brahe. And, as Newton used Kepler's laws to determine the gravitational force law for planetary motion, so Bohr used Eq. (11-45) for the hydrogen spectrum.

First, Bohr wrote the total energy of the electron in a circular orbit around the proton, $E = K + U = 0.5mv^2 - ke^2/r$, using the Coulomb potential energy with force constant $k = 1/4\pi\varepsilon_o \approx 9 \times 10^9\ \mathrm{N{\cdot}m^2/C^2}$ (*not* Boltzmann's constant). Then he used Newton's second law with the Coulomb force and the centripetal acceleration, $ke^2/r^2 = mv^2/r$. With these two equations he could write $K = 0.5mv^2 = 0.5ke^2/r$. In other words, the kinetic energy is half of the potential energy for a circular orbit.

It follows immediately that the total energy is given by $E(r) = -0.5ke^2/r$ for this bound state. With the audacity of youth, Bohr assumed that there must be no radiation from a charge in a circular orbit, contrary to classical electromagnetic theory in which accelerated charges radiate continuously. Following Planck and Einstein, Bohr assumed that $E_f - E_i = hf = hc/\lambda$, so that the electron would emit or absorb radiation discontinuously, in a quantum jump, not continuously. Now he could write

$$E_f - E_i = hc\left(\frac{1}{\lambda_f} - \frac{1}{\lambda_i}\right) = -0.5ke^2\left(\frac{1}{r_f} - \frac{1}{r_i}\right), \tag{11-46}$$

but this does not involve the squares of integers and there seemed little hope of measuring the actual radii of the electron orbits. There was only one thing left to do, and that was to think what no one in the world had ever thought before: Angular momentum *must be* quantized as $mvr = nh/2\pi$. This was necessary and sufficient.

Now Bohr had enough equations to eliminate unobservable quantities, because he could eliminate the velocity between the equation for kinetic energy and angular momentum to obtain

$$r_n = \frac{n^2h^2}{mke^2}\ . \tag{11-47}$$

Substituting Eq. (11-47) into Eq. (11-46) produces Eq. (11-45) and the numerical value of R. This brilliant achievement, shown here in less than one page, using only a bit of elementary algebra, was praised by Einstein. We can now see that this was semi-classical, a transition between the classical and the full quantum description.

What's Next?

Despite all of these successes with blackbody radiation, the photoelectric effect, and the spectrum of atomic hydrogen, there is a sense of incompleteness. The Wilson-Sommerfeld-Ishiwara quantization rule did not lead to any major new discoveries, and it could not treat non-periodic systems or even a free particle.

Clearly, a piece of the puzzle was missing. Maybe more than one was missing. Relativity was not included in Bohr's theory. Arnold Sommerfeld and others attempted to patch things up with relativity and elliptical orbits. For a while it looked like they were going in the right direction but they were in uncharted waters with an uncertain compass. It turned out that one of Sommerfeld's students was a bright young fellow named Werner Heisenberg, and he was destined to steer a new course.

What was needed was new information as strange as the quantization of energy and angular momentum appeared to be when these concepts were first suggested. This new information inspired new methods and new concepts that are still being explored theoretically and experimentally (because they are not yet completely understood). Despite the lack of full understanding, quantum mechanics is the basis for multi-billion-dollar industries, providing jobs for millions of people throughout the world in semiconductors, computers, lasers, telecommunications, medical diagnostics, and, of course, it has produced more Nobel prizes.

Next, let's examine a missing piece of the puzzle and see how it changes the whole picture. We will see the harmonic oscillator from an entirely new perspective, one that may be closer to the true state of Nature.

When Bohr's paper on the hydrogen atom was published in 1913, Sommerfeld immediately saw the importance of this new idea. I happened to be in his office when he opened the issue of the Philosophical Magazine, *which had just arrived; he glanced through it and told me: "There is a most important paper here by N. Bohr, it will mark a date in theoretical physics." And soon after, Sommerfeld started applying his own "quantum of action" method to rebuild a consistent theory of Bohr's atom. This is how the first quantized mechanics was born, and why it progressed so fast. It was definitely Sommerfeld's discovery of the importance of the $\int pdq$ integrals that paved the way and these integrals still are at the basis of the whole quantum theory. Everybody wondered (and still wonders) why the Stockholm committee systematically ignored Sommerfeld's pioneer work in modern physics. Such an omission is actually impossible to understand.*

Lèon Brillouin
in the Preface to *Wave Propagation and Group Velocity,*
(Academic Press, New York, 1960)

Brillouin's reference here is to Bohr's paper in *Philosophical Magazine* **26**, 1 (1913).

Chapter 12

Quantum Oscillators

The more I ponder about the physical part of Schrödinger's theory, the more disgusting ("desto abscheulicher") it appears to me.
Werner Heisenberg, in a letter to Wolfgang Pauli

In Chapter 11 it appeared that we lacked some information. Prince Louis Victor de Broglie provided the missing clue. In the seven chapters of his Ph.D. dissertation he presented arguments suggesting that particles may have wave characteristics, a duality with blackbody radiation and the photoelectric effect which had been interpreted to infer that electromagnetic waves have particle characteristics. Based on known results of classical mechanics and optics he proposed that there could be a connection between the momentum p of a particle and a wavelength λ associated with that particle. According to de Broglie, the wavelength may be expressed as

$$\lambda = \frac{h}{p} . \tag{12-1}$$

12.1 Wave-Like Property of Momentum

It was known from classical electromagnetic theory and experiments that electromagnetic waves have momentum $p = E/c$, where E is the energy and c is the speed of light. With the new information that $E = hf$, this could be written $p = hf/c$. Since we know that frequency, speed, and wavelength are related by $f/c = 1/\lambda$ we can now write $p = h/\lambda$ for photons. But photons travel with the speed of light and have no mass. According to relativity, particles with mass cannot travel with the speed of light, because their energy would be infinite. Could $p = h/\lambda$ apply to particles with mass? And, if so, what kind of wave is this particle wave?

The Ph.D. examining committee praised the originality of de Broglie's work, but did not believe there was any physical content. The chairman, Jean Baptiste Perrin (1926 Nobel Prize for experimental study of Brownian motion) asked if these waves could be verified by experiment. Prince de Broglie responded that it might be possible using electron diffraction by crystals, in the same way that x-ray diffraction can be used to measure photon wavelength. He apparently suggested an experiment to Alexandre Dauvillier, a colleague of his brother Maurice de Broglie, but Dauvillier was busy working on the newly emerging technology of television.

Drs. Clinton Joseph Davisson and Lester H. Germer at Bell Laboratories quickly confirmed Eq. (12-1) by their experiments on electron diffraction. The experimental conditions were favorable; electrons accelerated through 54 V had just about the right wavelength for easily measured diffraction by nickel crystals. They were assisted by the accidental breakage of a liquid air dewar, which cracked their vacuum system and oxidized their polycrystalline nickel target. After repairs they heated the target to remove the oxide, and this heating and cooling formed large nickel crystals in their target. Lucky accident. Prince de Broglie received his Ph.D. degree, and in 1929 received the Nobel Prize. Dr. Davisson received the Nobel Prize in 1937.

If we accelerate electrons (charge e, mass m) through a potential difference V volts, then the decrease in potential energy appears as an increase in kinetic energy,

$$eV = \frac{p^2}{2m} = \frac{h^2}{2m\lambda^2} \ . \tag{12-2}$$

So, the de Broglie wavelength can be expressed as

$$\lambda = \frac{1.226}{\sqrt{V}} \text{ nm.} \tag{12-3}$$

It was known from x-ray diffraction that the atomic spacing in nickel crystals was about 0.215 nm, so voltages less than 100 V would produce electrons with comparable wavelengths. Equation (12-3) is shown in Figs. 12-1 and 12-2. The Home screen of the worksheet DEBROGLI is shown in Fig. 12-3.

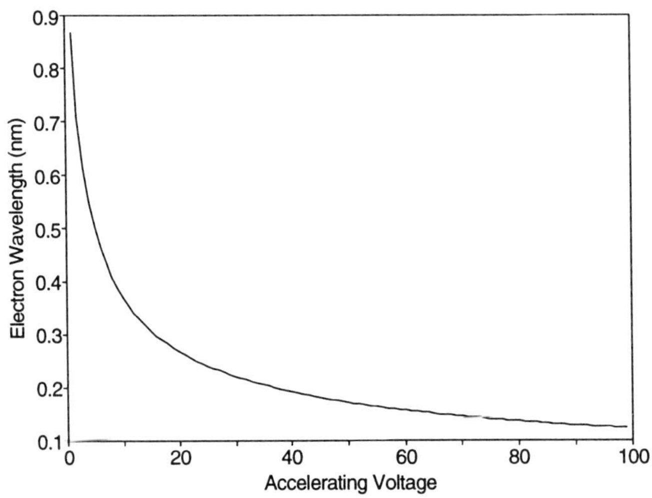

Figure 12-1. Electron de Broglie wavelength *vs.* accelerating voltage.

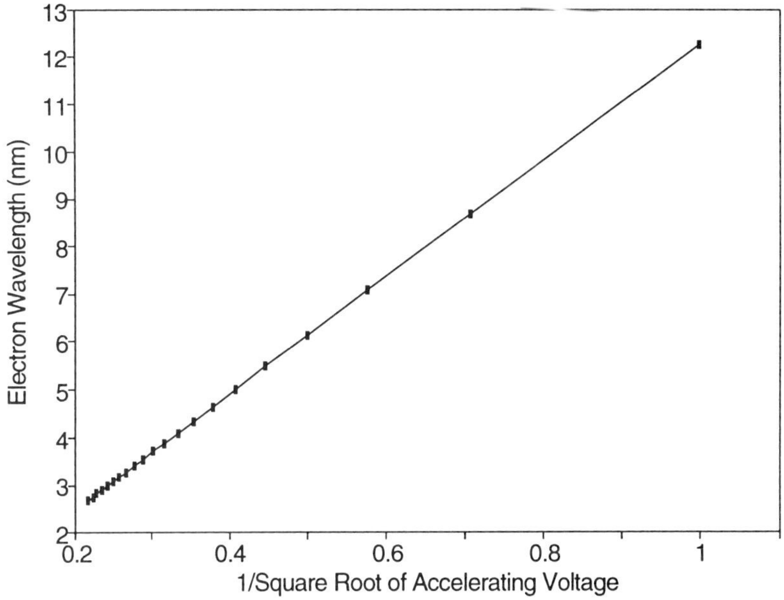

Figure 12-2. The de Broglie relation is particularly simple when viewed as a function of the reciprocal of the square root of the accelerating voltage.

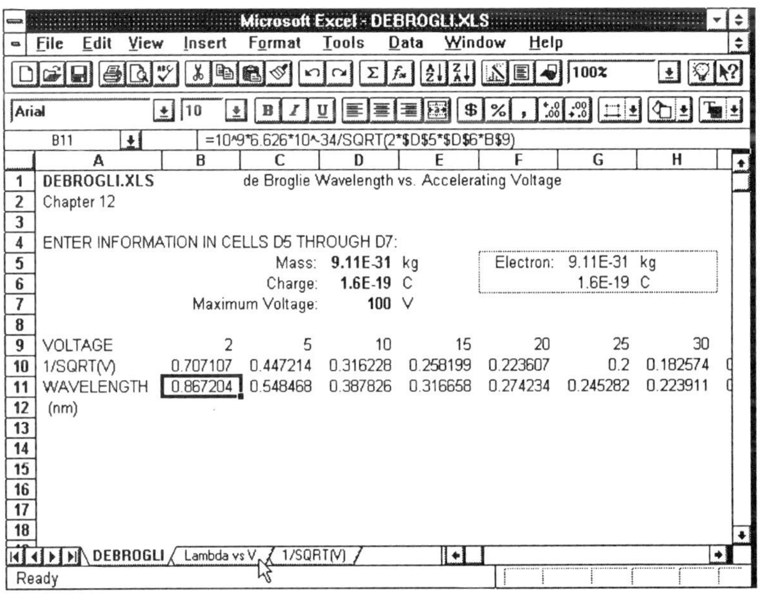

Figure 12-3. Home screen of DEBROGLI worksheet. You can change the charge and mass of the particle and the accelerating voltage by entering information in Cells D5 through D7. The highlighted cell shows formula for wavelength.

Because the electron wavelength could be tuned by the accelerating voltage, it soon became apparent that an electron microscope could be developed with a resolution far better than that of visible light microscopes. Dr. Ernst Ruska was awarded the Nobel Prize in 1986 for his invention of the electron microscope.

12.2 Schrödinger Equation

The Schrödinger equation has never been "derived" satisfactorily; its validity can only be justified by its results. Here we are going to show that the Schrödinger equation is plausible. As a start let's list some of the things we now know.

- Energy is quantized according to $E = hf = \hbar\omega$
- Momentum is associated with a wavelength; $p = \frac{h}{\lambda} = \hbar k$. (Here k is the wavenumber, $k = 2\pi/\lambda$. Don't confuse this with Boltzmann's constant k_B.)
- The total energy is $E = \frac{p^2}{2m} + U$.
- Phase space has dimensions of h, with units of J·s or eV·s ("action").
- The quantum domain is $hf > k_BT$. (Here k_B is Boltzmann's constant.) Results should approximate the classical domain for $hf < k_BT$. See Fig. 11-4.
- If there is a wave there must be a wave equation. What is it? When we find it, it should tell us everything we can know about the system. The wave equation must be consistent with what we now know.

We are usually on safe ground if we use energy arguments. The total energy is

$$\frac{p^2}{2m} + U = E. \tag{12-4}$$

Now, let's write the energy in terms of $E = hf$ and the momentum in terms of $p = h/\lambda = (h/2\pi)k = \hbar k$,

$$\frac{(\hbar k)^2}{2m} + U = \hbar\omega. \tag{12-5}$$

A wave is something that wiggles in space and time. One of the simplest equations that wiggles in space and time is an analytic signal representation of a plane wave,

$$\Psi(x,t) = A\exp[j(\omega t - kx)]. \tag{12-6}$$

Why not try something even simpler? For example, we could try,

$$\Psi(x,t) = A\cos(\omega t - kx). \tag{12-7}$$

Equation (12-7) is the real part of Eq. (12-6). It turns out that purely real or purely imaginary functions are not solutions of Schrödinger's equation. Only complex functions are solutions. This is often a conceptual difficulty, but recall from Chapter 6 that analytic signals contain *only positive frequencies* so $E = hf$ will be positive.

Even Eq. (12-6) looks pretty complicated, because it oscillates in both space and time, as waves do. We can stop the motion in space by keeping x constant (observing at one point) while we watch the oscillation in time. Then we can stop the motion in time by observing for an instant while we watch the oscillation in space. It might be helpful to think of using an oscilloscope probe, which has a very sharp point. The probe does not occupy much space, but it lets you watch the waveform over a considerable interval in time at the point where the probe is located.

Similarly, you can think of the space oscillation as being a photograph taken with a fast camera shutter and a wide-angle lens. The picture doesn't occupy much time but you can view the waveform oscillating over a wide interval in space.

The time derivative brings down a factor of $j\omega$ in Eq. (12-6). Each time the space derivative is applied it brings down a factor of $-jk$.

The factor $\hbar\omega$ in the total energy is suggestive of a time derivative because, if we ask how Ψ moves in time (keeping x constant), we need to take the partial derivative of Ψ with respect to t,

$$\frac{\partial \Psi}{\partial t} = j\omega\Psi . \tag{12-8}$$

The factor k^2 in the total energy is suggestive of a second partial derivative with respect to x (this keeps t constant and just looks at the oscillation in space),

$$\frac{\partial^2 \Psi}{\partial x^2} = jk \cdot jk\Psi = -k^2\Psi. \tag{12-9}$$

To ensure that the energy E is real and positive we must multiply the time derivative by $-j\hbar$ to obtain $\hbar\omega$ in Eq. (12-8). To obtain positive kinetic energy, as it must be, it is necessary to multiply Eq. (12-9) by $-\hbar^2/(2m)$. When we do this we obtain Schrödinger's equation, for which he was awarded the Nobel Prize in 1933,

$$-\frac{\hbar^2}{2m}\frac{\partial^2 \Psi}{\partial x^2} + U\Psi = j\hbar\frac{\partial \Psi}{\partial t} . \tag{12-10}$$

To check this out, plug our analytic signal into Eq. (12-10). The result is: The total energy is the sum of the kinetic and potential energies, just what we started with. This wave equation may look funny (especially the j factor). What is almost as bad is the first time derivative, which makes it look like a diffusion or heat transfer equation instead of a wave equation. Recall that the classical wave equation of mechanics, acoustics, and electromagnetics has a *second* time derivative. Obviously this thing requires further examination because it is not relativistically correct.

If you try $\text{Re}\,\Psi = \cos(\omega t - kx)$ or $\text{Im}\,\Psi = \sin(\omega t - kx)$ for $U(x) = 0$ you will find they are not solutions, but $\cos(\omega t - kx) + j\sin(\omega t - kx)$ is a solution. A complex solution is a fundamental requirement of Schrödinger's equation.

We have obtained this wave equation *only* from energy considerations plus the experimental results $E = hf$ and $p = h/\lambda$. Now, since

$$-j\hbar\,\frac{\partial\Psi}{\partial t} = -\hbar\, jj\omega\Psi = \hbar\omega\Psi = E\Psi, \tag{12-10}$$

We can rewrite the wave equation without the time,

$$-\frac{\hbar^2}{2m}\frac{\partial^2\Psi}{\partial x^2} + U\Psi = E\Psi. \tag{12-11}$$

However, now that our equation is independent of time we should modify Ψ, because Ψ is a function of x and t and we no longer need t. Let's use a wave function that only depends on x, and call it $\psi(x)$. Since we have gotten rid of t, we can do without the ∂'s, because all we have is an ordinary differential equation. Also, since we know there is more than one energy level, we should identify the ψ and E for each energy level with a label n,

$$-\frac{\hbar^2}{2m}\frac{d^2\psi_n}{dx^2} + U\psi_n = E_n\psi_n\,. \tag{12-12}$$

Now things are beginning to look pretty simple. Let's play with our new toy and see what it will do. We can try it on three easy steps of an Aristotelian Incline:

- A free particle. (Recall that we couldn't describe the free particle with the Wilson-Sommerfeld-Ishiwara quantization rule.)
- A particle in a box. (The particle is "free" until it hits the walls.)
- A simple harmonic oscillator.

But first, let's see another interpretation of what we have just done in building Schrödinger's equation.

12.3 Dynamic Variables and Their Operators

Recall that we started with the total energy,

$$K + U = E\,. \tag{12-13}$$

Then we developed a way of expressing this in terms of a wave function $\Psi(x,t)$,

$$\left[-\frac{\hbar^2}{2m}\frac{\partial^2}{\partial x^2} + U\right]\Psi = j\hbar\frac{\partial\Psi}{\partial t}. \tag{12-10}$$

If Eq. (12-10) is correct, then the kinetic energy, potential energy, total energy, momentum, and Hamiltonian must correspond to the operators shown in Table 12-1.

To see this, note that Eq. (12-10) can be expressed in operator form as

$$H\Psi = E\Psi. \tag{12-14}$$

What does all of this mean? Simply this: If we know $\Psi(x,t)$ and we want to know something about the momentum p_x, just operate on $\Psi(x,t)$ with thc operator, $-j\hbar\,\partial/\partial x$. If we want the energy, use the Hamiltonian operator.

Similarly, whatever quantity we want to know, we operate on $\Psi(x,t)$ with the appropriate operator and we get the answer. The wavefunction $\Psi(x,t)$ contains *all of the information* that anyone can get about the system.

Table 12-1 Some Dynamic Variables and Their Operators

Dynamic Variable	*Operator in Coordinate Form*
Kinetic energy $\frac{p_x^2}{2m}$	$-\frac{\hbar^2}{2m}\frac{\partial^2}{\partial x^2}$
Kinetic energy $\frac{p^2}{2m}$	$-\frac{\hbar^2}{2m}\nabla^2$
Potential energy $U(x)$	$U(x)$
Linear momentum p_x	$-j\hbar\frac{\partial}{\partial x}$
Linear momentum p	$-j\hbar\nabla$
Hamiltonian H	$-\frac{\hbar^2}{2m}\nabla^2 + U(x)$

Equation (12-14) is a well-known form, called an *eigenvalue* equation. Solutions exist only for certain values, the eigenvalues. The operator H, operating on the function Ψ, produces eigenvalue E associated with Ψ. The function Ψ is called an *eigenfunction* or *characteristic function*. This is reminiscent of a matrix equation.

This may, at first, appear to be unusual and exciting, and it is. However, it should also be somewhat familiar. After all, didn't the solution $x(t)$ for the classical harmonic oscillator contain all of the information that we could get about the system? If we wanted to know the momentum p we operated on $x(t)$ with d/dt and multiplied it by the mass m. If we wanted to know the kinetic energy we calculated $0.5mv^2$, and so on. The only unfamiliar parts are the new knowledge about the granular nature of energy ($E = hf$) and the wave-like nature of momentum ($p = h/\lambda$).

12.4 Free Particle

Now it appears that there is no reason why we cannot treat a free particle by letting $U(x) = 0$. This is something we couldn't do before. This looks inviting, so let's write,

$$-\frac{\hbar^2}{2m}\frac{d^2\psi_n}{dx^2} = E_n\psi_n \,, \tag{12-15}$$

because the total energy is in kinetic form,

$$\frac{p^2}{2m} = \frac{(\hbar k)^2}{2m} = E. \tag{12-16}$$

Then, from Eqs. (12-15) and (12-16),

$$\frac{d^2\psi_n}{dx^2} = -k_n^2\psi_n \,. \tag{12-17}$$

"Deja vu all over again," in the immortal words of Yogi Berra. (Dr. Berra received the honorary Doctor of Humane Letters in May, 1996, probably for his contributions to the French and English languages.) Equation (12-17) looks just like our old classical harmonic oscillator equation from Chapter 1, except x has replaced t. It's a one-dimensional harmonic "oscillator" in space. Clearly it has a familiar solution,

$$\psi_n = A_n \exp\left[\pm j(k_n x + \varphi_n)\right] . \tag{12-18}$$

Now it appears that n, k and E can have continuous values, because the particle is free (if there are no boundary conditions to impose discrete values on n). Based on what some people now believe, the universe is finite and expanding. This distant, but finite, boundary makes n, k and E discrete, even though for a free particle the energy levels are so dense that they appear to be almost continuous (recall Chapter 6, boundary conditions on Laplace's equation). For normalization the wavefunction ψ_n has to die away at the boundaries. Equation (12-18) is not normalizable for infinite boundaries. Is this lack of normalizability an argument for a finite universe or for the non-existence of free particles (Mach's principle), or both?

Presumably, using a Fourier series or Fourier integral we could construct almost any kind of waveform we wanted, using enough of the Ψ_n in superposition, with the right amount of each one selected by means of Fourier coefficients C_n,

$$\psi(x) = \sum_{n=1}^{N} C_n \, \psi_n(x) \, . \qquad (12\text{-}19)$$

So it's deja vu all over again, *once more*. This is looking pretty good; let's try it on something a little fancier, like a particle in a box. But first, let's see what all of this says about the velocity of a free particle.

Group, Phase, and Classical Velocities

According to de Broglie,

$$p = \frac{h}{\lambda} = \hbar k, \qquad (12\text{-}20)$$

where k is the wavenumber. According to Planck and Einstein,

$$E = hf = \hbar\omega. \qquad (12\text{-}21)$$

For a nonrelativistic free particle,

$$E = \frac{p^2}{2m} = \frac{\hbar^2 k^2}{2m} \, . \qquad (12\text{-}22)$$

So,

$$\hbar\omega = \frac{\hbar^2 k^2}{2m} \, , \qquad (12\text{-}23)$$

or

$$\omega(k) = \frac{\hbar k^2}{2m} \, . \qquad (12\text{-}24)$$

A relationship between frequency and wave number is known as a *dispersion relation*. Equation (12-24) is the dispersion relation for a free nonrelativistic particle. The group velocity $\partial\omega/\partial k$ is commonly associated with energy, momentum, and information transfer. For the free particle this is

$$v_g = \frac{\partial\omega}{\partial k} = \frac{\hbar k}{m} = \frac{p}{m} = v. \qquad (12\text{-}25)$$

This is a comforting result:

- The velocity of a classical particle is associated with the group velocity of its quantum wavefunction.
- This implies that kinetic energy and momentum are transported with the group velocity of the quantum wavefunction.

This is consistent with Paul Ehrenfest's theorem concerning quantum expectation values and their classical analogs. What about the phase velocity? This is

$$v_p = \frac{\omega}{k} = \frac{\hbar k}{2m} = \frac{p}{2m} = 0.5v. \tag{12-26}$$

The phase velocity is half of the velocity of a classical free particle. This may not be so surprising because, classically, the phase velocity and group velocity are different in electromagnetic waveguides, plasmas, and water waves.

What is surprising is that we know the group velocity and phase velocity in a vacuum are the same for classical electromagnetic waves; both equal c, the speed of light in vacuum. For the vacuum, the dispersion relation is $\omega = ck$, so $\omega/k = c$ and $\partial\omega/\partial k = c$. All electromagnetic waves travel with the same velocity in a vacuum, regardless of frequency. In contrast, Eq. (12-24) says that the vacuum is dispersive for a quantum wavefunction, even if there are no potentials in the space.

Figure 12-4 shows this parabolic form of dispersion relation on the "ω–k" or Brillouin diagram. In this figure there are no data in the neighborhood of $k = 0$, which corresponds to infinite wavelength, because there may be a cut-off wavelength and frequency if the universe is finite; if there is no infinite wavelength, then k cannot go to zero. Also, there must be a high-frequency cut-off, because a particle cannot have infinite energy. Figure 12-5 shows the electron group velocity and phase velocity corresponding to the dispersion relation of Fig. 12-4.

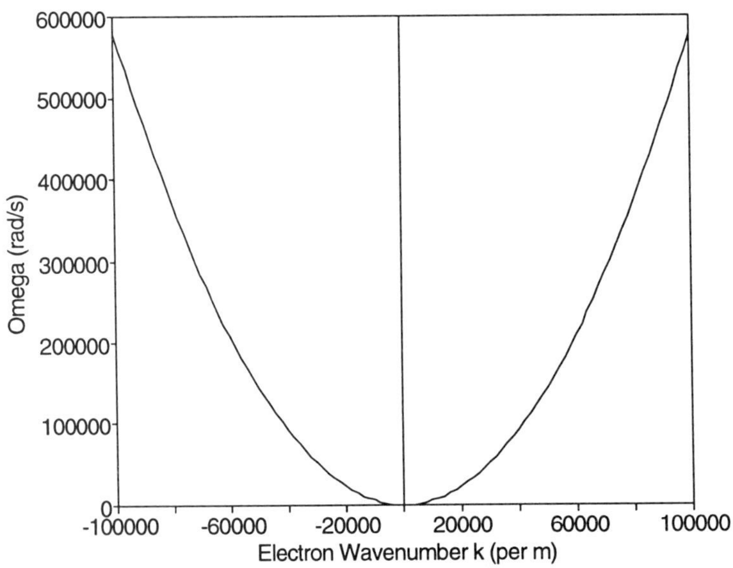

Figure 12-4. Quantum dispersion relation for a free nonrelativistic electron, Eq. (12-24). Slope of the curve is the group velocity, ω/k is the phase velocity.

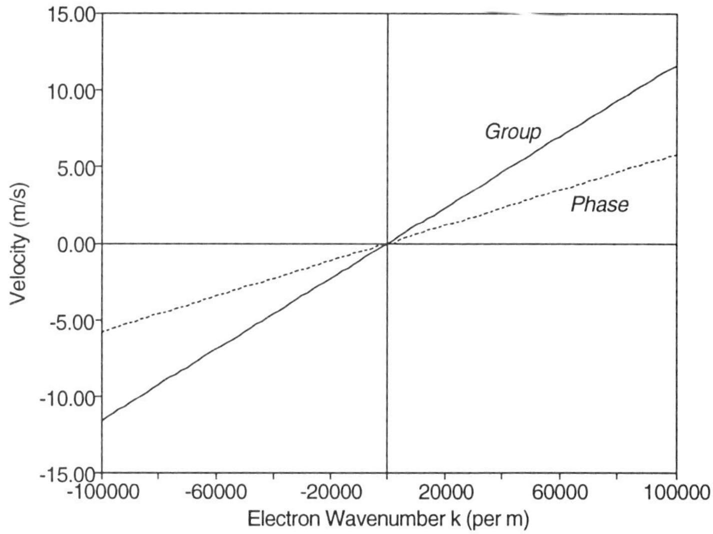

Figure 12-5. Nonrelativistic free electron group and phase velocities in vacuum.

For classical electromagnetic waves in a collisionless plasma, Eqs. (4-18) – (4-24), there is a cut-off at the plasma frequency where the electromagnetic group velocity is zero and the phase velocity is infinite. From Maxwell's equations and Newton's laws, the dispersion relation for a collisionless plasma is

$$\omega(k) = (c^2k^2 + \omega_p^2)^{1/2}, \tag{12-27}$$

where the square of the plasma frequency is $\omega_p^2 = ne^2/m\varepsilon_o$ and n is the electron number density. Equation (12-27) is shown in Fig. 12-6. From Eq. (12-27), v_p is

$$v_p = \frac{\omega}{k} = c\left(1 + \frac{\omega_p^2}{c^2k^2}\right)^{1/2}. \tag{12-28}$$

Equations (12-28) and (12-29) are shown in Fig. 12-7. The group velocity is

$$v_g = \frac{\partial\omega}{\partial k} = \frac{c^2}{v_p} . \tag{12-29}$$

The relativistic energy of a free electron is

$$E = \hbar\omega = (p^2c^2 + m_o^2c^4)^{1/2} = (\hbar^2k^2c^2 + m_o^2c^4)^{1/2} \tag{12-30}$$

and, curiously, this leads to the same mathematical form as Eq. (12-27),

$$\omega(k) = \left(c^2k^2 + \frac{m_o^2c^4}{\hbar^2}\right)^{1/2}. \tag{12-31}$$

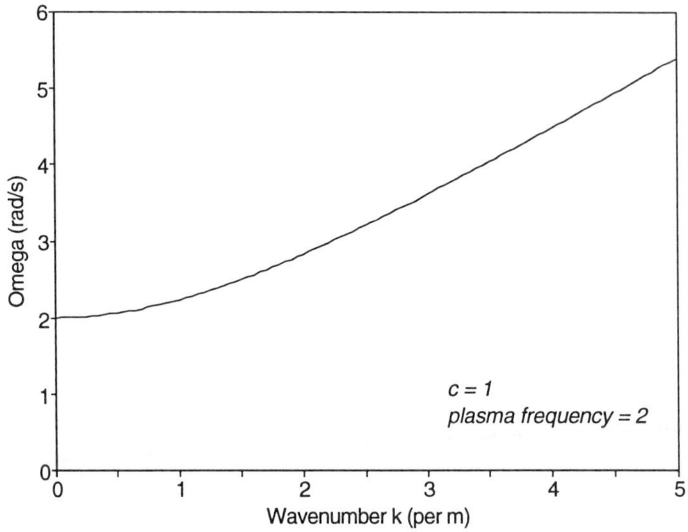

Figure 12-6. Dispersion relation for classical electromagnetic waves in a collisionless plasma, for the right half plane of the Brillouin diagram. Note the cut-off at the plasma frequency where ω remains finite as *k* goes to zero.

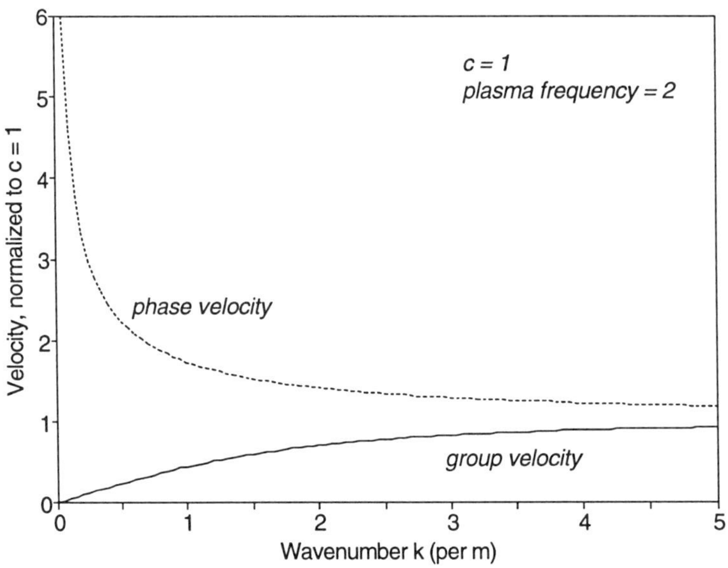

Figure 12-7. Group velocity and phase velocity for classical electromagnetic waves in a collisionless plasma. Both velocities approach *c* for large wavenumber. This is the same qualitative behavior as a free relativistic particle in the quantum domain in a vacuum.

Equation (12-31) has an interesting interpretation in terms of the Compton electron wavelength $\lambda_c = h/m_o c = 0.00243$ nm. (Arthur Holly Compton received the Nobel Prize in 1927 for his work in discovering that x-rays suffer a change in wavelength when they collide with electrons.) We can express Eq. (12-31) as

$$\omega(k) = \left(c^2k^2 + \frac{4\pi^2c^2}{\lambda_c^2}\right)^{1/2} = \left(c^2k^2 + \omega_c^2\right)^{1/2}, \tag{12-32}$$

where $\omega_c = 2\pi c/\lambda_c$. The Compton wavelength is associated with x-ray (*electromagnetic*) scattering from electrons, but this applies to any charged particle. (Each type of particle has its own Compton wavelength.) It is curious that λ_c appears in the *quantum* dispersion relation for a free relativistic particle.

The dispersive property of the vacuum for quantum wavefunctions leads to spreading of wave packets, increasing uncertainty both forward and backward in time. Our crystal ball is cloudy in the past and in the future. This property, which seems so strange at first glance, is consistent with experience. Historical events and memory become more blurred, the longer we look back in time. Predictions about the future are better for short times than for long times. To examine dispersion we need to superpose at least two states, so we will postpone wavefunction spreading until we discuss superposition of states in Section 12.6.

Even more mysterious than vacuum dispersion is the apparent dispersion in the classically-forbidden tunneling process, measured in recent experiments. Someday we will know what all of this means but it may require modification of concepts.

12.5 A Particle in a Box

Before getting down to the job of treating the harmonic oscillator with Schrödinger's equation let's look at a particle in a box. We've already examined this problem classically in Section 2.9 and with the old, or semi-classical, quantum theory in Section 11.4. A particle in a box is just a bit more complicated than a free particle and we will get some exercise to prepare for the harmonic oscillator, which is more complicated than a particle in a box.

Let's start with the complete time-dependent equation so we can see the general way we separate a partial differential equation into ordinary differential equations,

$$-\frac{\hbar^2}{2m}\frac{\partial^2\Psi}{\partial x^2} + U\Psi = j\hbar\frac{\partial\Psi}{\partial t}. \tag{12-10}$$

Separation of Variables

1. Let $\Psi(x,t)$ be represented as a product $\psi(x)T(t)$.
2. Next, substitute this product into the wave equation.

3. Divide the wave equation by this product. This will make the left side a function only of x, and the right side a function only of t.

4. How can a function of x be equal to a function of t? This can be true only if both functions are equal to a constant. This constant is called the *separation constant*.

5. Solve each ordinary differential equation, obtaining $\psi(x)$ and $T(t)$.

6. Obtain the full solution $\Psi(x,t) = \psi(x)T(t)$.

When we perform steps one and two on Eq. (12-10) we obtain,

$$-T\frac{\hbar^2}{2m}\frac{\partial^2\psi}{\partial x^2} + U\psi T = \psi j\hbar\frac{\partial T}{\partial t}. \tag{12-33}$$

Now we can change the ∂'s to d's, because each derivative operates on a function of only one variable. Divide Eq. (12-33) by ψT,

$$-\frac{1}{\psi}\frac{\hbar^2}{2m}\frac{d^2\psi}{dx^2} + U = \frac{1}{T}j\hbar\frac{dT}{dt}. \tag{12-34}$$

Notice the left-hand side is a function of x only and the right-hand side is a function of t only, so both sides must be equal to a constant. We know what's coming, so let's call this separation constant E. (It will turn out to be the energy.) Now we have two ordinary differential equations, The separated equation for x is,

$$-\frac{1}{\psi}\frac{\hbar^2}{2m}\frac{d^2\psi}{dx^2} + U(x) = E. \tag{12-35}$$

Let the box be from $x = 0$ to $x = L$, and recall $k^2 = 2mE/\hbar^2$. In the box $U(x) = 0$ so,

$$\frac{d^2\psi}{dx^2} = -\frac{2mE}{\hbar^2}\psi = -k^2\psi. \tag{12-36}$$

But this is just Eq. (12-17) again, and we know its solution.

The separated equation for t is

$$\frac{1}{T}j\hbar\frac{dT}{dt} = E \tag{12-37}$$

or

$$\frac{dT}{T} = \frac{E}{j\hbar}dt. \tag{12-38}$$

Now, since $E = \hbar\omega$, we can express this as

$$\frac{dT}{T} = -j\omega\, dt\,. \tag{12-39}$$

The time equation is certainly easy to solve,

$$\ln(T) = -j\omega t + C, \tag{12-40}$$

where C is a constant of integration that could be complex. Exponentiate both sides of Eq. (12-40),

$$T = T_o e^{j\omega t}. \tag{12-41}$$

So, we have the complete solution,

$$\Psi(x,t) = \psi(x)T(t) = [A_o \cos(kx) + B_o \sin(kx)]\, T_o e^{j\omega t}. \tag{12-42}$$

Let's clean this up by combining constants,

$$\Psi(x,t) = [A \cos(kx) + B \sin(kx)]\, e^{j\omega t}, \tag{12-43}$$

where A and B are constants.

Now we'll see that imposition of boundary conditions produces discrete solutions with integer quantum numbers. The boundary conditions for all time are

$$\Psi(0,t) = 0 \quad \text{and} \quad \Psi(L,t) = 0.$$

At $x = 0$, $\Psi(0,t) = 0$, so we must have $A = 0$. At $x = L$, $\Psi(L,t) = 0$. So,

$$B \sin(kL) = 0. \tag{12-44}$$

It would not be very interesting if B were zero, so let's see what happens when $\sin(kL)$ is zero. This is zero whenever

$$kL = n\pi, \tag{12-45}$$

where n is an integer, so,

$$k_n = \frac{n\pi}{L} = \frac{2\pi n}{\lambda_n}, \tag{12-46}$$

where we have put a subscript on k, because it is now a discrete set of values.

- Integers (quantum numbers) appear due to the imposition of boundary conditions on the wavefunction.

We should also discretize and label B, ω, and the wavefunction itself,

$$\Psi_n(x,t) = B_n \sin\left(\frac{n\pi x}{L}\right) \exp[j\omega_n t]. \tag{12-47}$$

This is a standing wave composed of n half-waves, with amplitude B_n and frequency ω_n. From Eq. (12-10) we can get the energy levels,

$$E_n = \frac{p^2}{2m} = \frac{(\hbar k_n)^2}{2m} = \frac{(\hbar n\pi/L)^2}{2m} = \frac{n^2h^2}{8mL^2}, \tag{12-48}$$

which is exactly in agreement with the results of using the Wilson-Sommerfeld-Ishiwara quantization rule, Eq. (11-31). This should give us a little confidence that we are on the right track. Table 12-2 summarizes some of this information.

Table 12-2. Characteristics of a Particle in a Box

Energy Level n	*Wavelength*	*Symmetry* [See Eq. (12-47) and Fig. 12-8]	$\frac{E_n}{E_1}$
1	$2L$	Even	1
2	L	Odd	4
3	$2L/3$	Even	9
4	$L/2$	Odd	16

The probability of finding the particle in the box is 1, so we should adjust B_n in Eq. (12-47) to normalize the probability,

$$\int_0^L P(x)\,dx = \int_0^L \Psi_n^*\Psi_n\,dx = 1. \tag{12-49}$$

The wavefunctions in Eq. (12-47) comprise a complete orthogonal set so normalization is quick and easy. Equation (12-49) becomes

$$\int_0^L B_n^*\,B_n \sin^2\left(\frac{n\pi x}{L}\right)dx = 1, \tag{12-50}$$

from which we find,

$$B_n^*\,B_n = |B_n|^2 = \frac{2}{L}. \tag{12-51}$$

Now we can calculate the probability of finding the particle in any range Δx in the box, for any state. The normalized wavefunction that contains *all of the information we can get* about the particle in the box is given by,

$$\Psi_n(x,t) = \sqrt{2/L}\,\sin\left(\frac{n\pi x}{L}\right)\exp[j\omega_n t]. \tag{12-52}$$

The first three wavefunctions are shown in Fig. 12-8 and their associated probability densities are shown in Fig. 12-9. The lowest energy state is very nonuniform, located mostly at the center, but at the third level things begin to take on a more uniform appearance, on the average, consistent with a classical particle in a box.

Probability of Finding the Particle

Suppose we want to know the probability of finding the particle in the lowest energy level ($n = 1$) near the center of the box, in a small range from $0.499L$ to $0.501L$. In this narrow range near the center of the box, the wavefunction is substantially constant (see Fig. 12-8). In this range $\sin(\theta) \approx \sin(\pi/2) = 1$ and Eq. (12-52) tells us,

$$\frac{2}{L}\int_{0.499L}^{0.501L} \sin^2\left(\frac{\pi \cdot 0.5L}{L}\right) dx \approx \frac{2}{L}\int_{0.499L}^{0.501L} dx = 2(0.501 - 0.499) = 0.004. \tag{12-53}$$

(In Fig. 12-8 the box is centered on $x = 0$.) Notice that the probability increases as the spatial interval increases, as we would expect. If we are interested in finding the probability over a larger range, or where the wavefunction is rapidly changing, we would have to perform the integration with the sine function in the integral, instead of approximating it at the center of this narrow range, as we did in Eq. (12-53).

For $n > 1$ the wavefunctions have zero-crossings, so the probability density goes to zero at those points. Nevertheless, the probability does not go to zero, because the probability density must be integrated over an interval. Also, if we try to pin the particle down by specifying a narrow spatial interval, we will lose information about its momentum. This is consistent with what we know about uncertainty principles.

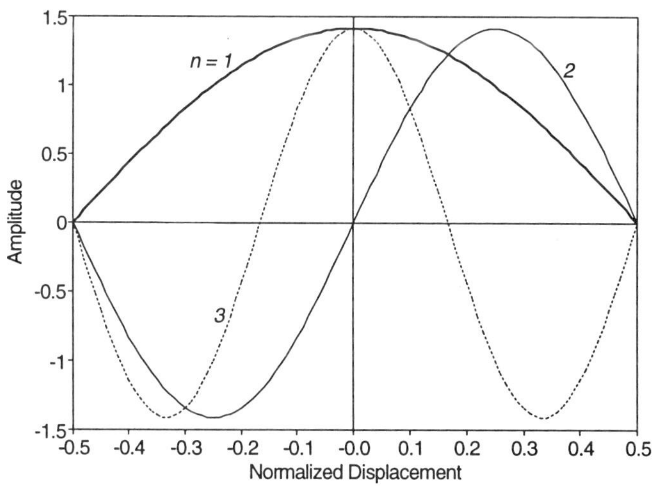

Figure 12-8. Three lowest wavefunctions for a particle in a one-dimensional box. Observe the alternating symmetry, maxima, and zeros of wavefunctions. $-L/2 \le x \le +L/2$. Center of box has been shifted from $x = 0.5$ to $x = 0$.

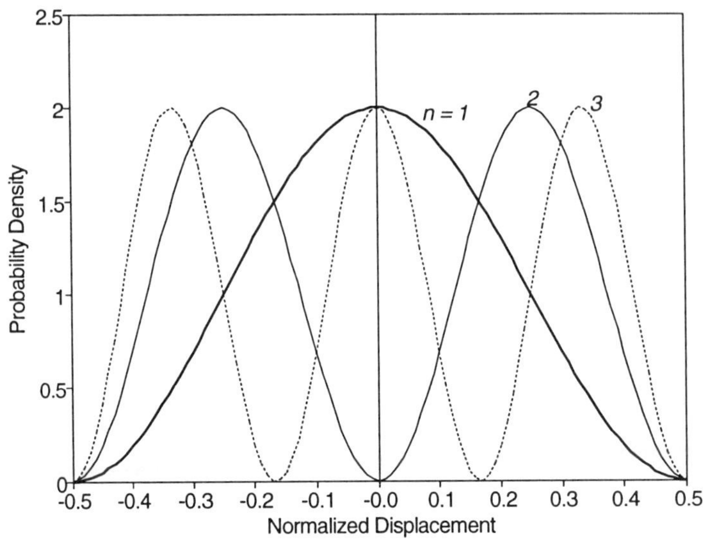

Figure 12-9. Probability densities for the wavefunctions in Fig. 12-8.

Now we have enough information to calculate the probability of a transition from one energy level to another, and to find the "selection rules" that relate what transitions are allowed. We are going to do this for electric dipole transitions, the strongest, and therefore the most common, type of transition. From what we have seen about classical harmonic and square-well oscillators in Section 2.9 we might expect some similarities and some differences in the quantum domain.

12.6 Superposition of States

For simplicity let's assume there there are only two energy levels in the square well and we have a superposition of the two states,

$$\Psi = \sum_{n=1}^{2} \Psi_n = \Psi_1 + \Psi_2 . \tag{12-54}$$

We can calculate the probability density,

$$|\Psi|^2 = |\Psi_1 + \Psi_2|^2, \tag{12-55}$$

which can be expanded as,

$$|\Psi|^2 = |\Psi_1|^2 + |\Psi_2|^2 + \Psi_1^*\Psi_2 + \Psi_2^*\Psi_1 . \tag{12-56}$$

The first two terms are independent of time, because the complex exponential goes to unity, but the last two terms need further examination,

$$\psi_1 \exp(-j\omega_1 t)\, \psi_2 \exp(j\omega_2 t) \;+\; \psi_1 \exp(+j\omega_1 t)\, \psi_2 \exp(-j\omega_2 t)\,. \qquad (12\text{-}57)$$

Let's collect terms,

$$\psi_1\psi_2 \exp[j(\omega_2 - \omega_1)t] \;+\; \psi_1\psi_2 \exp[j(\omega_1 - \omega_2)t], \qquad (12\text{-}58)$$

and rewrite this as

$$2\,\psi_1\psi_2 \cos[(\omega_2 - \omega_1)t]. \qquad (12\text{-}59)$$

Finally,

$$|\Psi|^2 = |\psi_1|^2 + |\psi_2|^2 + 2\,\psi_1\psi_2 \cos[(\omega_2 - \omega_1)t]. \qquad (12\text{-}60)$$

- The probability density oscillates at the frequency $\omega = \omega_2 - \omega_1$. If this probability density is associated with a charged particle, then this oscillator would absorb and emit photons at this frequency.

A classical oscillator will absorb and emit electromagnetic radiation at this frequency. Classical electromagnetic fields have energy proportional to the square of the electric field, so $|\psi|^2$ is proportional to $N\hbar\omega$, where N is the number of photons at this frequency. The electric field structure radiated by a classical harmonic oscillator is shown in Fig. 12-10. The transverse part of the electric field is the part that radiates; the photons are most dense where the electric field is strongest. Note that the radiation field is maximum perpendicular to the direction of motion of the oscillator, and there is no radiation from the ends of the oscillator.

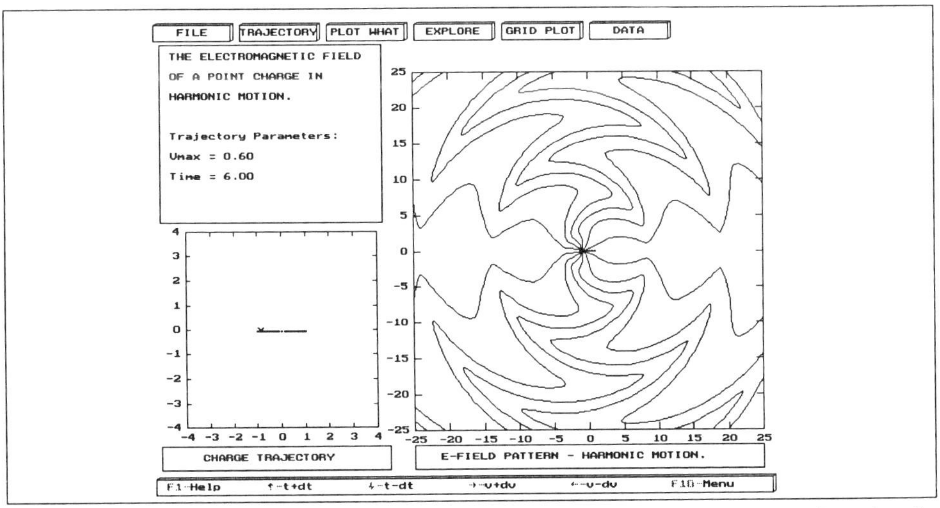

Figure 12-10. Classical electric field structure from a charge executing simple harmonic motion. The oscillator trajectory is the horizontal line in the box at the lower left. This screen was generated with *Electricity and Magnetism Simulations* (see Software section in Bibliography).

Schrödinger's Cat and Her Kitten

In 1926 Schrödinger proposed a *gedanken* (thought) experiment that contains the seeds of a possible paradox in quantum theory, related to the superposition of states. Suppose a cat is put in a sealed box with a weak radioactive source. Atoms in the radioactive source decay at random. Suppose that there are detectors in the box, and the detectors are connected to a device that kills the cat painlessly if a detector is triggered by a radioactive decay. This is a kinder, gentler, version of the *gedanken* experiment, which used cyanide in a "hellish device" (Schrödinger's words) but even this is not kind enough. Contemporary experiments use atoms instead of cats.

The cat is alive (state 1) when it is sealed in the box. Later, we do not know if the cat is dead or alive. When we open the box, the cat is either alive (state 1) or dead (state 2). Clearly, there must have been a superposition of the two states in the time interval when the cat was sealed in the box. Measurement is required to determine which state the cat is in, and measurement shows that the cat is in one state or the other when we open the box and observe the system.

In the May 24, 1996, issue of *Science*, a group at the National Institute of Standards and Technology (Boulder, Colorado) reported a Schrödinger-cat experiment with a supercooled beryllium atom instead of a cat. The atom was forced into harmonic oscillations as a superposition of two coherent-state wave functions. The two states were separated by 80 nm; the atom appeared to be in two places.

Now, a classical cat is a macroscopic object, composed of many particles, and such a cat can be described quite well by classical mechanics. A single supercooled atom is better described by quantum mechanics, so perhaps the beryllium atom study is a Schrödinger-kitten experiment. Experiments with groups of atoms may eventually clarify the transition between quantum and classical descriptions.

Wave Packet Spreading

The natural spreading of a quantum wave packet in a vacuum can be demonstrated in a simple way by using the superposition of only two waves. Clearly, a single wave will not work, because the momentum is well defined, but the position is completely unknown. To make a packet we need at least two waves of different wavelengths, so we will have a somewhat localized particle. Analytically, the wave packet is usually constructed using a Gaussian distribution, because we can get results in closed form, but a Gaussian requires an infinite number of wave components.

Use your spreadsheet to build up various wave packets and let them propagate via the non-relativistic dispersion relation, Eq. (12-24) or the relativistic dispersion relation, Eq. (12-31). Let all of the components start at $t = 0$. The shorter wavelengths will move faster and the packet will spread.

Worksheet tip: Use the IF function in conjunction with the MAX function to find the location and magnitude of the maximum of the wave packet. The IF function is a logical function analogous to the exclusive-or gate used in digital circuits. Set the MAX function to display the maximum value of the wave packet. Then set the IF function to read out the time value on the time axis, corresponding to the MAX value. Consult your manual or on-screen Help for the syntax of these functions.

What sort of wave packet would not spread? This has been investigated in some detail by many people. A particle experiencing a particular constant linear acceleration will not spread, because the shorter wavelengths start later. Can wave packets become compressed, that is, become more localized? Wave packets can be compressed if they "chirp" properly [S. C. Bloch, *Am. J. Phys.* **61**, 789–798 (1993)]. All of this is analogous to spread-spectrum techniques long used in some types of radar, and now becoming popular in digital cellular telephones and other telecommunications applications. *Solitons* are nonlinear waves that retain their shape, and it appears that soliton-like solutions may exist in Schrödinger's equation.

12.7 Electric Dipole Radiation

A particle in a box is not alone in the universe. It is immersed, as you and I are, in the sea of photons left over from the Big Bang. There is a plethora of stray radiation from natural and human activity on this planet. In our galaxy the largest local source is the Sun, but if we take our box (or our eyes) out at night, the moon, planets, stars and other galaxies provide more than enough photons to perturb the oscillator (or our eyes). You can see the stars at night; there are so many photons that many of them will match the energy jump necessary to be absorbed, or they may cause stimulated emission of radiation from an oscillator.

You can shield a particle in an electrically conducting box, but there will still be photons associated with blackbody (thermal) radiation from the inner surface. Make the box superconducting; there will still be photons. Even if absolute zero could be achieved there would still be the zero-point energy and vacuum fluctuations of the electromagnetic field. As in Chapter 9, oscillators will always be perturbed in some way by coupling (photon exchange) with other oscillators in the field.

Now let's calculate the electric dipole moment (Sections 4.1 and 4.2) for a particle in a box. We can write the dipole moment connecting two states as

$$(qx)_{1,2} = \int_0^L \Psi_1^* \, qx \, \Psi_2 \, dx. \tag{12-61}$$

The first thing we notice about Eq. (12-61) is its symmetry. If the integrand is an odd function, then the integral is zero. The dipole moment qx is an odd function, so the wavefunctions must be of *different symmetry* to make the entire integrand an even function. This simple example is the origin of "selection rules" which tell what transitions are allowed and which ones are forbidden for a particular multipole. As in a dipole antenna, the square of the dipole moment determines the power radiated.

12.8 Dipole Transitions, Particle in a Box

Let's state this fundamental fact again. The dipole moment qx is an odd function, so one wavefunction must be even and one must be odd if the integral in Eq. (12-61) is to be non-zero. In other words, a dipole moment only connects states of different symmetry. Clearly, a dipole transition starting and ending in the same state is forbidden. The system must *change symmetry* during a dipole transition in emission or absorption. The dipole moment is definitely a moment to remember.

- Electric dipole radiation occurs between two states of different symmetry. The charge distribution must change is some non-symmetrical way.

In Table 12-2 we saw that, for the square well, adjacent states have opposite symmetry, so we would expect that dipole radiation would be associated with adjacent energy levels, but not between two levels with one level between them. Since a harmonic oscillator potential function can be regarded as a softened infinite square well, we would expect the same sort of selection rules to apply, even though the wavefunctions for the square well and the harmonic oscillator are different.

The calculation for the particle in a box is easier than the harmonic oscillator. We can use Eq. (12-47) in Eq. (12-56), for a transition between $n = 1$ and $n = 2$,

$$(qx)_{1,2} = \frac{2}{L}\int_0^L \sin\left(\frac{\pi x}{L}\right) qx \sin\left(\frac{2\pi x}{L}\right) dx. \tag{12-62}$$

The dipole moment qx is written in standard form between the wavefunctions, because x is regarded as an operator in this representation, but this is purely formal, because x has no effect on the function that follows it. For exercise, you can do this integral for the infinite square well. Equation (12-62) is called a matrix element; it is labeled by two numbers. The numerical value of the matrix element is the expectation value of the dipole moment connecting the two states and this is a measure of the intensity of the radiation from that transition (or how probable it is).

The *finite* square well brings in some entirely new features, barrier penetration and unbound states. The bound-state wavefunction is not required to go to zero at the boundaries, but we must match the wavefunction and its first derivative at the boundaries, like an electromagnetic transmission line at an impedance mismatch. This leads to a wavefunction that is exponentially decaying in the classically forbidden region, so its probability density decays rapidly with distance, like a microwave in a waveguide operated below cut-off frequency.

This non-classical region for particles is a source of continuing debate and experiment. Tunneling through the non-classical region is generally accepted as the process of alpha-particle emission from the nucleus. Charles H. Townes (Columbia University) received the Nobel Prize in 1964 for his work in quantum electronics, which included constructing oscillators and amplifiers based on the tunneling of the nitrogen nucleus through the plane of the protons in the NH_3 molecule; this is the *maser* (microwave amplification by stimulated emission of radiation). In 1973, Leo Esaki received the Nobel Prize for discovering tunneling in semiconductors, Ivar Giaever received the Prize for discovering tunneling in superconductors, and Brian Josephson (at age 23) received the Prize for the theoretical prediction of a supercurrent through a tunnel barrier. The negative resistance property of tunnel diodes makes them useful in electronic circuits. You can purchase tunnel diodes at your local electronics store, and Josephson junctions have a wide range of applications.

12.10 Harmonic Oscillator

When we insert the one-dimensional simple harmonic oscillator potential function in Schrödinger's time-independent equation, Eq. (12-35), we have,

$$-\frac{\hbar^2}{2m}\frac{d^2\psi}{dx^2} + \frac{1}{2}\kappa x^2\,\psi = E\psi, \tag{12-63}$$

and we can restate this as,

$$\frac{d^2\psi}{dx^2} + \frac{2m}{\hbar^2}(E - 0.5\,m\omega_o^2x^2)\psi = 0, \tag{12-64}$$

or

$$\frac{d^2\psi}{dx^2} + \left(\frac{2mE}{\hbar^2} - \frac{m^2\omega_o^2x^2}{\hbar^2}\right)\psi = 0. \tag{12-65}$$

From our experience with the infinite square well, we know some general properties of the solutions to this equation.

- The lowest energy level will have a maximum at the center.
- Adjacent energy levels will have opposite symmetry.
- Wavefunctions will be orthogonal, and must be normalizable so total probability = 1.
- In contrast with the square well, the wavelength will change with position, because the oscillator potential is proportional to x^2; momentum changes.
- The time-dependent Schrödinger equation should show a moving wavepacket (a superposition of states) bouncing between the walls.

Let's simplify the notation in Eq. (12-65) using normalized dimensionless quantities. For the dimensionless energy we can write

$$\varepsilon = 2E/\hbar\omega_o$$

and for the dimensionless displacement,

$$\chi = (m\omega_o/\hbar)^{1/2}\, x\,.$$

With these, Eq. (12-65) becomes

$$\frac{d^2\psi}{d\chi^2} + (\varepsilon - \chi^2)\psi = 0. \tag{12-66}$$

This type of differential equation with eigenvalues ε was studied by Hermite in 1864 and we will not repeat the somewhat lengthy process of solution. (Also, in 1873 Hermite proved that *e*, the base of natural logarithms, is a transcendental number.) When you try a series solution you find that the series must terminate, so the solutions are polynomials, either even or odd. To terminate the series for a particular solution *n*, you find $\varepsilon_n = 2n + 1$. Consequently, the energy levels are $E_n = (n + 0.5)\hbar\omega_o$.

- Energy levels are evenly spaced. The spacing is $\hbar\omega_o$.
- There is a non-removable zero-point energy, $0.5\hbar\omega_o$.

The solutions involve Gaussians and the Hermite polynomials $H_n(\chi)$, which are related to the Gaussian. Perhaps this is not surprising, considering the universality of the Gaussian and the harmonic oscillator. It turns out that the lowest energy level is a pure Gaussian, the minimum uncertainty state in any uncertainty relation. It's really beautiful how this works out; it must have been exciting for Schrödinger when he discovered this. At the time, he was already a mature, highly experienced physicist, and in his first papers he covered all the bases remarkably well.

Zero-Point Energy

Notice how naturally the zero-point energy $E_o = 0.5\hbar\omega_o$ appears. This is more than a curiosity; it has measurable consequences in subtle effects. Today we think of the vacuum as a polarizable spacetime of virtual particles, not merely an empty volume. The electrostatic polarizability of the fluctuations of virtual particles is measurable in the Lamb shift of the fine structure of hydrogen (1955 Nobel Prize for Willis E. Lamb, Jr.) and in the scattering of photons by photons (Delbrück scattering). The Casimir effect, the mutual attraction of two uncharged conducting plates in vacuum, is accounted for in terms of zero-point energy of virtual particles. The force in the Casimir effect is inversely proportional to the fourth power of the plate separation. It also appears that the vacuum may be unstable with regard to positron emission.

Harmonic Oscillator Wavefunctions

It is not difficult to show that the solutions to Eq. (12-66) comprise a complete orthogonal set as

$$\psi_n(\chi) = C_n H_n(\chi) \exp[-0.5\chi^2], \tag{12-67}$$

where $H_o = 1$. Equation (12-67) is a polynomial in a Gaussian window (recall data windows in Chapter 7). The Hermite polynomials have a recursion relation

$$H_{n+1} = 2\chi H_n - 2nH_{n-1},$$

by which you can generate all others (see Tables 12-3 and 12-4). The normalization factor C_n is given in the worksheet QUANTOSC; in general form C_n is given by

$$C_n = \frac{1}{\pi^{1/4}(2^n n!)^{1/2}}. \tag{12-68}$$

Table 12-3 lists the lowest wavefunctions; the first two are shown in Fig. 12-11. The first three probability densities are shown in Fig. 12-12. Normalization factors and Hermite polynomials for $n = 0$ to $n = 10$ are collected in Table 12-4.

Table 12-3. Normalized Harmonic Oscillator Wavefunctions

n	$H_n(\chi)$	$\psi_n(\chi)$
0	1	$0.751126 \exp[-0.5\chi^2]$
1	2χ	$0.531126\ H_1(\chi) \exp[-0.5\chi^2]$
2	$4\chi^2 - 2$	$0.265563\ H_2(\chi) \exp[-0.5\chi^2]$
3	$8\chi^3 - 12\chi$	$0.108416\ H_3(\chi) \exp[-0.5\chi^2]$

Table 12-4. Normalization Factors and Hermite Polynomials

n	*Normalization*	*Hermite Polynomial*
0	0.751126	1
1	0.531126	2χ
2	0.265563	$4\chi^2 - 2$
3	0.108416	$8\chi^3 - 12\chi$
4	0.038331	$16\chi^4 - 48\chi^2 + 12$
5	0.012121	$32\chi^5 - 160\chi^3 + 120\chi$
6	0.003499	$64\chi^6 - 480\chi^4 + 720\chi^2 - 120$
7	0.000935	$128\chi^7 - 1344\chi^5 + 3360\chi^3 - 1680\chi$

continued next page

Table 12-4 (continued). Normalization Factors and Hermite Polynomials

n	*Normalization*	Hermite Polynomial
8	0.000234	$256\chi^8 - 3584\chi^6 + 13440\chi^4 - 13440\chi^2 + 1680$
9	5.51E–05	$512\chi^9 - 9216\chi^7 + 48384\chi^5 - 80640\chi^3 + 30240\chi$
10	1.23E–05	$1024\chi^{10} - 23040\chi^8 + 161280\chi^6 - 403200\chi^4 + 302400\chi^2 - 30240$

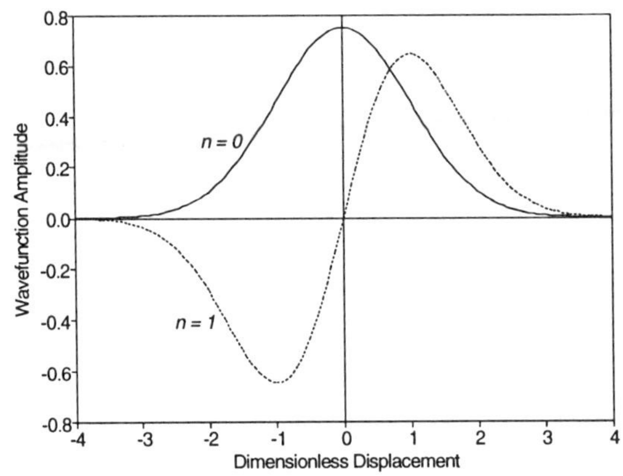

Figure 12-11. The two lowest wavefunctions ψ_n for the harmonic oscillator. Note even and odd symmetry. The Gaussian, $n = 0$, is the minimum uncertainty state.

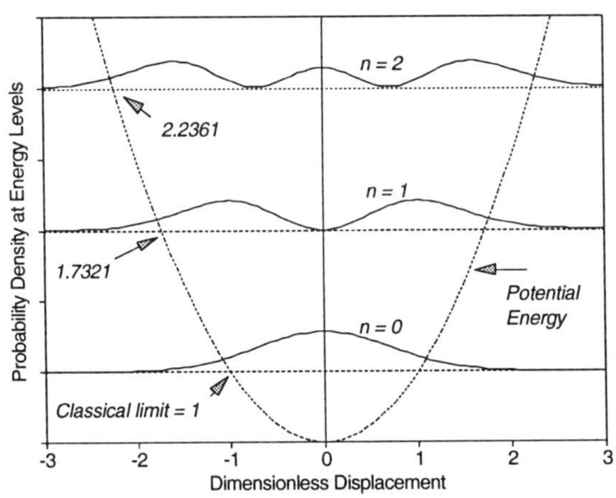

Figure 12-12. Probability densities $|\psi_n|^2$ associated with the three lowest energy levels of the harmonic oscillator. Compare these with the classical harmonic oscillator, Figs. 10-4 and 10-5. Note tunneling into the classically forbidden region (outside of classical turning point). Energy is in units of $0.5\hbar\omega_o$.

Let's examine some of the lowest states in detail. The ground state probability density is shown in Fig. 12-13. Note penetration into the non-classical region, $\chi > 1$.

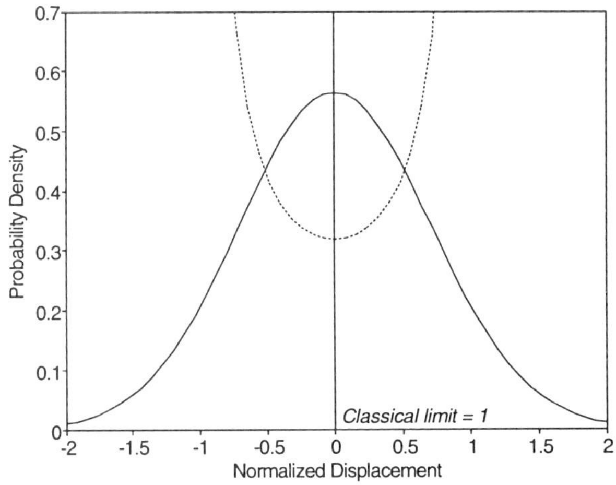

Figure 12-13. *Solid line:* Ground state probability density of the harmonic oscillator. *Dotted line:* Classical probability density at this energy.

Ground state, $n = 0$

The ground state of the harmonic oscillator is not an excited state, but it is an exciting state, in the emotional sense of the word. The ground state, shown in Fig. 12-13, is a pure Gaussian function. The classical probability density for the same energy is also shown in this Figure, but it can be argued that this is not a valid comparison. If you measure the positions of an ensemble of identical classical oscillators at "rest" at a finite temperature, you would get something approximating a Gaussian distribution because of random thermal motions. This is the thermal agitation (Brownian motion) at the rest position. You would *not* find the classical probability density at energy kT which is associated with a *coherent* sinusoidal oscillation at this energy. The non-removable energy $0.5\hbar\omega$ of the ground state has experimentally-observable consequences.

- The ground state has the only wavefunction that does not have a zero-crossing, so it has the only probability density that does not have a zero.
- Every higher energy level is modulated or "windowed" by the Gaussian of the ground state. See Eq. 12-67 and Table 12-3.
- The central limit theorem tells us that the Gaussian is the end result of an infinte number of self-convolutions of almost every realizable probability density (Chapters 7 and 10). Actually, even a rectangular

- Penetration or "tunneling" into the classically-forbidden region beyond the turning point provides a small (but finite) probability of coupling to other oscillators. This is yet another clue that all oscillators are connected to, and eventually influence, all other oscillators.
- The ground state of the harmonic oscillator is the *minimum uncertainty state* in which $\Delta p_x \, \Delta x = 0.5\hbar$ and $\Delta E \, \Delta t = 0.5\hbar$. This is the gound state of the electromagnetic field in vacuum, and probably the ground state of the universe.

Using the ground state wavefunction, verify the $\Delta p_x \, \Delta x$ uncertainty relation.

First excited state, $n = 1$

Attention is often directed to the differences between the probability densities of the states of the quantum oscillator and the classical oscillator. For example, above the ground state, the quantum oscillator probability density has zeros; the classical density has no zeros for any energy state. The quantum oscillator has a larger probability density before the classical turning points; the classical oscillator has maximum probability density at the turning points. The quantum oscillator has a probability density that has a different form for each state; the classical oscillator has the same mathematical form for every energy. The wavefunction for $n = 1$ is shown in Fig. 12-11 and the probability density is shown in Fig. 12-14.

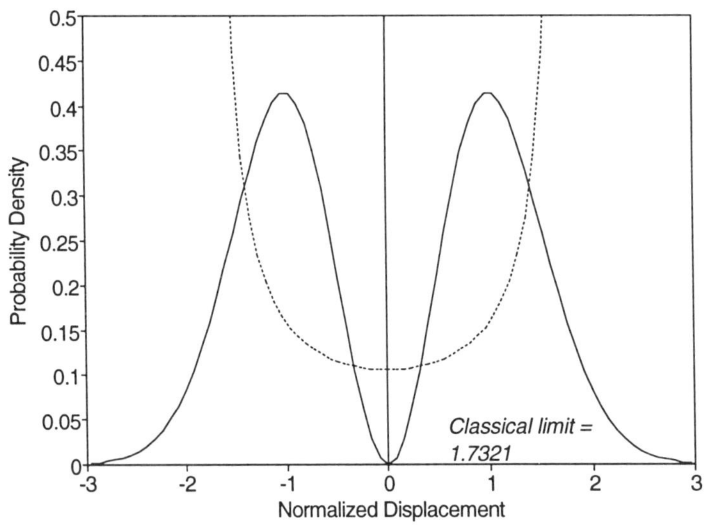

Figure 12-14. *Solid line:* First excited state, $n = 1$, for the quantum harmonic oscillator. *Dotted line:* Classical probability density for this energy. Note the zero probability density at equilibrium displacement, (0).

The similarity of the classical and quantum oscillator is as remarkable as the differences. Even the first excited state has some of the characteristics of the classical oscillator. Both oscillators have smaller probability densities at the equilibrium point than near the turning points. This corresponds to the fact that all oscillators have greater momentum as they pass through the equilibrium point than they have near the turning points. All quantum oscillators, above the ground state, have shorter wavelengths near the equilibrium point than near the turning points, because momentum is greater near the equilibrium point.

Second excited state, $n = 2$

The probability density for this state is shown in Fig. 12-15. The quantum probability density is approximating the classical density more closely.

Tenth excited state, $n = 10$

The probability density for $n = 10$ is shown in Fig. 12-16, The average is rapidly approaching the probability density for the classical harmonic oscillator (Figs. 10-4 and 10-5).

12.11 Harmonic Oscillator Dipole Transitions

By integrating the time-averaged Poynting vector over a sphere of radius r enclosing the dipole we obtain the classical result for the total time-averaged power

$$\langle P \rangle = \frac{8\pi^3 f^4}{3\varepsilon_o c^3} d_{rms}^2 = 3.466 \times 10^{-13} f^4 d_{rms}^2 \text{ watts.} \tag{12-69}$$

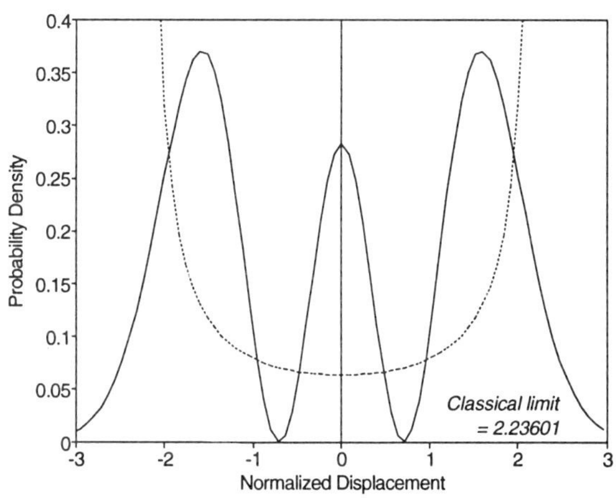

Figure 12-15. Second excited state, $n = 2$, for the quantum harmonic oscillator. Classical probability density for this energy is shown as dotted line.

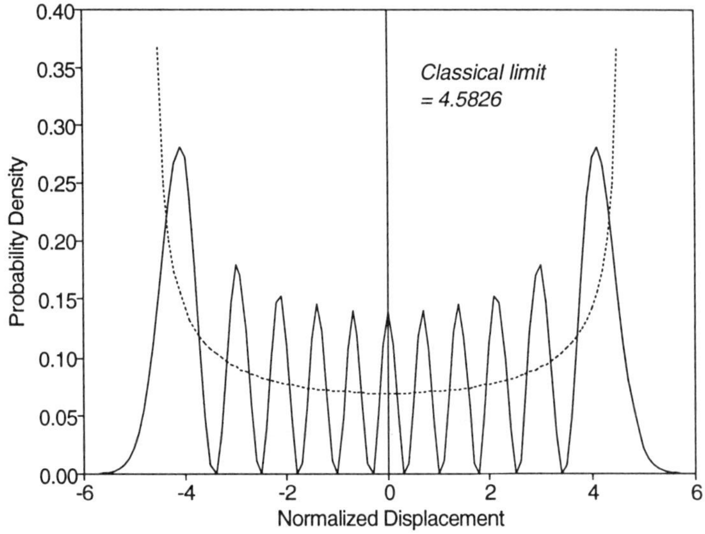

Figure 12-16. *Solid line:* Harmonic oscillator probability density for $n = 10$. *Dotted line:* Classical probabilty density.

In Eq. (12-69), d_{rms} is the r.m.s. value of the classical dipole moment and f is in Hz. We can calculate the quantum analog of the classical dipole moment as,

$$<q\chi>_{0,1} \; = \; C_0 C_1 \int_{-\infty}^{+\infty} H_0(\chi) \exp[-0.5\chi^2] q\chi H_1(\chi) \exp[-0.5\chi^2] \, d\chi \qquad (12\text{-}70)$$

where q is the charge, $H_0(\chi) = 1$ and $H_1(\chi) = 2\chi$. So, Eq. (12-70) simplifies to a standard integral,

$$<q\chi>_{0,1} \; = \; 2qC_0 C_1 \int_0^{+\infty} 2\chi^2 \exp[-\chi^2] \, d\chi. \qquad (12\text{-}71)$$

where the normalization constants are given in Eq. (12-68) and Table 12-3. The factor of 2 before the integral is due to the change in limits from Eq. (12-70) to Eq. (12-71).

12.12 QUANTOSC, Quantum Oscillator Worksheet

This worksheet contains methods of generating Hermite polynomials and normalizing factors, which are then used to compute wavefunctions and probability densities for several of the lowest states. The methods can be used to compute any desired higher state. Figure 12-17 shows the Home screen. Press PgDn for more.

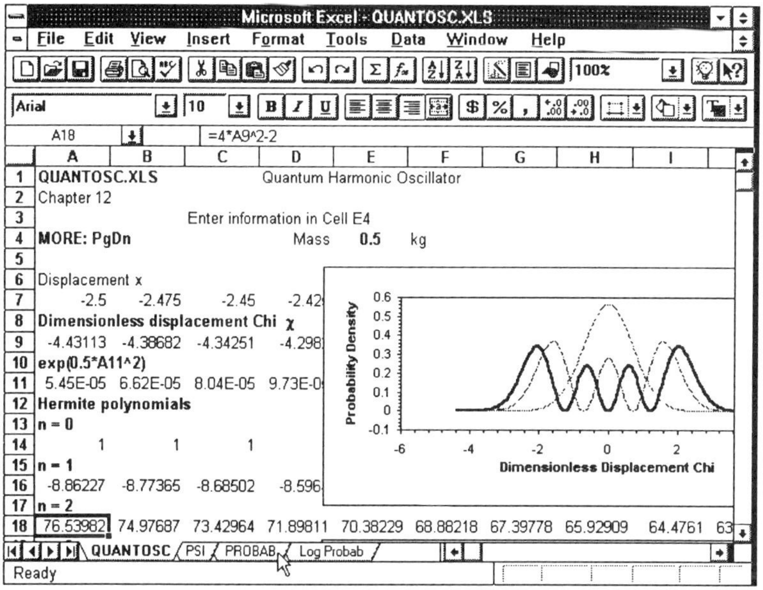

Fig. 12-17. Home screen of the QUANTOSC worksheet, used for computing oscillator wavefunctions and probability densities.

12.13 Revival of Wavefunctions

If Schrödinger's cat were a particle in a box, it would have many more than nine lives, because of the periodic collapse and revival properties associated with this potential. Oscillator wavefunctions do not collapse so they never need to revive. (See Section 10.12).

12.14 Discovery of Classical Oscillators

In the plausibility argument for the existence of Schrödinger's equation (Section 12.2) we started with an algebraic equation for the total energy expressed in a form consistent with the quantization of energy and the wave-like property of matter. Then, following a process that was essentially inverse to the phasor method (Sections 1.4 and 7.7) we transformed an algebraic equation into a partial differential equation.

Suppose that, once upon a time, long ago and far away in one of the spiral arms of a distant galaxy, there lived some people who had discovered quantum mechanics before they knew anything about classical mechanics. This state of affairs came about because they had always studied oscillators at low energy levels. Then, suppose that about 300 orbits after their discovery of quantum mechanics they found some strange data in routine measurements of harmonic oscillators in high energy levels. One of the people found that a simple way of fitting the experimental data was to assume that

$$m\omega_o^2 = \kappa, \text{ a constant.} \tag{12-72}$$

Later, another person conceived the idea that ω_o^2 suggested an eigenvalue equation of the form

$$D\psi = j\omega_o\psi \tag{12-73}$$

where D is the operator d/dt. Then,

$$D^2\psi = -\omega_o^2\psi. \tag{12-74}$$

In view of the assumption based on experimental data, Eq. (12-74) could be written as

$$\frac{d^2\psi}{dt^2} = -\frac{\kappa}{m}\psi \tag{12-75}$$

or,

$$m\frac{d^2\psi}{dt^2} = -\kappa\psi. \tag{12-76}$$

There were great debates over this equation because it had the strange property that a purely real or a purely imaginary ψ could be a solution, as well as a complex ψ.

The meaning of ψ remained a mystery until a young physicist suggested that, although ψ itself was indeed mysterious, it was possible to interpret the function

$$\frac{1}{2}(\psi^* + \psi) \tag{12-77}$$

for $\psi = C\exp(-j\omega_o t)$ and

$$\frac{j}{2}(\psi^* - \psi) \tag{12-78}$$

for $\psi = C\exp(+j\omega_o t)$. The interpretation that Eq. (12-77) and (12-78) represented a distribution of the system displacement in the time domain was reluctantly accepted, and not with universal approval. For compact notation these solutions were written as $x(t)$. It soon became clear that $x(t)$ could provide all information about a macroscopic system. For example, momentum could be obtained by using the operator $m\,d/dt$ on $x(t)$. And everyone lived happily ever after.

12.15 Meanwhile, Back in Heligoland . . .

Werner Heisenberg's formulation actually preceded Schrödinger's equation by several months. I have reversed the historical sequence because Schrödinger's equation is usually easier to understand. Schrödinger's methods may seem more understandable for two reasons. First, people are often more comfortable with differential equations than with matrices. Second, Heisenberg's methods are very abstract, as Schrödinger noted in *Annalen der Physik*, **70** 735 (1926): ". . . I was discouraged, if not repelled, by what appeared to me a rather difficult method of transcendental algebra, defying any visualization (Anscheulicheit)." Heisenberg was not kinder to Schrödinger; see the quote at the beginning of this chapter.

Heisenberg believed that a physical theory should be based only on observables. It is difficult to argue with this belief, although later some people tried to introduce "hidden variables". Ironically, Heisenberg's concrete belief in observables led to a very abstract formalism that was (even more ironically) later proven to be mathematically consistent with Schrödinger's formulation. The two methods, which appeared so different, were merely different ways of doing the same things. It is said that David Hilbert, the great mathematician, had a good laugh because physicists were slow to recognize that the two methods were connected. Heisenberg was awarded the Nobel Prize in 1932 "for the creation of quantum mechanics".

There is a curious parallel between Newton and Heisenberg. Both were young men in their twenties, each man invented his own brand of mathematics because he needed it, and for health reasons both sought a period of isolation during which they produced their masterpieces. Newton went to a farm in the countryside to escape the plague that was ravaging the cities of England and continental Europe; in 1925 Heisenberg went to an island named Heligoland to find relief from a severe case of hay fever. Before we begin, please return your seat to the full upright and locked position, and fasten your seat belt.

Quantum Extension of Classical Formalism

With Max Born, one of Heisenberg's key discoveries in his mathematical development was the non-commutation of canonically conjugate coordinates and momenta,

$$p_r q_r - q_r p_r = -j\hbar. \tag{12-79}$$

This is the quantum analog of classical Poisson brackets (Section 8.5).

In his Nobel Prize lecture of December 11, 1954, Max Born recalled that "It was easy to guess what the remaining quantities must be, namely, zero; and at once there stood before me the peculiar formula, $pq - qp = h/2\pi j$. This meant that coordinates q and momenta p cannot be represented by figure values but by symbols, the product of which depends upon the order of multiplication – they are said to be non-commuting."

Born continued, "I was as excited by this result as a sailor would be who, after a long voyage, sees from afar, the longed-for land, and I felt regret that Heisenberg was not there. I was convinced from the start that we had stumbled on the right path. Even so, a great part was only guess-work, in particular, the disappearance of the non-diagonal elements in the above-mentioned expression."

In 1956 Born wrote, "I shall never forget the thrill I experienced when I succeeded in condensing Heisenberg's ideas on quantum conditions in the mysterious equation $pq - qp = h/2\pi j$."

Olivia Newton-John must have been very proud of her grandfather.

Harmonic Oscillator

The harmonic oscillator has the distinction of being easier to analyze in matrix mechanics than by means of the Schrödinger equation. Let's start with the Hamiltonian operator for the one-dimensional harmonic oscillator,

$$H(p,x) = \frac{1}{2m} p^2 + \frac{1}{2} \kappa x^2. \tag{12-80}$$

Our task is to determine the matrix elements of the matrices **p** and **x** in the H representation in such a way that the commutation relation

$$\mathbf{px} - \mathbf{xp} = -j\hbar \tag{12-81}$$

is valid, and that the Hamiltonian matrix,

$$\mathbf{H} = \frac{1}{2m} \mathbf{p}^2 + \frac{1}{2} \kappa \mathbf{x}^2 \tag{12-82}$$

is diagonal. As before, let E_n be the desired eigenvalues, that is, the diagonal elements H_{nn} of the Hamiltonian matrix. Let's call the lowest energy level E_o, the next energy level E_1, and so on.

Now let's write the matrix equations in terms of the corresponding matrix elements,

$$\sum_{r=0}^{\infty} (p_{ir}x_{rk} - x_{ir}p_{rk}) = -j\hbar\delta_{ik} \tag{12-83}$$

$$E_k\delta_{ik} = \frac{1}{2m} \sum_r p_{ir}p_{rk} + \frac{1}{2} \kappa \sum_r x_{ir}x_{rk}\,. \tag{12-84}$$

In view of the commutation rules we have,

$$\mathbf{pH} - \mathbf{Hp} = -j\hbar\kappa\mathbf{x} \tag{12-85}$$

$$\mathbf{xH} - \mathbf{Hx} = j\frac{\hbar}{m}\mathbf{p}. \tag{12-86}$$

These matrix equations imply the following relations among the matrix elements,

$$\sum_r (p_{ir}H_{rk} - H_{ir}p_{rk}) = -j\hbar\kappa x_{ik}$$

$$\sum_r (x_{ir}H_{rk} - H_{ir}x_{rk}) = j\frac{\hbar}{m}p_{ik} \tag{12-87}$$

or, in other words,

$$p_{ik}(E_k - E_i) = -j\,\hbar\,\kappa\,x_{ik}$$

$$x_{ik}(E_k - E_i) = j\frac{\hbar}{m}p_{ik}. \tag{12-88}$$

When we eliminate p_{ik} we find,

$$x_{ik}\left[1 - \left(\frac{2\pi}{h}\right)^2\frac{m}{\kappa}(E_k - E_i)^2\right] = 0. \tag{12-89}$$

The oscillator frequency is $f_o = (1/2\pi)\sqrt{\kappa/m}$ so Eq. (12-89) becomes simplified to,

$$x_{ik}\left[1 - \left(\frac{E_k - E_i}{hf_o}\right)^2\right] = 0. \tag{12-90}$$

If $E_i = E_k$ then $x_{ik} = 0$, so it is clear that all of the matrix elements x_{ik} are zero, except for the elements with indices i and k so that,

$$E_k - E_i = \pm hf_o. \tag{12-91}$$

So, we are led inevitably to the constant spacing of the energy levels of the harmonic oscillator, but this abstract formalism does not speak of wavefunctions. If we continued, we would soon find Heisenberg's famous uncertainty principles (in momentum and its conjugate coordinate, and energy and time) that incited and excited philosophers at the time they were enunciated. The words relativity and uncertainty warped a generation of philosophers who didn't understand physics, but who thought that physics gave them justification for relativism and uncertainty.

12.16 Single-Electron Oscillator

In 1989, Professor Hans Dehmelt (University of Washington) shared the Nobel Prize with Professors Norman Ramsey (Harvard University) and Wolfgang Paul (University of Bonn). The Royal Swedish Academy presented the prize for "the development of atomic precision spectroscopy" that led to the development of atomic clocks and precision timing. Ramsay developed the separated oscillatory field method used in the hydrogen maser and other atomic clocks, and Paul developed the trap that held the particles that Dehmelt and his colleagues studied.

One of Dehmelt's objectives was to measure the gyromagnetic ratio of an electron and a positron, that is, the ratio of the spin to the cyclotron frequency. In 1987, this was measured to 4 parts in 10^{12}. An important part of the experimental method was "sideband cooling" with an electromagnetic field, because the excitation and observation frequencies are separated. (A similar technique uses a laser to cool neutral atoms.) Single-ion spectroscopy may eventually be the basis for a frequency standard. A charged particle in a Paul trap is a pseudo-atom with well-defined energy levels; Dehmelt calls this pseudo-atom "geonium." Electrons have been confined in Paul traps, and Penning traps, for about a year, and a positron was observed continuously for three months. The methods should enable atomic clocks to be improved by a factor of 100,000. Such clocks may permit tests of the Hubble expansion of the universe, believed to be about 1 part in 10^{18} per second.

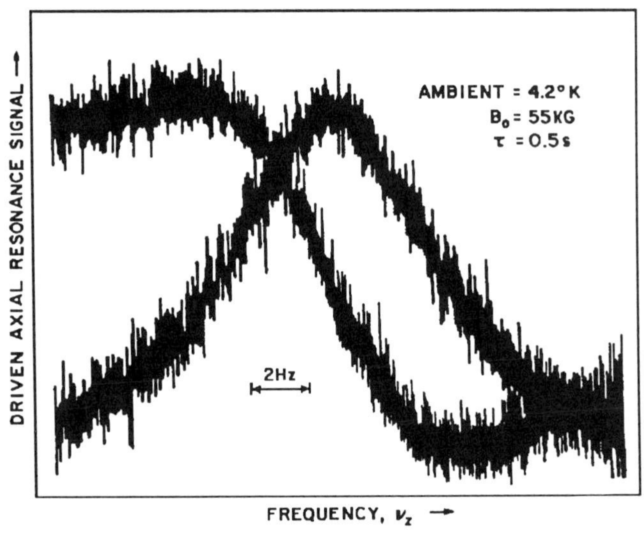

Fig. 12-18. Dispersion (S-shape) and absorption modes of a single electron oscillator. Q is about 10^7. (Courtesy of Prof. Hans G. Dehmelt.)

12.17 Trouble in Paradise

Schrödinger's equation was a remarkable leap forward. It was, and is, extraordinarily successful when applied to atomic, molecular, and condensed matter, but Eq. (12-10) could not be completely correct, because it does not treat space and time on an equal basis, as required by relativity.

Equation (12-10) has a second space derivative and a first time derivative; it looks like a diffusion or heat-transfer equation. From classical mechanics and classical electromagnetics we know that the classical wave equation can be written as

$$\nabla^2\xi - \frac{1}{v^2}\frac{\partial^2\xi}{\partial t^2} = 0, \tag{12-92}$$

where ξ is a physical quantity such as electric field, magnetic field, displacement, pressure, and so on, and v is the speed of the wave described by this wave equation. The electromagnetic wave equation is derivable from Maxwell's equations *without approximation*. Equation (12-92) involves four coordinates, x, y, z, t. We can put all of the coordinates in the same units by changing the time scale $t \rightarrow jvt$. With this change of scale we can write the classical wave equation as

$$\nabla^2\xi + \frac{\partial^2\xi}{\partial(jvt)^2} = 0. \tag{12-93}$$

Equation (12-93) says that the sum of four second partial derivatives is zero, so at least one partial derivative must be of opposite sign from the others. Recall Laplace's equation from Chapter 6. This is the same thing, but now it is in four dimensions. We know that this is relativistically correct, because this wave equation is invariant under the Lorentz transform. For the special case of electromagnetic waves, Maxwell's equations agree with experiments in the classical domain $hf << k_bT$ and they lead to Eq. (12-92) without approximation.

So the question was: Are relativity and quantum mechanics incompatible, or is there something we don't know? Two such magnificent concepts could not be completely separate; shouldn't one imply the existence of the other? Sommerfeld and his colleagues had attempted to apply relativistic modifications to the semiclassical Bohr theory, which they extended to elliptic orbits, but their success was limited by lack of information (de Broglie wavelength).

If you look back at Eq. (12-4) you can see the reason for the first time derivative and the incompatibility of Schrödinger's equation with relativity. Equation (12-4) is not the relativistic energy. The relativistic form of Eq. (12-4) is

$$E = [(cp)^2 + (m_oc^2)^2]^{1/2} + U. \tag{12-94}$$

In the meantime, experimental information had come to light in the form of another quantum number, suggested by the atomic beam experiments of Stern and Gerlach. Otto Stern won the Nobel Prize in 1943 for development of the beam method and his discovery of the magnetic moment of the proton. The new quantum number was called *spin,* because there did not seem to be any other type of motion left for particles, except spin. As early as 1921, Arthur Holly Compton suggested that electrons might be spinning, to account for magnetism. Compton did not pursue this, though, and he won the Nobel Prize in 1927 for the Compton effect, essentially playing billiards with photons and electrons.

The existence of the new quantum number implied that Schrödinger's equation was incomplete. The spin quantum number and its peculiar properties were later incorporated *ad hoc* into Schrödinger's equation but this repair job was not a satisfying solution. Starting with the relativistic energy in Schrödinger's equation resulted in a theoretical description of spin-zero particles, but experiment showed that electrons were spin-½ particles. Once more, it seemed that a new approach and/or more information was needed. It remained for Paul Adrien Maurice Dirac to show the way, and the electron spin (and antimatter) made their debut in a natural four-dimensional formalism. Dirac shared the 1933 Nobel Prize with Erwin Schrödinger. At present we do not think of spin in a literal sense; it is just a name for a quantum number.

It is often said that Dirac's belief that physical laws should have mathematical beauty led him to predict the existence of antimatter. Antimatter was confirmed experimentally in 1932 with the discovery of the first antiparticle, the positron (1936 Nobel Prize for Carl David Anderson). In addition to its fundamental importance for our understanding of the universe, positron emission tomography (PET) is a useful technique in medical diagnosis.

History was made at 6 p.m. on November 13, 1995, when Sir Michael Atiyah, president of the Royal Society, unveiled a plaque in Westminster Abbey to commemorate the scientific achievements of Dirac (1902–1984). The memorial carries Dirac's celebrated equation of 1928, the first mathematical equation in history to be enshrined in the Abbey.

Dirac's memorial is situated next to the tomb of Sir Isaac Newton. The small group of Cambridge scientists honored at the Abbey include George Green (1841), James Clerk Maxwell (1879), and Lord Kelvin (1907).

What's Next?

Walter Huckel composed a little poem to Erwin Schrödinger in the German language. The following is a translation of the poem into the English language by Felix Bloch (Stanford University), who was awarded the Nobel Prize in 1952, not for translating this poem, but for nuclear magnetic resonance (NMR). NMR is useful in physics, chemistry, and biology. Magnetic resonance imaging (MRI) is extraordinarily important in non-invasive medical diagnosis.

Erwin with his Ψ can do
Calculations quite a few.
But one thing has not been seen
Just what does Ψ really mean?

Statistical Interpretation

"Just what does Ψ really mean?" Max Born gave the statistical interpretation that is now generally accepted. Yet the curious person may ask, since Schrödinger's equation appears to be a deterministic prescription for the nonrelativistic spacetime evolution of Ψ, why is Ψ probabilistic?

The foundations of Schrödinger's equation are the experimentally-based and confirmed quantization of energy $E = hf$ and the wave-like nature of momentum $p = h/\lambda$, and these led to the deterministic equation for Ψ. The interpretation of $\Psi^*\Psi$ as a probability density is not as unusual as it appears to be at first glance. In Chapter 6 we obtained the power spectral density $F^*(\omega)F(\omega)$, which can be interpreted as a probability density for power in the frequency domain. The PSD was based entirely on the Fourier transform, which has a classical uncertainty relation for every pair of variables related by the transform. In Chapter 10 we obtained a displacement probability density $P(x)$ for the classical harmonic oscillator, and this was based entirely on Newton's laws, often regarded as deterministic. Probabilities arise naturally in deterministic systems (where they are useful) and random systems (where they are necessary).

Expansion of the Universe

There are several clues that the universe is expanding. These include Olber's paradox, the cosmic red shift, entropy increase, and the cosmic microwave background radiation. Feynman mentioned a clue (page 271) to the connection between the expansion of the universe, adiabatic invariance, and Planck's constant, and this eventually goes back to Galileo's adjustable-length pendulum.

It is interesting to speculate that energy is quantized and $E = hf$ because of the adiabatic invariance of field oscillators, as the universe expands. This would imply that Planck's constant has truly cosmic importance, from the subatomic domain to atomic and molecular structure, to the boundary of spacetime.

Superstrings

According to Michael Green (Cambridge), the most fundamental paradox of twentieth century physics is the apparent conflict between general relativity (gravitation) and quantum mechanics. Green and John Schwarz (California Institute of Technology) originated the theory of superstrings, which appears to remove this paradox and also to unify fundamental particle theory. If this turns out to be a viable theory, it will be quite an accomplishment.

Could spacetime be filled with little harmonic oscillators? Huygens may have thought so in 1678; the Huygens construction is successful in describing electromagnetic wave propagation in terms of every point on a wavefront acting as an oscillator, a source of spherical waves. Green and Schwarz's theory describes the fundamental components of matter (electrons, photons, quarks) as one-dimensional strings which vibrate as a violin string might vibrate at a fundamental frequency and its harmonics. In this theory, which sounds simplistic, the harmonics of vibrating superstrings correspond to the various types of fundamental paarticles.

The strings in this theory are very small. How small? As the nucleus compares with the sun, superstrings compare with the nucleus. Are these superstrings the harmonic oscillators that fill spacetime? Recall that the Planck length is $(Gh/c^3)^{1/2} \approx 10^{-35}$m and the Planck time is $(G/c^5)^{1/2} \approx 10^{-43}$s. At present these dimensions appear to be beyond our limits, but only recently the atomic domain seemed beyond our limits and now we can arrange a group of atoms to spell IBM, make lasers with quantum dots, and trap a single electron to make an oscillator. With present technology, quantum dots are assembled with an uncertainty of about 20 atoms; we must do better in order to sharpen the spectra of these little lasers.

With Paul Townsend (Cambridge) and others, Green is attempting to describe properties of spacetime as well as fundamental particles. Perhaps quantization of spacetime will enable us to derive the values of the fundamental constants (G, h, k_B, c, e, and m) from first principles. Then we will know why the fine-structure constant α is approximately 1/137. There are still many mysteries.

S. G. Djorgovski (California Institute of Technology) said "The universe is very, very strange." Disregarding Gödel's proof, is the universe strange, or is our preconception strange, because of our limited abilities of perception?

How does it come about then, that great scientists such as Einstein, Schrödinger, and De Broglie are nevertheless dissatisfied with the situation? Of course, all these objections are levelled not against the correctness of the formulae, but against their interpretation. Two closely knitted points of view are to be distinguished: the question of determinism and the question of reality.

Max Born

"The statistical interpretation of quantum mechanics"

Nobel Lecture, December 11, 1954,

in *Nobel Lectures, Physics,* Vol. 3, 1942–1962 (Elsevier, Amsterdam 1964)

In quantum mechanics people discovered that they could find no way of imagining the meaning of the theory. This was brought out most clearly and consistently by Niels Bohr. I'm not sure that any other physicist really understands exactly what Bohr meant to say.

David Bohm
in *A Question of Physics: Conversations in Physics and Biology*
(Routledge and Kegan Paul, London, 1979)

Perhaps Bohm had in mind Bohr's statement that "One must always do what one really cannot."

Words of caution:

This double nature of radiation (and of material corpuscles) is a major property of reality, which has been interpreted by quantum-mechanics in an ingenious and amazingly successful fashion. This interpretation, which is looked upon as essentially final by almost all contemporary physicists, appears to me as only a temporary way out.

Albert Einstein
in *Albert Einstein: Philosopher–Scientist*
(Harper, New York, 1951)

. . . The eternal surge
of time and tide rolls on, and bears afar
our bubbles, as the old burst, new emerge. . .

George Gordon, Lord Byron
in *Don Juan*, Canto xv, Stanza 99

Appendix 1

Worksheets

Installing the Files from the Diskette

1. Assuming you are using drive A: as the floppy drive for your diskette, at the A:> prompt type `INSTALL` and press Enter ↵ .
2. You may also type `A:INSTALL` and press Enter ↵ at the C:> prompt.
3. Follow the instructions displayed by the installation program. The default directory path is C:\OSCILLAT, but you may enter a different directory name.

If you intend to use the FFT and IFFT tools available in Microsoft *Excel*, check under the Tools menu for *Data Analysis* to ensure that the *Analysis Tools* are installed in your spreadsheet, and *Fourier Analysis* is present. If these are not installed, run *Excel* Setup and install them. If you have any questions about installing the files from your diskette, call Wiley Technical Support Services at (212) 850-6194. Browse the world wide web site at http://www.wiley.com.

Name	*Chapter*	*Application*
ABC-OF-Q	5	*Q* in time and frequency
COHERENC	7	MSC and SNR functions
CONVOLVE	7	Convolution using FFT, and non-FFT
COUPLOSC	9	Magnetically coupled *LCR* resonators
COUPLOS2	9	Coupled oscillators, time and frequency (FFT)
CRITDAMP	3	Critically-damped classical oscillator
CROSSCOR	7	Cross-correlation using FFT
DAMPMASS	4	Resistive force and external force
DATWINDO	7	Data windows for spectral resolution enhancement
DEBROGLI	12	de Broglie wavelength of a particle

Name	*Chapter*	*Application*
DECONVOL	7	Deconvolution using FFT
FFT-IFFT	7	Tutorial on use of FFT and IFFT
FORCEOSC	4	Forced oscillator
FORCE-RV	3	Mass with velocity-dependent resistive force
FOURIER	6	Synthesis using Fourier series
FREEMASS	4	Free mass with driving force
FREE-OSC	1	Free harmonic oscillator
OVERDAMP	3	Over-damped classical oscillator
PHASE-SP	1	Phase space for undamped oscillator
PHASECOR	10	Phase-space correlation and covariance
PLANCK	11	Planck distribution, Rayleigh-Jeans, more
PROBABIL	10	Oscillator x-distribution; Poisson, Gauss, exp
Q-IN-T-F	5	Q in the time and frequency domains
Q-OF-LCR	5	Q of LCR resonator
QUANTOSC	12	Quantum harmonic oscillator
SYST-I-D	7	System identification using FFT
THERMAL	10	Thermal velocity vs. mass for 2 temperatures
THERMAL2	10	Thermal velocity vs. temperature for 2 masses
UNDAMPED	2	Undamped classical oscillator
UNDRDAMP	3	Under-damped classical oscillator
Y-AND-Z	5	Admittance and impedance

Useful Spreadsheet Functions

In *Excel* on-screen Help, many of the functions in the following list are named *Engineering Functions.* Additional functions for number base conversions are also available. The CONVERT function is especially useful because it relieves the tedium and mistakes of unit conversion. CONVERT contains a vast array of units. You should check it out, and also explore the statistical tools.

AVG
Returns the arithmetic mean of a specified range of values
BESSELI
Returns the modified Bessel function In(x)
BESSELJ
Returns the Bessel function Jn(x)
BESSELK
Returns the modified Bessel function Kn(x)
BESSELY
Returns the Bessel function Yn(x)
COMPLEX
Converts real and imaginary coefficients into a complex number
CONVERT
Converts a number from one measurement system to another
DELTA
Tests whether two values are equal
ERF
Returns the error function
ERFC
Returns the complementary error function
GESTEP
Tests whether a number is greater than a threshold value
IF
Returns a specified value depending on whether a given statement is true or false
IMABS
Returns the absolute value (modulus) of a complex number
IMAGINARY
Returns the imaginary coefficient of a complex number
IMARGUMENT
Returns the argument theta, an angle expressed in radians
IMCONJUGATE
Returns the complex conjugate of a complex number
IMCOS
Returns the cosine of a complex number
IMDIV
Returns the quotient of two complex numbers
IMEXP
Returns the exponential of a complex number
IMLN
Returns the natural logarithm of a complex number
IMLOG10
Returns the base-10 logarithm of a complex number

IMLOG2
Returns the base-2 logarithm of a complex number
IMPOWER
Returns a complex number raised to an integer power
IMPRODUCT
Returns the product of two complex numbers
IMREAL
Returns the real coefficient of a complex number
IMSIN
Returns the sine of a complex number
IMSQRT
Returns the square root of a complex number
IMSUB
Returns the difference of two complex numbers
IMSUM
Returns the sum of complex numbers
MAX
Returns the maximum value of a specified range of values
MIN
Returns the minimum value of a specified range of values
SQRTPI
Returns the square root of (number * PI)
STD
Returns the population standard deviation of a specified range of values
STDS
Returns the sample standard deviation of a specified range of values
SUM
Returns the algebraic sum of a range of real values
VAR
Returns the population variance of a specified range of values
VARS
Returns the sample variance of a specified range of values

Caution: The FFT and IFFT operations return complex values that cannot be graphed directly. Use IMREAL, IMAGINARY, IMABS, IMARGUMENT, and so on, to separate the part that you want to graph. See examples of these operations in the worksheets. FFT and IFFT do not recalculate automatically. You must invoke these operations manually, or write a macro to automate them.

Set the recalculation mode in your spreadsheet to your requirements. Remember, if you set this mode to Manual, you must recalculate by pressing the appropriate Function Key, usually F9.

Appendix 2

The Hilbert Transform

We mentioned repeatedly that causality has important consequences not only in our treatment of data in the time domain (where it is obvious) but also in the frequency domain (where it is less obvious).

The Hilbert transform, a manifestation of causality in the frequency domain, relates the real and imaginary parts of an analytic function. One of the practical applications of this transform is relating the real and imaginary parts of a frequency-dependent linear response function, such as the system function $H(\omega)$ or the complex permittivity $\varepsilon(\omega)$. It is also useful in digital signal processing.

The universality of this transform leads to its rediscovery in different contexts. It is sometimes known as the Kramers-Kronig relations in electromagnetic scattering (especially with x-rays), the Bode gain-phase relations or the Wiener-Lee transform in electronics, or the dispersion relations in fundamental particle theory.

Although the Hilbert transform may appear complicated, its meaning is as simple as the relation between the refractive and absorptive properties of a lens, or reactance and resistance. It may seem curious that causality is manifested in such mundane properties that do not explicitly involve time, but that's part of the mystique of Nature and this transform.

The Hilbert transform is useful because it can save a lot of time and effort (see Ref. 10 in this Appendix). Since the transform relates the real and imaginary parts of the response function, it is not necessary to measure both parts. Measuring both is redundant, so you can measure whichever is easier. Knowing one, you can in principle compute the other (but you have to be very careful; see Refs. 3 and 16).

Derivations of the Hilbert transform, named for the great mathematician David Hilbert (Universities of Königsberg and Göttingen), often involve analytic continuation of $H(\omega)$ to complex frequencies, and utilization of the analyticity in the upper half plane. In this Appendix we will show a quick and easy derivation using one of our favorite test signals, a unit step function. Ernst Guillemin (M.I.T.) may have been the first to do it this way.

A2.1 Derivation via Unit Step

Let $u(t)$ be the unit step or Heaviside function with the usual properties,

$$u(t) = 0, \quad t < 0 \quad \text{and} \quad u(t) = 1, \quad t > 0.$$

We will use the Fourier transform of this step function together with the convolution theorem to derive the Hilbert transform. This is particularly clear not only because it is short and sweet but also because it shows that the Hilbert transform is simply a convolution of the parts of a linear response function with $1/\omega$.

The Fourier transform of the step function is given by,

$$F(\omega) = \int_{-\infty}^{+\infty} u(t)\, e^{-j\omega t}\, dt = \int_{0}^{+\infty} e^{-j\omega t}\, dt\,.$$

This famous integral that perturbed mathematicians has the well-known result,

$$F(\omega) = \pi\delta(\omega) - P\left(\frac{j}{\omega}\right) \tag{A2-1}$$

where P indicates the Cauchy principal value and $\delta(\omega)$ is the Dirac delta function. The convolution theorem for the Fourier transform tells us that,

$$\int_{-\infty}^{+\infty} f(t)\, g(t)\, e^{-j\omega t}\, dt = \frac{1}{2\pi}\int_{-\infty}^{+\infty} F(\omega-\omega')\, G(\omega')\, d\omega'. \tag{A2-2}$$

These are definite integrals so the variable of integration ω' is just a dummy variable. As in Chapter 7, let's discuss a linear impulse response $h(t)$ and its associated system function $H(\omega) = \mathrm{Re}H(\omega) + j\mathrm{Im}H(\omega)$. Causality requires that $h(t)$ has the form,

$$h(t) = u(t)y(t), \text{ where } y(t) = h(t) \text{ for } t > 0.$$

- Since there is no condition on $y(t)$ for $t < 0$ *we are free to choose this part*, and we can choose it to be even or odd.

Taking the Fourier transform of $h(t)$, and using Eq. (A2-1) and (A2-2) we have,

$$H(\omega) = \frac{1}{2\pi} P \int_{-\infty}^{+\infty} \frac{jY(\omega')}{\omega-\omega'}\, d\omega' + \frac{1}{2}\, Y(\omega). \tag{A2-3}$$

The second term in Eq. (A2-3) comes from the delta function and the first term comes from j/ω. We now exercise our freedom to choose $y(t)$ for $t < 0$.

- First, let's choose $y(-|t|) = y(|t|)$, an even function.

This makes $Y(\omega)$ purely real, so $Y(\omega) = 2\,\mathrm{Re}\,H(\omega)$. Substitution of this into the integrand of Eq. (A2-3) yields,

$$\text{Im}\, H(\omega) \;=\; \frac{-1}{\pi}\, P \int_{-\infty}^{+\infty} \frac{\text{Re}H(\omega')}{\omega - \omega'}\, d\omega' \,. \tag{A2-4}$$

- Next, let's choose $y(-|\,t\,|) \;=\; -y(|\,t\,|)$, an odd function.

This makes $Y(\omega)$ purely imaginary, so $jY(\omega) = -2\,\text{Im}\,H(\omega)$. Substituting this into the integrand of Eq. (A2-3) gives the result,

$$\text{Re}\, H(\omega) \;=\; \frac{1}{\pi}\, P \int_{-\infty}^{+\infty} \frac{\text{Im}H(\omega')}{\omega - \omega'}\, d\omega' . \tag{A2-5}$$

Equations (A2-4) and (A2-5) comprise the Hilbert transform pair, which establishes a relation between the real and imaginary parts of $H(\omega)$.

A2.2 Constructing an Analytic Function

Analytic functions play many important roles in science and technology. One of their most significant attributes is that their spectra contain only positive frequencies. To construct an analytic function from a real function use Eq. A2-4. If the integral exists you will obtain the imaginary part which, when joined with the real function, will form an analytic function.

Conversely, if you have an imaginary function use Eq. A2-5 to find the real part of the associated analytic function (if the integral exists).

A2.3 Fourier Transform of Unit Step

We placed a great burden on the Fourier transform of $u(t)$ in the derivation of the Hilbert transform, so we should examine it carefully. It is a bit tricky because $u(t)$ is a member of the class of *generalized functions*, which includes the delta function introduced in quantum theory by Dirac in 1927. (The Dirac delta function is certainly not a function in the Dirichlet sense, but the terminology persists.)

Although some of these "functions" were invented because they were needed, there was always a suspicious and somewhat repulsive mathematical odor about them until around 1950, when Laurent Schwartz (École Polytechnique) showed, by means of distribution theory, that they had validity. Of course, the delta function only has meaning inside an integral. Dirac, who used as a guide what he perceived to be mathematical beauty, knew it must be true and Schwartz proved it. Schwartz held the same chair in mathematics that Fourier held a century before.

First, let's see how a seemingly safe and standard method for normal functions gives wrong results in this case. The Fourier transform of the unit step is

$$F(\omega) \;=\; \int_{-\infty}^{+\infty} u(t)\, e^{-j\omega t}\, dt \;=\; \int_{0}^{+\infty} e^{-j\omega t}\, dt \,. \tag{A2-6}$$

so,

$$F(\omega) = \frac{e^{-j\omega t}}{-j\omega}\Big|_0^{\infty} \tag{A2-7}$$

The lower limit is no problem, but there is trouble at the upper limit. Let's go back and try the usual epsilonic approach, adding a small real part that we will let go to zero. Poisson and Lord Rayleigh used this in difficult situations. We can rewrite Eq. (A2-6) as

$$F(\omega) = \int_0^{+\infty} e^{-j\omega t - \varepsilon t}\, dt\,. \tag{A2-8}$$

so,

$$F(\omega) = \frac{e^{-j\omega t - \varepsilon t}}{-j\omega - \varepsilon}\Big|_0^{\infty} = \frac{0 - 1}{-j\omega - \varepsilon} = \frac{1}{j\omega + \varepsilon}\,. \tag{A2-9}$$

As ε goes to zero, Eq. (A2-9) goes to

$$?\quad F(\omega) = \frac{1}{j\omega}\quad ? \qquad \text{(A2-10).}$$

but this cannot be correct, because it is purely imaginary and $u(t)$ is purely real and uncentered (asymmetric) in time. Compare with our known results for a real exponential decay, a purely real, uncentered function, with a transform that contains both real and imaginary parts [see Eq. (6-63)]. For a real exponential decay, the real part of the transform is an even function and the imaginary part is an odd function. The real part of Eq. (A2-10) is missing. The inverse Fourier transform of Eq. (A2-10) should bring back the unit step, but it doesn't.

Let's try another approach, because the Fourier transform should be a complex function with an inverse transform that has a value of zero for $t < 0$ and a magnitude of 1 for $t > 0$. Our past experience will guide us. The unit step has a "dc" component so this should give a real, even part that is proportional to a delta function at zero frequency. Numerical computation with the FFT can serve as another guide. Try the FFT on a unit step in your spreadsheet.

We are going to use a trick. The signum function sgn(t) has a value of -1 for $t < 0$ and a value of $+1$ for $t > 0$. We can make a unit step from a signum by adding the value 0.5 to half of a signum, $u(t) = 0.5\,\mathrm{sgn}(t) + 0.5$. The advantage of this is that the signum is an odd function so its Fourier transform is odd and imaginary. The additive constant 0.5 will give the spike at zero frequency, and the delta function is real and even. This separates the unit step and its transform into their odd and even parts. We know that a delta function has a uniform spectrum with an amplitude of unity, so the constant 0.5 must be the inverse transform of a delta function times π,

$$0.5 = \frac{1}{2\pi}\int_{-\infty}^{+\infty} \pi\delta(\omega)\, e^{+j\omega t}\, d\omega = \frac{1}{2\pi} \times \pi$$

because the integral has a value of π at $\omega = 0$, by the properties of the delta function. Finally, if we join the real and imaginary parts we have the spectrum of the unit step,

$$F(\omega) = \pi\delta(\omega) - P\left(\frac{j}{\omega}\right).$$

This is Eq. (A2-1).

A2.4 Application of the Hilbert Transform

Let's see how to use the Hilbert transform to compute the imaginary part of a system function or spectrum, knowing the real part. The same sort of procedure applies to the inverse transform. We will do this slowly and carefully so the complex operations will be clear.

Suppose we know the real part of the Fourier transform of a causal function, the real exponential decay,

$$f(t) = e^{-Kt} \text{ for } t > 0 \Leftrightarrow \operatorname{Re} F(\omega) = \frac{K}{\omega^2 + K^2}. \tag{A2-11}$$

Here $f(t)$ could be a causal impulse response $h(t)$ and $F(\omega)$ would then correspond to its system function $H(\omega)$. This is applicable to the damped harmonic oscillator by means of the complex modulation or frequency-shift property of the Fourier transform, Table 6-3. Inserting Re $F(\omega)$ from Eq. (A2-11) in Eq. (A2-4) gives

$$\operatorname{Im} F(\omega) = \frac{-1}{\pi} P\int_{-\infty}^{+\infty} \frac{K}{(\omega'^2 + K^2)} \frac{1}{(\omega - \omega')}\, d\omega'. \tag{A2-12}$$

This convolution in the frequency domain clearly has a pole at $\omega = \omega'$ so we have to regard ω as a complex variable and perform a contour integration in the complex plane, using Cauchy's residue theorem. However, there are two more poles lurking in this integral, and they can be seen by factoring the first denominator,

$$\operatorname{Im} F(\omega) = \frac{-1}{\pi} P\int_{-\infty}^{+\infty} \frac{K}{(\omega' + jK)(\omega' - jK)} \frac{1}{(\omega - \omega')}\, d\omega'. \tag{A2-13}$$

We are free to choose the path of integration, as long as we don't go through any poles. Let's choose a Bromwich contour along the real axis and close it with an infinite semi-circle in the upper half-plane. As we integrate along the real axis from negative infinity there is no problem until we come to the pole at $\omega = \omega'$. Let's go around this pole clockwise with an infinitesimal semi-circle and continue on to positive infinity. Then we loop back counterclockwise to negative infinity along the Bromwich contour, enclosing the pole at $\omega' = jK$ in the upper half-plane. We can ignore the pole in the lower half-plane, because it is outside of our integration area.

By the residue theorem, the pole on the real axis contributes $-\pi j$ times the residue at this pole because it is a semi-circle and we traversed it in the negative (clockwise) direction. The pole in the upper half-plane contributes $+2\pi j$ times the residue at this pole, because it is fully enclosed by a counterclockwise path. Now we can collect the contributions of the residues to this integral,

$$\text{Im}\, F(\omega) = -\frac{K}{\pi}\left[-\pi j\frac{1}{\omega^2+K^2} + 2\pi j\frac{1}{(jK+jK)}\frac{1}{(\omega-jK)}\right] \tag{A2-14}$$

Let's simplify the algebra in Eq. (A2-14),

$$\text{Im}\, F(\omega) = -K\left[-j\frac{1}{\omega^2+K^2} + 2j\frac{1}{(2jK)}\frac{1}{(\omega-jK)}\right] \tag{A2-15}$$

so

$$\text{Im}\, F(\omega) = j\frac{K}{\omega^2+K^2} - 2j\frac{K}{(2jK)}\frac{1}{(\omega-jK)}\ . \tag{A2-16}$$

Finally, we have

$$\text{Im}\, F(\omega) = j\frac{K}{\omega^2+K^2} - \frac{1}{\omega-jK} = j\frac{K}{\omega^2+K^2} - \frac{\omega+jK}{\omega^2+K^2} \tag{A2-17}$$

and the imaginary parts cancel giving

$$\text{Im}\, F(\omega) = -\frac{\omega}{\omega^2+K^2} \tag{A2-18}$$

in agreement with Eq. (6-63). This is remarkable. It seems almost magical that the entire contribution from the pole on the real axis cancelled the imaginary part of the contribution from the pole on the imaginary axis, to produce the expected result.

- The Hilbert transform and the Fourier transform have the global property that *all points* inside the integral contribute to *each point* in the transformed domain.

In other words, every single data point in one domain depends on *all* data points in the other domain. These integral transforms are not point-to-point transforms like cartesian to polar representation. This is why truncation of data can have serious consequences.

Hilbert and Vitamin B12

When David Hilbert became ill with a liver disease, his life was saved by a new medicine sent from the United States. This liver preparation led to the discovery of vitamin B12.

A2.6 References for the Hilbert Transform

Engineering and scientific databases show that a hundred or so works involving the Hilbert transform are published every year. Most are applications. The following selection provides the fundamentals to help you apply this transform.

1. B. D. O. Anderson and M. Green, "Hilbert transform and gain/phase error bounds for rational functions," *IEEE Trans. on Circuits and Systems* **35**, No. 5, 528–535 (1988).
2. R. N. Bracewell, *The Fourier Transform and Its Applications* (McGraw-Hill, New York, 1986) pp. 267–272
3. R. N. Bracewell, "Numerical transforms," *Science* **248**, 697-704 (1990).
4. D. C. Champeney, *Fourier Transforms and Their Physical Applications* (Academic Press, New York, 1973) pp. 244–246.
5. B. Gold, A. V. Oppenheim, and C. M. Rader, "Theory and implementation of the discrete Hilbert transform," in *Digital Signal Processing*, L. R. Rabiner and C. M. Rader, Eds., (IEEE Press, New York, 1972) pp. 94–109.
6. E. A. Guillemin, *Theory of Linear Physical Systems* (John Wiley, New York, 1963) Chap. 18.
7. B. Y.-K. Hu, "Kramers-Kronig in Two Lines," *American Journal of Physics* **57**, 821 (1989).
8. J. D. Jackson, *Classical Electrodynamics*, Second Edition (John Wiley, New York, 1975), pp. 306–412.
9. Y. W. Lee, *Statistical Theory of Communication* (John Wiley, New York, 1960) pp. 491–500.
10. J. M. Maisel, "Hilbert transform works with Fourier transforms to dramatically lower sampling rates," *Personal Engineering & Instrumentation News* **7**, No. 2, 83–86, February (1990).
11. P. M. Morse and H. Feshbach, *Methods of Theoretical Physics* (McGraw-Hill, New York, 1953) pp. 372, 944.
12. A. V. Oppenheim and R. W. Schafer, *Digital Signal Processing* (Prentice Hall, Englewood Cliffs, NJ, 1975) Chap. 7.
13. W. K. H. Panofsky and M. Phillips, *Classical Electricity and Magnetism*, Second Edition (Addison-Wesley, Reading, MA, 1962) pp. 416–419.
14. A. Papoulis, *The Fourier Integral and Its Applications* (McGraw-Hill, New York, 1962) Chap. 10.

15. A. B. Pippard, *The Physics of Vibration* (Cambridge University Press, Cambridge, 1989) pp. 107–113.

16. J. R. Reitz, F. J. Milford, and R. W. Christy, *Foundations of Electromagnetic Theory*, Fourth Edition (Addison-Wesley, Reading, MA, 1993) pp. 484–494, 613–618.

17. H. S. Robertson, *Statistical Thermophysics* (Prentice Hall, Englewood Cliffs, NJ, 1993) pp. 509–513.

18. H. E. Rowe, *Signals and Noise in Communication Systems* (D. Van Nostrand Co., Princeton, NJ, 1965) pp. 13–20, 47–49, 59–60, 98–99.

19. E. B. Saff and A. D. Snider, *Fundamentals of Complex Analysis for Mathematics, Science, and Engineering*, Second Edition (Prentice Hall, Englewood Cliffs, NJ, 1993) pp. 415–425,

20. M. Sharnoff, "Validity conditions for the Kramers-Kronig relations," *American Journal of Physics* **32**, 40–44 (1964).

21. A. Yariv, *Quantum Electronics*, Third Edition (John Wiley, New York, 1989) pp. 651–652.

We must know and we shall know.

David Hilbert (as quoted by Jean Dieudonnè)
in *Great Currents of Mathematical Thought,* Vol. 1
(Dover Publications, New York, 1971)
a translation of *Les Grandes Courants de la Pensèe Mathèmatique*
(Librairie Scientifique et Technique, Paris, 1962)

Bibliography

Books

Introduction to Classical Mechanics, A. P. Arya (Allyn and Bacon, Needham Heights, Massachusetts, 1990).

Systems with Small Dissipation, V. B. Braginsky, V. P. Mitrofanov, and V. I. Panov (University of Chicago Press, Chicago, 1985).

Quantum Measurement, V. B. Braginsky and F. Y. Khalili, K. S. Thorne, Ed. (Cambridge University Press, New York, 1992).

Classical Mechanics, Second Ed., H. C. Corben and P. Stehle (John Wiley, New York, 1960).

Classical Mechanics, T. L. Chow (John Wiley, New York, 1995).

Classical Mechanics, Volumes 1 and 2, E. A. Desloge (John Wiley, New York, 1982).

Statistical Physics, E. A. Desloge (Holt, Rinehart and Winston, New York, 1966).

An Introduction to the Physics of Vibrations and Waves, N. Feather (Aldine Publishing Co., Chicago, 1961).

Analytical Mechanics, Fifth Ed., G. R. Fowles and G. L. Cassidy (Harcourt Brace, New York, 1993).

Vibrations and Waves, A. P. French (W. W. Norton, New York, 1971).

Classical Mechanics, Second Ed., H. Goldstein (Addison-Wesley, Reading, Massachusetts, 1980).

Theory of Linear Physical Systems, E. A. Guillemin (John Wiley, New York, 1963).

Phase-Space Dynamics of Particles, A. J. Lichtenberg (John Wiley, New York, 1969).

Quantum Statistical Properties of Radiation, W. H. Louisell (John Wiley, New York, 1973).

Classical Dynamics of Particles and Systems, Third Ed., J. B. Marion and S. T. Thornton (Harcourt Brace Jovanovich, Orlando, Florida, 1988).

Theory of Vibrations, N. W. McLachlan (Dover Publications, New York, 1951).

Methods of Theoretical Physics, P. M. Morse and H. Feshbach (McGraw-Hill, New York, 1953).

The Harmonic Oscillator in Modern Physics: From Atoms to Quarks, Second Edition, M. Moshinsky and Y. F. Smirnov (Gordon and Breach, Buffalo, New York, 1996).

Classical Dynamics and Its Quantum Analogues, Second Edition, D. Park (Springer-Verlag, Berlin, 1990).

The Physics of Vibration, A. B. Pippard (Cambridge University Press, Cambridge, 1989).

Statistical Thermophysics, H. S. Robertson (Prentice-Hall, Englewood Cliffs, New Jersey, 1993).

Mechanics, Second Ed., K. R. Symon (Addison-Wesley, Reading, Massachusetts, 1960).

"Inertial reference frames," E. F. Taylor and J. A. Wheeler, in *A Universe of Physics*, J. B. Marion, Ed. (John Wiley, New York, 1970).

Waves and Oscillations, R. A. Waldron (D. Van Nostrand, New York, 1964).

Software

Classical Mechanics Simulations, B. Hawkins and R. S. Jones (John Wiley, New York, 1995).

Electricity and Magnetism Simulations, R. Ehrlich, J. Tuszynski, L. Roelofs, and R. Stoner (John Wiley, New York, 1995).

Quantum Mechanics Simulations, J. R. Hiller, I. D. Johnston, and D. F. Styer (John Wiley, New York, 1995).

Modern Physics Simulations, D. Brandt, J. R. Hiller, and M. J. Moloney (John Wiley, New York, 1995).

Selected Papers from American Journal of Physics

"Ehrenfest's theorem and the particle-in-a-box," D. S. Rikhsar, *Am. J. Phys.*, Vol. 64, No. 11, November 1996, pages 1416–1418.

"Why are the energy levels of the quantum harmonic oscillator equally spaced?," A. Royer, *Am. J. Phys.*, Vol. 64, No. 11, November 1996, pages 1393–1399.

"Ripples in the energy of a damped oscillator: The experimental point of view," L. Basano, P. Ottonello, and V. Palestini, *Am. J. Phys.*, Vol. 64, No. 10, October 1996, pages 1326–1329.

"The evolution and revival structure of localized quantum wave packets," R. Bluhm, V. A. Kostelecky, and J. A. Porter, *Am. J. Phys.*, Vol. 64, No. 7. July 1996, pages 944-953.

"Nonlinear coupled oscillators and Fourier transforms: An advanced undergraduate laboratory," P. A. DeYoung, D. LaPointe, J. Levy, and W. Lorenz, *Am. J. Phys.*, Vol. 64, No. 7. July 1996, pages 898–902.

"On the electromagnetically damped mechanical harmonic oscillator," L. McCarthy, *Am. J. Phys.*, Vol. 64, No. 7, July 1996, pages 885–891.

"Nonspreading wave packets," C. W. Wong, *Am. J. Phys.*, Vol. 64, No. 6, June 1996, pages 792–799.

"Coupled oscillators: A laboratory experiment," G. Hansen, O. Harang, and R. J. Armstrong, *Am. J. Phys.*, Vol. 64, No. 5. May 1996, pages 656–660.

"Visualizing the solutions for the circular infinite well in quantum and classical mechanics," R. W. Robinett, *Am. J. Phys.*, Vol. 64, No. 4.April 1996, pages 440–446.

"The libration limits of the elastic pendulum," D. M. Davidovic, B. A. Anicin, and V. M. Babovic, *Am. J. Phys.*, Vol. 64, No. 3, March 1996, pages 338–342.

"Solving one-dimensional Schrödinger-like equations using a numerical matrix method," J. F. Van der Maelen Uría, S. García-Granda, and A. Menéndez-Velázquez, *Am. J. Phys.*, Vol. 64, No. 3, March 1996, pages 327–332.

"Photon states made easy: A computational approach to quantum radiation theory," I. D. Johnston, *Am. J. Phys.*, Vol. 64, No. 3, March 1996, pages 245–255.

"Resource Letter CD-1: Causality and determinism in physics," R. Jones, *Am. J. Phys.*, Vol. 63, No. 3, March 1996, pages 208–215.

"Smooth transitions between the three damping cases for the harmonic oscillator," F. J. McCormack, *Am. J. Phys.*, Vol. 63, No. 12, December 1995, page 1151.

"Sudden expansion or squeezing of a harmonic oscillator," C. Aslangul, *Am. J. Phys.*, Vol 63, No. 11, November 1995, pages 1021–1025.

"Quantum and classical probability distributions for position and momentum," R. W. Robinett, *Am. J. Phys.*, Vol. 63, No. 9, September 1995, pages 823–832.

"The two-dimensional harmonic oscillator at finite temperature and nonzero chemical potential," H. Haugerud and F. Ravndal, *Am. J. Phys.*, Vol. 63, No. 9, September 1995, pages 839–844.

"Asymptotic form of the penetration probability of the quantum harmonic oscillator into the classically forbidden region," D. E. Atems and J. M. Wadhedra, *Am. J. Phys.*, Vol. 63, No. 5, May 1995, pages 443–446.

"Time-dependent perturbation and exact results for the periodically driven quantum harmonic oscillator," R. Akridge, *Am. J. Phys.*, Vol. 63, No. 2, February 1995, pages 141–147.

"Hellmann-Feynman theorem in a linear superposition of energy eigenstates," S. Balasubramanian, *Am. J. Phys.*, Vol. 62, No. 12, December 1994, pages 1116–1117.

"Quantum spectrum of a chain of oscillators," D. H. Zanette, *Am. J. Phys.*, Vol. 62, No. 5, May 1994, pages 404–407.

"Ripples in the energy of a damped harmonic oscillator," E. A. Karlow, *Am. J. Phys.*, Vol. 62, No. 7, July 1994, pages 634–636.

"Simple mechanical forced damped oscillator with electronic output," G. I. Moore, *Am. J. Phys.*, Vol. 62, No. 2, February 1994, pages 140–143.

"Soliton-like solutions of the Schrödinger equation for simple harmonic oscillator," C. C. Yan, *Am. J. Phys.*, Vol. 62, No. 2, February 1994, pages 147–151.

"Experimental determination of normal frequencies in coupled mechanical oscillator systems using fast Fourier transforms: An advanced undergraduate laboratory," B. J. Weigman and H. F. Perry, *Am. J. Phys.*, Vol. 61, No. 11, November 1993, pages 1022–1027.

"Adiabatic invariance derived from the Rund-Trautman identity and Noether's theorem," D. E. Neuenschwander and S. R. Starkey, *Am. J. Phys.*, Vol. 61, No. 11, November 1993, pages 1008–1013.

"Kepler's third law and the oscillator's isochronism," V. M. Gorringe and P. G. L. Leach, *Am. J. Phys.*, Vol. 61, No. 11, November 1993, pages 991–995.

"The envelopes of some families of fixed-energy trajectories," A. P. French, *Am. J. Phys.*, Vol. 61, No. 9, September 1993, pages 805–811.

"An alternative approach for the propagator of a charged harmonic oscillator in a magnetic field," C. Farina, M. Maneschy, and C. Neves, *Am. J. Phys.*, Vol. 61, No. 7, July 1993, pages 636–641.

"Heisenberg picture in quantum mechanics and linear evolutionary systems," P. Nardone, *Am. J. Phys.*, Vol. 61, No. 3, March 1993 , pages 232–237.

"Time evolution of minimum uncertainty states of a harmonic oscillator," H. A. Gersch, *Am. J. Phys.*, Vol. 60, No. 11, November 1992, pages 1024–1030.

"Classically forbidden behavior of the quantum harmonic oscillator for large quantum numbers," J. J. Diamond, *Am. J. Phys.*, Vol. 60, No. 10, October 1992, pages 912–916.

"Experiments utilizing an ultrasonic range finder," I. R. Gatland, R. Kahlscheuer, and H. Menkara. *Am. J. Phys.*, Vol. 60, No. 5, May 1992, pages 451–454.

"Time dependence in quantum mechanics – Study of a simply decaying system," D. S. Onley and A. Kumar, *Am. J. Phys,*. Vol. 60, 1992 , pages 432–439.

"Molecular-dynamics simulator for optimal control of molecular motion," M. Husman, C. Schwieters, M. Littman, and H. Rabitz, *Am. J. Phys.*, Vol. 59, No. 11, pages 1012–1017.

"Finite eigenfunctions in the WKB approximation," U. Sukhatme, A. Pagnamenta, *Am. J. Phys.*, Vol. 59, No. 10, October 1991, pages 944–947.

"On the use of direct variational methods to study confined quantum systems," J. L. Marin and S. A. Cruz, *Am. J. Phys.*, Vol. 59, No. 10, October 1991, pages 931–935.

"Zero-point energy in early quantum theory," P. W. Milonni and M.-L. Shih, *Am. J. Phys.*, Vol. 59, 1991, pages 684–698.

"Comment on 'An exotic harmonic oscillator' by J. B. Sztrajman," W. T. Doyle, *Am. J. Phys.*, Vol. 59, No. 4, April 1991, pages 373–374.

"New quantum mechanical perturbation technique using an 'electronic scratchpad' on an inexpensive computer," R. Lynch and H. A. Mavromatis, *Am. J. Phys.*, Vol. 59, No. 3, March 1991, pages 270–273.

"Theory of a nonharmonic oscillator," I. R. Gatland, *Am. J. Phys.*, Vol. 59, No. 2, February 1991, pages 155–158.

"Test of the adiabatic approximation in quantum mechanics: Forced harmonic oscillator," Y. Nogami, *Am. J. Phys.*, Vol. 59, No. 1, January 1991, pages 64–68.

"Quantum mechanical study of particles in 'softened' potential boxes and wells, " L. S. Salter, *Am. J. Phys.*, Vol. 58, No. 10, October 1990, pages 961–967.

"Measuring small vibrations with interferometry," J. B. Diamond, D. P. Donnelly, J. D. Breault, and M. E. McCarthy, *Am. J. Phys.*, Vol. 58, No. 10, October 1990, pages 919–922.

"The influence of distant boundaries on quantum mechanical energy levels," G. Barton, A. J. Bray, and A. J. McKane, *Am. J. Phys.*, Vol. 58, No. 8, August 1990, pages 751–755.

"The motion of wave packets through their expectation values and uncertainties," D. F. Styer, *Am. J. Phys.*, Vol. 58, No. 8, August 1990, pages 742–744.

"The pumping of a swing from the seated position," W. B. Case and M. A. Swanson, *Am. J. Phys.*, Vol. 58, No. 5, May 1990, pages 463–467.

"Canonical transformation in quantum mechanics," Y. S. Kim and E. P. Wigner, *Am. J. Phys.*, Vol. 58, 1990, pages 4394–48. Note: This excellent, clearly written paper discusses the phase-space picture in quantum mechanics and its use in describing coherent states, squeezed states, and spreading of wave packets.

"Elementary examples of adiabatic invariance," F. S. Crawford, *Am. J. Phys.*, Vol. 58, No. 4, April 1990, pages 337–344.

"An exotic harmonic oscillator: The frequency depends on initial conditions," J. B. Sztrajman and C. B. Comun, *Am. J. Phys.*, Vol. 58, No. 2, February 1990, pages 159 –160.

"Tunneling through a truncated harmonic oscillator potential barrier," J. D. Chalk, *Am. J. Phys.*, Vol. 58, No. 2, February 1990, pages 147–151.

"Less is more: Experiments with an individual atomic particle at rest in free space," H. Dehmelt, *Am. J. Phys.*, Vol. 58, No. 1, January 1990, pages 17 - 27. This remarkable paper describes Dehmelt's work which led to his Nobel Prize. Single electrons, and other single particles, were trapped in macroscopic electromagnetic fields for months at a time, and clearly demonstrated harmonic oscillations and the quantization of energy in these macroscopic environments.

"Extensions of the Feynman-Hellman theorem and applications," S. B.Singh and C. A. Singh, *Am. J. Phys.*, Vol. 57, No. 10, October 1989, pages 894–899.

"Correlations and squeezing of two-mode oscillations," A. K. Ekert and P. L. Knight, *Am. J. Phys.*, Vol. 57, No. 8, August 1989 , pages 692–697.

"An elementary introduction to solitons," J. Gratton and R. Delellis, *Am. J. Phys.*, Vol. 57, No. 8, August 1989, pages 683–687.

"Applications of Bohr's correspondence principle," F. S. Crawford, *Am. J. Phys.*, Vol. 57, No. 7, July 1989, pages 621–628.

"Laplace vectors for the harmonic oscillator," J. Sivardiere, *Am. J. Phys.*, Vol. 57, No. 6, June 1989, pages 524–525 .

"An approximate analytic solution of the spectral inverse problem," R. Engelke, *Am. J. Phys.*, Vol. 57, No. 4, April 1989, pages 339–344.

"The quantum mechanical few-body problem," J. W. Norbury, *Am. J. Phys.*, Vol. 57, No. 3, March 1989, pages 264–266.

"Effective Lagrangians and quantum mechanics: The index of refraction," B. R. Holstein, *Am. J. Phys.*, Vol. 57, No. 2, February 1989, pages 142–148.

"The stretched harmonic oscillator: An analytical test of semiclassical approximations," M. Casas, H. Krivine, and J. Martorell, *Am. J. Phys.*, Vol. 57, No. 1, January 1989, pages 35–39.

"On the harmonic oscillator inside an infinite potential well," J. L. Marin and S. A. Cruz, *Am. J. Phys.*, Vol. 56, No. 12, December 1988, pages 1134–1136.

"The harmonic oscillator at finite temperature using path integrals," A. Larsen and F. Ravndal, *Am. J. Phys.*, Vol. 56, No. 12, December 1988, pages 1129–1133.

"Critique and correction of the textbook comparison between classical and quantum harmonic oscillator probability densities," C. Leubner, M. Alber, and N. Schupfer, *Am. J. Phys.*, Vol. 56, No. 12, December 1988 , pages 1123–1129.

"The effect of the mass of the center spring in one-dimensional coupled harmonic oscillators," D. L. Wallach W. Beatty, K. Beisler, P. Chronowski, M. Holloway, R. Hutchinson, D. Kacmarik, R. Lesko, J. Lutz, D. Miller, and M. Sivak, *Am. J. Phys.*, Vol. 56, No. 12, December 1988, pages 1120–1123.

"Numerical path integration without Monte Carlo," C. C. Gerry and J. Kiefer, *Am. J. Phys.*, Vol. 56, No. 11, November 1988, pages 1002–1005.

"Semiclassical approximation and exact evaluation of the propagator for a harmonic oscillator with time-dependent frequency," C. P. Natividade, *Am. J. Phys.*, Vol. 56, No. 10, October 1988, pages 921–922.

"An aging harmonic oscillator," C. F. Lo, *Am. J. Phys.*, Vol. 56, No. 9, September 1988, pages 827–828.

"Walsh functions and the solution of nonlinear differential equations," H.F. Ahner, *Am. J. Phys.*, Vol. 56, No. 7, July 1988, pages 628–633.

"Moments of probability distribution, wavefunctions, and their derivatives at the origin of N-dimensional central potentials," A. Ray, K. Mahata, and P. P. Ray, *Am. J. Phys.*, Vol. 56, No. 5, May 1988, pages 462–464.

"A squeezed-state primer," R. W. Henry and S. C. Glotzer, *Am. J. Phys.*, Vol. 56, No. 4, April 1988, pages 318–328.

"The harmonic oscillator via functional techniques," J. F. Donoghue and B. R. Holstein, *Am. J. Phys.*, Vol. 56, No. 3, March 1988 , pages 216–222.

"Supersymmetry, shape invariance, and exactly solvable potentials," R. Dutt, A. Khare, and U. P. Sukhatme, *Am. J. Phys.*, Vol. 56, No. 2, February 1988, pages 163 –168.

"Electromagnetically driven resonance apparatus," P. J. Ouseph and J. P. Ouseph, *Am. J. Phys.*, Vol. 55, No. 12, December 1987, pages 1126–1129.

"Coherent states of a harmonic oscillator," S. Howard and S. K. Roy, *Am. J. Phys.*, Vol. 55, No. 12, December 1987, pages 1109 - 1117.

"The physicist's toolkit," M. H. Krieger, *Am. J. Phys.*, Vol. 55, No. 11, November 1987, pages 1033–1038.

"A general expression for the multipole moments of the quantum harmonic oscillator," V. Namias, *Am. J. Phys.*, Vol. 55, No. 11, November 1987, pages 1008–1009.

"Resource Letter IQM-2: Foundations of quantum mechanics since the Bell inequalities," L. E. Ballantine, *Am. J. Phys.*, Vol. 55, 1987, pages 785–792.

"Using videotapes to study underdamped motion of a pendulum: A laboratory project," M. S. Greenwood, *Am. J. Phys.*, Vol. 55, No. 7, July 1987, pages 645–648.

"The human body's response to glucose and three physical models," M. J. Ruiz, *Am. J. Phys.*, Vol. 55, No. 7, July 1987, pages 641–645.

"Coulomb-Kepler problem and the harmonic oscillator," A. C. Chen, *Am. J. Phys.*, Vol. 55, No. 3, March 1987, pages 250–252.

"The total probability current and the quantum period," J.-M. Levy-Leblond, *Am. J. Phys.*, Vol. 55, No. 2, February 1987, pages 146–149.

"Quantizing the damped harmonic oscillator," B. Yurke, *Am. J. Phys.*, Vol. 54, No. 12, December 1986, pages 1133–1139.

"Using the virial theorem," J. Sivardiere, *Am. J. Phys.*, Vol. 54, No. 12, December 1986, pages 1100–1103.

"Predicting charmonium and bottomonium spectra with a quark harmonic oscillator," J. W. Norbury, F. F. Badavi, and L. W. Townsend, *Am. J. Phys.*, Vol. 54, No. 11, November 1986, pages 1031–1034.

"Lagrangians for dissipative systems," D. H. Kobe, G. Reali, and S. Sieniutycz, *Am. J. Phys.*, Vol. 54, No. 11, November 1986, pages 997–999.

"Damped harmonic oscillator: A correction in some standard textbooks," A. Soares de Castro, *Am. J. Phys.*, Vol. 54, No. 8, August 1986, pages 741–742.

"Classical dynamics and Lie groups," J. D. Louck and N. Metropolis, *Am. J. Phys.*, Vol. 54, No. 6, June 1986, pages 558–564.

"Argand diagrams, harmonic oscillators, and record-playing tonearms," R. D. Piccard, *Am. J. Phys.*, Vol. 54, No. 4, April 1986, pages 342–345.

"A variational principle and the classical and quantum mechanics of the damped harmonic oscillator," L. Herrera, V. L. Nunez, A. Patino, and H. Rago, *Am. J. Phys.*, Vol. 54, No. 3, March 1986, pages 273–277.

"Energy analysis of the conical-spring oscillator," F. C. Grant, *Am. J. Phys.*, Vol. 54, No. 3, March 1986, pages 227–233.

"Computer-aided experiments with the damped harmonic oscillator," M. F. McInerney, *Am. J. Phys.*, Vol. 53, No. 10, October 1985, pages 991–996.

"The hydrogenic atom and the periodic table of the elements in two spatial dimensions," F. J. Asturias and S. R. Aragon, *Am. J. Phys.*, Vol. 53, No. 9, September 1985, pages 893–899.

"Forced harmonic oscillator: A path integral approach," B. R. Holstein, *Am. J. Phys.*, Vol. 53, No. 8, August 1985, pages 723–725.

"Forced and damped harmonic oscillator experiment using an accelerometer," J. L. Hunt, *Am. J. Phys.*, Vol. 53, No. 3, March 1985, pages 278–279.

"Conservative model for the damped harmonic oscillator," B. Yurke, *Am. J. Phys.*, Vol. 52, No. 12, December 1984, pages 1099–1102.

"Motion of a harmonic oscillator with variable sliding friction," I. R. Lapidus, *Am. J. Phys.*, Vol. 52, No. 11, November 1984, pages 1015–1016.

"Precession of elliptic orbits," J. Sivardiere, *Am. J. Phys.*, Vol. 52, No. 10, October 1984, pages 909–912.

"Inexpensive demonstration of an anharmonic oscillator," H. J. T. Smith and K. A. Woolner, *Am. J. Phys.*, Vol. 52, No. 9, September 1984, pages 800–801.

"Resolution of the Schrödinger equation through a simple algebraic technique," F. M. Fernandez and E. A. Castro, *Am. J. Phys.*, Vol. 52, No. 4, April 1984, pages 344 –346.

"Quantum solutions of the damped harmonic oscillator," G. J. Milburn and D. F. Walls, *Am. J. Phys.*, Vol. 51, No. 12, December 1983 , pages 1134 – 1136.

"Ladder operators in the Morse potential obtained from a related harmonic oscillator," R. Montemayor and L. Urrutia, *Am. J. Phys.*, Vol. 51, No. 7, July 1983, pages 641 – 644.

"Motion in a time-dependent, uniform electric field," S. Fallieros and J. L. Friar, *Am. J. Phys.*, Vol. 50, No. 11, November 1982, pages 1001–1005.

"Motion of a harmonic oscillator with sliding and viscous friction," A. Ricchiuto and A. Tozzi, *Am. J. Phys.*, Vol. 50, No. 2, February 1982, pages 176–179.

"Dirac's form of relativistic quantum mechanics," D. Han and Y. S. Kim, *Am. J. Phys.*, Vol. 49, No. 12, December 1981, pages 1157–1161.

"Generalized Lagrangian and conservation law for the damped harmonic oscillator," L. Y. Bahar and H. G. Kwatny, *Am. J. Phys.*, Vol. 49, No. 11, November 1981, pages 1062–1065.

"Sliding friction and the harmonic oscillator," C. Barratt and G. L. Strobel, *Am. J. Phys.*, Vol. 49, No. 5, May 1981, pages 500–501.

"Bohlin's and other integrals for the damped harmonic oscillator," W. E. Gettys, J. R. Ray, and E. Breitenberger, *Am. J. Phys.*, Vol. 49, No. 2, February 1981, pages 162 –164.

"Experiment on impulsive excitation, resonance, and Fourier analysis of a harmonic oscillator," H. K. Macomber, *Am. J. Phys.*, Vol. 49, No. 1, January 1981, pages 31–33.

"Applications of microcomputers in the teaching of physics – 6502 software," D. P. Marsh, *Am. J. Phys.*, Vol. 48, No. 11, November 1980, pages 933–937.

"Entropic approach to Brownian movement," R. M. Neumann, *Am. J. Phys.*, Vol. 48, No. 5, May 1980, pages 354–357.

"Methods of establishing the asymptotic behavior of the harmonic oscillator wave functions," M. Bowen and J. Coster, *Am. J. Phys.*, Vol. 48, No. 4, April 1980, pages 307–307.

"Hydrogen atom and relativistic pi-mesic atom in N-space dimensions," M. M. Nieto, *Am. J. Phys.*, Vol. 47, No. 12, December 1979, pages 1067–1072.

"A simple method for illustrating the difference between the homogeneous and inhomogeneous Lorentz groups," Y. S. Kim, M. E. Noz, and H. Oh, *Am. J. Phys.*, Vol. 47, No. 10, October 1979, pages 892–897.

"Canonical approach to the damped harmonic oscillator," N. A. Lemos, *Am. J. Phys.*, Vol. 47, No. 10, October 1979, pages 857–858.

"Lagrangians and systems they describe:How not to treat dissipation in quantum mechanics," J. R. Ray, *Am. J. Phys.*, Vol. 47, No. 7, July 1979, pages 626–629.

"Spatial Bose-Einstein condensation," R. Masut and W. J. Mullin, *Am. J. Phys.*, Vol. 47, No. 6, June 1979, pages 493–497.

"Linear collisions with harmonic oscillator forces: The inverse scattering problem," N. D. Newby Jr., *Am. J. Phys.*, Vol. 47, No. 2, February 1979, pages 161–165.

"Quantization of inequivalent classical Hamiltonians," I. K. Edwards, *Am. J. Phys.*, Vol. 47, No. 2, February 1979, pages 153–155.

"Comparison of classical and quantum mechanical uncertainties," J. Peslak, Jr., *Am. J. Phys.*, Vol. 47, No. 1, January 1979, pages 39–45.

"Note on the time-dependent damped and forced harmonic oscillator," P. G. L. Leach, *Am. J. Phys.*, Vol. 46, No. 12, December 1978, pages 1247–1249.

"String-coupled pendulum oscillators: Theory and experiment," M. J. Moloney, *Am. J. Phys.*, Vol. 46, No. 12, December 1978, pages 1245–1246.

"Forces giving no orbit precession," L. S. Brown, *Am. J. Phys.*, Vol. 46, No. 9, September 1978, pages 930–931.

"Elementary computer physics, a concentrated one-week course," G. Christiansen, *Am. J. Phys.*, Vol. 46, No. 7, July 1978, pages 748–751.

"Inertia demonstration revisited," F. G. Karioris, *Am. J. Phys.*, Vol. 46, No. 7, July 1978, pages 710–713.

"Ladder operators for some spherically symmetric potentials in quantum mechanics," J. D. Newmarch and R. M. Golding, *Am. J. Phys.*, Vol. 46, No. 6, June 1978, pages 658–660.

"Momentum autocorrelation functions from classical Green's functions," P. P. Lottici, *Am. J. Phys.*, Vol. 46, No. 5, May 1978, pages 507–508.

"Relativistic harmonic oscillators and hadronic structures in the quantum-mechanics curriculum," Y. S. Kim and M. E. Noz, *Am. J. Phys.*, Vol. 46, No. 5, May 1978, pages 484–488.

"Group theory of covariant harmonic oscillators," Y. S. Kim and M. E.Noz, *Am. J. Phys.*, Vol. 46, No. 5, May 1978, pages 480–483.

"Thermodynamic work on a harmonic oscillator," J. D. Edmonds, Jr., *Am. J. Phys.*, Vol. 46, No. 3, March 1978, pages 289–290.

"Quantum measurements on a computer: An interactive teaching program," A. S. Krass, *Am. J. Phys.*, Vol. 46, No. 2, February 1978, pages 152–158.

"The problem of several delta-shell potentials in the Lippmann-Schwinger formulation," A. K. Jain, S. K. Deb, and C. S. Shastry, *Am. J. Phys.*, Vol. 46, No. 2, February 1978, pages 147–151.

"Anharmonic vibrations of an 'ideal' Hooke's law oscillator," J. Thomchick and J. P. McKelvey, *Am. J. Phys.*, Vol. 46, No. 1, January 1978, pages 40–45.

"Derivation of noise formulas using Campbell's theorem," E. Mathieson, *Am. J. Phys.*, Vol. 45, No. 12, December 1977, pages 1184–1186.

"The reflecting galvanometer as a damped, driven simple harmonic oscillator with recording facilities," B. A. Eagle and P. J. Jackson, *Am. J. Phys.*, Vol. 45, No. 11, November 1977, pages 1113–1115 .

"Forced harmonic oscillator in the interaction picture," R. A. Ferrell, *Am. J. Phys.*, Vol. 45, No. 5, May 1977, pages 468–469.

"Cranking model calculation of the moment of inertia using the method of Dalgarno and Lewis," S. M. Harris, *Am. J. Phys.*, Vol. 45, No. 4, April 1977, pages 373–375.

"Canonical transformation to the free particle," E. N. Glass and J. J. G. Scanio, *Am. J. Phys.*, Vol. 45, No. 4, April 1977, pages 344–346.

"Retardation in coupled dipole-oscillator systems," P. W. Milonni, *Am. J. Phys.*, Vol. 44, No. 8, August 1976, pages 741–743.

"How good is Hartree-Fock? Calculation of the static electric polarizability in an exactly soluble model," J. B. Krieger and S. E. Kaufman, *Am. J. Phys.*, Vol. 44, No. 6, June 1976, pages 571–573.

"Conditions on wave functions derived from operator domains," T. F. Jordan, *Am. J. Phys.*, Vol. 44, No. 6, June 1976, pages 567–570.

"Classical particles in quantum mechanics," G. Patsakos, *Am. J. Phys.*, Vol. 44, No. 2, February 1976, pages 158–166.

"Conserved and piecewise-conserved Runge vectors for the isotropic harmonic oscillator," L.H. Buch and H. H. Denman, *Am. J. Phys.*, Vol. 43, No. 12, December 1975, pages 1046–1048.

"Solution of a simple inelastic scattering problem," S. K. Knudson, *Am. J. Phys.*, Vol. 43, No. 11, November 1975, pages 964–968.

"Quantum solution for the biharmonic oscillator," W. E. Gettys and H. W. Graben, *Am. J. Phys.*, Vol. 43, No. 7, July 1975, pages 626–629.

"WKB connection rules from the harmonic oscillator," H. Kroemer, Am. J. Phys., Vol. 43, No. 6, June 1975, pages 514 - 517.

"Statistical analysis of classical few-particle systems," R. J. Olness and W. G. Hoover, *Am. J. Phys.*, Vol. 42, No. 7, July 1974, pages 584–590.

"Remark on the solutions of the harmonic oscillator equation," H. A. Buchdahl, *Am. J. Phys.*, Vol. 42, No. 1, January 1974, pages 47–50.

"Time evolution of predictions in statistical mechanics," B. M. Casper, S. Freier, and D. Van Atta, *Am. J. Phys.*, Vol. 41, No. 12, December 1973, pages 1358–1360.

"Optical rotatory power in a classical one-electron model," G. E. Desobry and P. K. Kabir, *Am. J. Phys.*, Vol. 41, No. 12, December 1973, pages 1350–1357.

"Comparison of the Hartree-Fock and maximum overlap orbitals for a simple model," W. P. Latham, Jr., *Am. J. Phys.*, Vol. 41, No. 11, November 1973, pages 1258–1266.

"The simple harmonic oscillator: An alternative solution which avoids solving the second-order differential equation," R. T. Bush, *Am. J. Phys.*, Vol. 41, No. 5, May 1973, pages 738–739.

"Comment on energy quantization and the simple harmonic oscillator," R. L. Intemann, *Am. J. Phys.*, Vol. 41, No. 5, May 1973, pages 717–718.

"The forced harmonic oscillator and the zero-phonon transition of the Mössbauer effect," F. Ninio, *Am. J. Phys.*, Vol. 41, No. 5, May 1973, pages 648–649.

"The Brattsev-Epstein theorem applied to the ground state energy of a molecule," T. Inagaki, *Am. J. Phys.*, Vol. 41, No. 4, April 1973, pages 494–496.

"Comment on energy quantization and the simple harmonic oscillator, " A. L. Loeb, *Am. J. Phys.*, Vol. 41, No. 2, February 1973, pages 297–297.

"Plane pendulum in quantum mechanics," T. Pradhan and A. V. Khare, *Am. J. Phys.*, Vol. 41, No. 1, January 1973, pages 59–66.

"Exact and impulse scattering of weakly coupled probes by an harmonic oscillator," H. A. Gersch, *Am. J. Phys.*, Vol. 40, No. 12, December 1972, pages 1803–1810.

"Eigenvalues of one-dimensional systems by a matrix method," M. Nicola, *Am. J. Phys.*, Vol. 40, No. 11, November 1972, pages 1647–1652.

"Exponential decay with a memory," R. T. Robiscoe and J. C. Hermanson, *Am. J. Phys.*, Vol. 40, No. 10, October 1972, pages 1443–1448.

"Energy quantization and the simple harmonic oscillator," R. J. Swenson and J. C. Hermanson, *Am. J. Phys.*, Vol. 40, No. 9, September 1972, pages 1258–1260.

"The period of damped simple harmonic motion," M. Allen and E. J. Saxl, *Am. J. Phys.*, Vol. 40, No. 7, July 1972, pages 942–944.

"Uncertainty relations for nonsimultaneous measurements," J.-M. Levy-Leblond, *Am. J. Phys.*, Vol. 40, No. 6, June 1972, pages 899–901.

"Quantum theory of a damped two-level atom," T. von Foerster, *Am. J. Phys.*, Vol. 40, No. 6, June 1972, pages 854–862.

"Computer solutions to the Schrodinger equation," P. C. Chow, *Am. J. Phys.*, Vol. 40, No. 5, May 1972, pages 730–734.

"Generating function for a linear harmonic oscillator," G. S. Khandelwal, *Am. J. Phys.*, Vol. 40, No. 3, March 1972, pages 476–477.

"Evolution of the quantum states of a harmonic oscillator in a uniform time varying electric field," G. W. Parker, *Am. J. Phys.*, Vol. 40, No. 1, January 1972, pages 120 –125.

"Superradiance in radiatively coupled tuning forks," W. L. Lama, R. Jodoin, and L. Mandel, *Am. J. Phys.*, Vol. 40, No. 1, January 1972, pages 32–37.

"Resource Letter IQM-1 on the interpretation of quantum mechanics," B. DeWitt and R. Graham, *Am. J. Phys.*, Vol. 39, 1971, pages 724–738.

"Information theory and the approach to equilibrium," H. S. Robertson and M. A. Huerta, *Am. J. Phys.*, Vol. 38, 1970, pages 619 – 730.

"Laboratory exercises in classical electromagnetic field theory," H. A. Atwater, *Am. J. Phys.*, Vol. 36, No. 8, August 1968, pages 672–682. (Describes an experiment using eight coupled *LC* resonators with variable coupling.)

"Resonant detection of light pressure by a torsion pendulum in air—an experiment for underclass laboratories," R. E. Pollock, *Am. J. Phys.*, Vol. 31, No. 12, December 1963, pages 901–904.

Selected Papers from Other Journals

"Confinement of electrons to quantum corrals on a metal surface," M. F. Crommie, C. P. Lutz, and D. M. Eigler, *Science* Vol. 262, pages 218–220, 1993.

"Harmonic-Oscillator Transformation Coefficients," Davies, K. T. R. and Krieger, S. J., *Canadian Journal of Physics* Vol 69 No. 1, pages 62–69 1991. References: 38

"Recurrence Relations for Harmonic-Oscillator Brackets," Dobes, J., *Czechoslovak Journal of Physics* Vol 41 No. 1, pages 31–35 1991. References: 13

"On Connections Between the 4-Dimensional Harmonic-Oscillator and the Coulomb-Problem in the Representation with the Discrete Basis," Campigotto, C. and Smirnov, Y.F., *Helvetica Physica Acta* Vol 64 No. 1, pages 48–60 1991. References: 43

"Operation of a High Harmonic Cusptron Oscillator," Tiong, K. K. and Kuo, S. P., *International Journal of Electronics* Vol 70 No. 4, pages 815–825 1991. References: 18

"On the Harmonic-Oscillator and Its Invariance Superalgebras," Debergh, N., *Journal of Physics A-Mathematical and General* Vol 24 No. 1, pages 147–151 1991. References: 12

"The 2-Dimensional Harmonic-Oscillator Interacting with a Wedge," Cheng, B. K., *Journal of Physics A—Mathematical and General* Vol 23 No. 24, pages 5807–5814 1990. References: 17

"Harmonic-Oscillator on a Lattice in a Constant Force-Field and Associated Bloch Oscillations," Gallinar, J. P. and Chalbaud, E., *Physical Review B—Condensed Matter* Vol 43 No. 3, pages 2322–2333 1991. References: 20

"Equivalence of a Harmonic-Oscillator to a Free Particle," Takagi, S., *Progress of Theoretical Physics* Vol 84 No. 6, pages 1019-1024 1990. References: 26

"Harmonic-Oscillator Realization of the Canonical Q-Transformation," Filippov, A. T., Gangopadhyay, D., and Isaev, A. P., *Journal of Physics A—Mathematical and General* Vol 24 No. 2, pages L63–L68 1991. References: 29

"Hamiltonian-Mechanics of the Damped Harmonic-Oscillator," Nagem, R., Rhodes, B. A., and Sandri, G. V., *Journal of Sound and Vibration* Vol 144 No. 3, pages 536–538 1991. References: 4

"COBE satellite finds no hint of excess in the cosmic microwave spectrum," *Physics Today*, March 1990 Page 17.

"An Explicit Minimum-Time Solution for the Simple Harmonic-Oscillator," Choura, S. and Jayasuriya, S., *Journal of Dynamic Systems Measurement and Control—Transactions of the ASME* Vol 112, No. 4, pages 783–787 1990. References: 3

"The Exact Solution for the Generalized Time-Dependent Harmonic-Oscillator and Its Adiabatic Limit ," Gao, X. C., Xu, J. B., and Qian, T. Z., *Annals of Physics* Vol 204, No. 1, pages 235–243 1990. References: 25

"Applications of the Differentiability of Eigenvectors and Eigenvalues to a Perturbed Harmonic-Oscillator," Ifantis, E. K. and Panagopoulos, P. N., *Journal of Mathematical Physics* Vol 31, No. 10, pages 2410–2413 1990. References: 9

"Invariance of the Generalized Berry Phase Under Unitary Transformations - Application to the Time-Dependent Generalized Harmonic-Oscillator," Kobe, D. H., *Journal of Physics A—Mathematical and General* Vol 23, No. 19, pages 4249–4268 1990. References: 16

"Quantum-Theory of the Damped Harmonic-Oscillator," Harris, E. G., *Physical Review A* Vol 42, No. 7, pages 3685–3694 1990. References: 12

"Information-Theory Approach to the Variable-Mass Harmonic-Oscillator," Crespo, G., Proto, A. N., Plastino, A. , and Oter, D., *Physical Review A* Vol 42, No. 6, pages 3608–3617 1990. References: 31

"Light-Front Dynamics of the Covariant Harmonic-Oscillator," Fuda, M. G., *Physical Review D—Particles and Fields* Vol 42, No. 8, pages 2898–2910 1990. References: 68

"Effect of Nonconstant Gamma-N(Epsilon-J) on Neutron Resonance Broadening. 2. Harmonic-Oscillator," Kuwaifi, A., Summerfield, G. C., and Shamaoun, A., *Annals of Nuclear Energy* Vol 17, No. 10, pages 565–568 1990. References: 9

"Relativistic Harmonic-Oscillator Potential for Quarks in the Nucleon," Tegen, R., *Annals of Physics* Vol 197, No. 2, pages 439–455 1990. References: 41

"Statistical Mechanical Investigation of the Pseudo-Harmonic Oscillator," Buyukkilic, F. and Demirhan, D. A., *Chemical Physics Letters* Vol 166, No. 3, pages 272–274 1990. References: 4

"Analytical Expressions and Recursion-Relations of 2-Center Harmonic-Oscillator Matrix-Elements of Arbitrary Functions," Drallos, P. J. and Wadehra, J. M.,

International Journal of Quantum Chemistry Vol 37, No. 6, pages 797–809 1990. References: 16

"Time Evolution and Delocalization of Excitation in Models of a Harmonic-Oscillator Chain with an Impurity Mass," Yu, M. B., Kim, J. H., and Lee, M. H., *Journal of Luminescence* Vol 45, No. 1-6, pages 144–146 1990. References: 5

"Nonspurious Harmonic-Oscillator States in Single-Particle Coordinates," Novoselsky, A., and Katriel, J., *Journal of Mathematical Physics* Vol 31, No. 5, pages 1164–1166 1990. References: 9

"Untenability of a Proposed Constrained Dynamics for the Damped Harmonic-Oscillator," Lemos, N. A., *Journal of Mathematical Physics* Vol 31, No. 2, pages 348–350 1990 References: 30

"Variational and Perturbative Schemes for a Spiked Harmonic-Oscillator," Aguileranavarro, V. C., Estevez, G. A., and Guardiola, R., *Journal of Mathematical Physics* Vol 31, No. 1, pages 99–104 1990 References: 12

"Mass Renormalization and Squeezing of a Material Harmonic-Oscillator," Compagno, G., and Salamone, G. M., *Journal of Modern Optics* Vol 37, No. 2, pages 183–193 1990 References: 20

"Inconsistency of a Proposed Constrained Dynamics for the Damped Harmonic-Oscillator," Lemos, N. A., *Journal of Physics A—Mathematical and General* Vol 23, No. 17, pages 3929–3935 1990 References: 27

"Dynamic Parasuperalgebras of Parasupersymmetric Harmonic-Oscillator, Cyclotron Motion and Morse Hamiltonians," Durand, S. and Vinet, L., *Journal of Physics A—Mathematical and General* Vol 23, No. 16, pages 3661–3672 1990 References: 13

"Use of the Coherent States in Evaluating the Canonical Density-Matrix for a Harmonic-Oscillator," Yonei, K., *Journal of Physics A—Mathematical and General* Vol 23, No. 10, pages 1705–1715 1990 References: 13

"On (K=3/4) Coherent States for the Harmonic-Oscillator," Debergh, N., *Journal of Physics A—Mathematical and General* Vol 23, No. 2, pages 147–152 1990 References: 22

"Semiclassical Matrix-Mechanics. 1. The Harmonic-Oscillator," More, R. M., *Journal de Physique* Vol 51, No. 1, pages 47–58 1990 References: 8

"The Harmonic-Oscillator with Random Stiffness," Rehak, M. and Benaroya, H., *Mechanics of Structures and Machines* Vol 17, No. 4, pages 473–491 1989 References: 11

"Squeezing in Harmonic-Oscillator with Time-Dependent Mass and Frequency," Lo, C. F., *Nuovo Cimento della Societa Italiana di Fisica B-General Physics Relativity Astronomy and Mathematical Physics and Methods* Vol 105, No. 5, pages 497–506 1990 References: 6

"On the Harmonic-Oscillator with Nonzero Periodic Mass," Leaute, B., Marcilhacy, G., and Pons, R., *Nuovo Cimento della Societa Italiana di Fisica D-Condensed Matter Atomic Molecular and Chemical Physics Fluids Plasmas Biophysics* Vol 12, No. 8, pages 1051–1060 1990 References: 17

"An Electron in a Magnetic-Field As a Nonlocal Harmonic-Oscillator," Sayakanit, V., and Poulter, J., *Physics Letters A* Vol 144, No. 1, pages 31–34 1990 References: 5

"Nonzero-Temperature Coherent and Squeezed States for the Harmonic-Oscillator — The Time-Dependent Frequency Case," Aliaga, J., Crespo, G., and Proto, A. N., *Physical Review A* Vol 42, No. 1, pages 618–626 1990 References: 28

"Nonperturbative Expansion Method for a Quantum System Coupled to a Harmonic-Oscillator Bath," Tanimura, Y., *Physical Review A* Vol 41, No. 12, pages 6676–6687 1990 References: 29

"Application of the Generalized Feynman-Vernon Approach to a Simple System - The Damped Harmonic-Oscillator," Smith, C. M., and Caldeira, A. O., *Physical Review A* Vol 41, No. 6, pages 3103–3115 1990 References: 17

"Propagator for the Time-Dependent Harmonic-Oscillator," Macek, J., *Physical Review A* Vol 41, No. 3, pages 1361–1364 1990 References: 7

"Almost Sure Instability of the Random Harmonic-Oscillator," Feng, X. B., and Loparo, K. A., *SIAM Journal on Applied Mathematics* Vol 50, No. 3, pages 744–759 1990 References: 22

"A Gunn Diode Harmonic-Oscillator Using a Disk Resonator," Krotov, V. I., and Levterov, A. N., *Telecommunications and Radio Engineering* Vol 44, No. 6, pages 39–41 1989 References: 6

"A Harmonic-Oscillator," Pospelov, A. N., *Telecommunications and Radio Engineering* Vol 44, No. 3, pages 142–143 1989 References: 1

"The Construction of the Q-Analogs of the Harmonic-Oscillator Operators from Ordinary Oscillator Operators," Song, X. C., *Journal of Physics A—Mathematical and General* Vol 23, No. 16, pages L821–L825 1990 References: 7

"Magnetic Form-Factor of Li-6 Using Hard-Core Common Harmonic-Oscillator Potential," Lu, P., and Nigam, B. P., *Journal of Physics G—Nuclear and Particle Physics* Vol 16, No. 8, pages L135–L136 1990 References: 11

"Effect of Bound Self-Field Energy on Quantum Charged Harmonic-Oscillator," Sasabe, S., *Journal of the Physical Society of Japan* Vol 59, No. 8, pages 2607–2609 1990 References: 13

"Comment on the Q-Analogs of the Harmonic-Oscillator," Ng, Y. J., *Journal of Physics A—Mathematical and General* Vol 23, No. 6, pages 1023–1027 1990 References: 6

"Energy Characteristics and Stability of Harmonic-Oscillator with Intense Parameter Fluctuations," (Russian) Muzychuk, O. V., *Izvestiya Vysshikh Uchebnykh Zavedenii Radiofizika* Vol 33, No. 1, pages 43–48 1990 References: 7

"Non-Spurious Harmonic-Oscillator States with Arbitrary Symmetry," Novoselsky, A., and Katriel, J., *Annals of Physics* Vol. 196 No. 1, pages 135–149 (1989) References: 18

"Noether Theorem Invariants for a Time-Dependent Damped Harmonic-Oscillator with a Force Quadratic in Velocity," Gu, Z. Y., and Qian, S. W., *Europhysics Letters* Vol. 10 No. 7, pages 615–619 (1989) References: 14

"Study of Certain Aspects of Anharmonic, Time-Dependent and Damped Harmonic-Oscillator Systems," Bose, S. K., Dubey, U. B., and Varma, N., *Fortschritte der Physik—Progress of Physics* Vol. 37 No. 10, pages 761–818 (1989) References: 120

"Calculation of Resonant States Using an Optimal Path of Continuous Extension of the Bound-States – The Cubic and Quartic Energy Distortion of the Harmonic-Oscillator," Jolicard, G., and Perrin, M. Y., *Journal of Chemical Physics* Vol. 91 No. 12, pages 7780–7786 (1989) References: 35

"Multidimensional Harmonic-Oscillator Matrix-Elements," Fernandez, F. M., and Tipping, R. H., *Journal of Chemical Physics* Vol. 91 No. 9, pages 5505–5508 (1989) References: 15

"Free Green-Function in a Harmonic- Oscillator Basis," Silvestrebrac, B., Ayant, Y., and Gignoux, C., *Journal of Mathematical Physics* Vol. 30 No. 10, pages 2288–2290 (1989) References: 13

"On Q-Analogues of the Quantum Harmonic-Oscillator and the Quantum Group Su(2)Q," Macfarlane, A. J., *Journal of Physics A—Mathematical and General* Vol. 22 No. 21, pages 4581–4588 (1989) References: 40

"Non-Relativistic Spectrum of the Q2Qbar2 System in Large Harmonic-Oscillator Bases," Ceuleneer, R., and Semay, C., *Journal of Physics G—Nuclear and Particle Physics* Vol. 15 No. 12, pages 1783–1789 (1989) References: 11

"The Effect of Damping on a Pumped Harmonic-Oscillator," Colegrave-RK Stephens, D. B L., Wille, L. T., and Khosravi, A., J*ournal of Sound and Vibration* Vol. 135 No. 1, pages 53–61 (1989) References: 15

"Euler and Correlated Harmonic-Oscillator Wave-Functions for 3-Nucleon Systems," Kievsky, A., Viviani, M., and Rosati, S., *Nuclear Physics A* Vol. 501 No. 3, pages 503–512 (1989) References: 28

"Classical and Quantum Description of the Two-Dimensional Simple Harmonic-Oscillator in Elliptic Coordinates," Fendrik, A. J., and Bernath, M., *Physical Review A-General Physics* Vol. 40 No. 8, pages 4215–4223 (1989) References: 8

"New Canonic Variable Frequency RC Harmonic-Oscillator with 2 Similar Sections," Vosper, J. V., *Electronics Letters* Vol. 25 No. 20, pages 1374–1374 (1989) References: 0

"New Canonic Variable Frequency RC Harmonic-Oscillator with 2 Similar Sections - Reply," Ganguly, U. S., and Ganguly, S., *Electronics Letters* Vol. 25 No. 20, pages 1374–1375 (1989) References: 2

"Phase Operators of the Harmonic—Oscillator and Suppression of a Number State," Sukumar, C. V., *Physical Review A-General Physics* Vol. 40 No. 9, pages 5426–5429 (1989) References: 18

"Squeezing Caused by a Change of Mass in a Harmonic- Oscillator," Fan, H. Y., and Zaidi, H. R., *Canadian Journal of Physics* Vol. 67 No. 2-3, pages 152–154 (1989) References: 20

"Characterization of the Classical States of the Quantum Harmonic-Oscillator by Means of Definetti Theorem," Bach, A., and Srivastav, A. A., *Communications in Mathematical Physics* Vol. 123 No. 3, pages 453–462 (1989) References: 11

"Some Representations for Time-Dependent Harmonic-Oscillator,"Fan, H. Y., and Jing, S. C., *Communications in Theoretical Physics* Vol. 10 No. 3, pages 363–369 1988 References: 8

"A Complete Asymptotic of the Spectral-Function for Harmonic-Oscillator," Karadzhov, G. E., *Doklady Bolgarskoi Akademii Nauk* Vol. 41 No. 10, pages 9–11 1988 References: 0

"On Relativistic Singular Harmonic-Oscillator Potentials," Dominguezadame, F., and Macia, E., *Europhysics Letters* Vol. 8 No. 8, pages 711–715 (1989) References: 8

"A Simple Harmonic-Oscillator Model of the Intrinsic Barrier to Group Transfer," Lee, I., *Journal of the Chemical Society-Perkin Transactions II* No. 8, pages 943–950 (1989) References: 54

"Spectral-Analysis of a Quantum Harmonic-Oscillator Coupled to Infinitely Many Scalar Bosons," Arai, A., *Journal of Mathematical Analysis and Applications* Vol. 140 No. 1, pages 270–288 (1989) References: 15

"On Chains of Orthosymplectic Lie-Superalgebras and the N-Dimensional Quantum Harmonic-Oscillator," Beckers, J., and Cornwell, J. F., *Journal of Mathematical Physics* Vol. 30 No. 8, pages 1655–1661 (1989) References: 39

"On Generalized Coherent States with Maximal Symmetry for the Harmonic-Oscillator," Beckers, J., and Debergh, N., *Journal of Mathematical Physics* Vol. 30 No. 8, pages 1732–1738 (1989) References: 33

"Long-Time Behavior of 2-Point Functions of a Quantum Harmonic-Oscillator Interacting with Bosons," Arai, A., *Journal of Mathematical Physics* Vol. 30 No. 6, pages 1277–1288 (1989) References: 19

"On Topological Effects in Quantum-Mechanics — The Harmonic-Oscillator in the Pointed Plane," Doebner, H. D., Elmers, H. J., and Heidenreich, W. F., *Journal of Mathematical Physics* Vol. 30 No. 5, pages 1053–1059 (1989) References: 14

"Harmonic-Oscillator System with the Generalized Commutation Relations," Watanabe, S. A., *Journal of Mathematical Physics* Vol. 30 No. 2, pages 376–378 (1989) References: 19

"Recurrence Relations for Two-Center Harmonic-Oscillator Integrals and Their Solutions Using Generating-Functions," Rashid, M. A., *Journal of Mathematical Physics* Vol. 30 No. 2, pages 398–405 (1989) References: 4

"Supersymmetric Embedding of a Model of a Quantum Harmonic-Oscillator Interacting with Infinitely Many Bosons," Arai, A., *Journal of Mathematical Physics* Vol. 30 No. 2, pages 512–520 (1989) References: 43

"Symmetry Algebra of the Anisotropic Harmonic-Oscillator with Commensurate Frequencies," Rosenstee, G., and Draayer, J. P., *Journal of Physics A—Mathematical and General* Vol. 22 No. 9, pages 1323–1327 (1989) References: 13

"On Remarkable Properties of Invariance Superalgebras for the Harmonic-Oscillator," Beckers, J., and Cornwell, J. F., *Journal of Physics A—Mathematical and General* Vol. 22 No. 8, pages 925–929 (1989) References: 12

"An Sp(1,R) Calculation in a Deformed Harmonic-Oscillator Basis," Avancini, S. S., and Depassos, E. J. V., *Journal of Physics G–Nuclear and Particle Physics* Vol. 15 No. 2, pages 161–172 (1989) References: 13

"A Kinetic Discrete Model of a Gas Under an Harmonic-Oscillator Potential," Alves, A. S., *Mechanics Research Communications* Vol. 16 No. 4, pages 235–240 (1989) References: 4

"The Evolution of the Harmonic-Oscillator in Quantum-Mechanics with Spontaneous Localization," Benatti, F., and Weber, T., *Nuovo Cimento della Societa Italiana di Fisica B—General Physics Relativity Astronomy and Mathematical Physics and Methods* Vol. 103 No. 5, pages 511–536 (1989) References: 7

"Specific-Heat of the Harmonic-Oscillator Within Generalized Equilibrium Statistics," Ito, N., and Tsallis, C., *Nuovo Cimento della Societa Italiana di Fisica D—Condensed Matter Atomic Molecular and Chemical Physics Fluids Plasmas Biophysics* Vol. 11 No. 6, pages 907–911 (1989) References: 1

"The Stanley Decomposition of the Harmonic-Oscillator," Billera, L. J., Cushman, R., and Sanders, J. A., *Proceedings of the Koninklijke Nederlandse Akademie van Wetenschappen Series A—Mathematical Sciences* Vol. 91 No. 4, pages 375–393 (1989) References: 13

"Exact Wave-Functions and Coherent States of a Damped Driven Harmonic-Oscillator," Oh, H. G., Lee, H. R., George, T. F., and Um, C. I., *Physical Review A—General Physics* Vol. 39 No. 11, pages 5515–5522 (1989) References: 26

"Dynamics of SU(1,1) Coherent States Driven by a Damped Harmonic-Oscillator," Gerry, C. C., Ma, P. K., and Vrscay, E. R., *Physical Review A—General Physics* Vol. 39 No. 2, pages 668–674 (1989) References: 20

"Harmonic-Oscillator Driven by a Quantum Current," Slosser, J. J., Meystre, P., and Braunstein, S. L., *Physical Review Letters* Vol. 63 No. 9, pages 934–937 (1989) References: 8

"Measurement of the Coherent Harmonic Emission from a Free-Electron Laser-Oscillator," Bamford, D. J., and Deacon, D. A. G., *Physical Review Letters* Vol. 62 No. 10, pages 1106–1109 (1989) References: 28

"Quantum-Mechanics of Rotating and Translating Molecules. 2.Far Infrared-Spectrum of the Harmonic-Oscillator Rigid Rotor," Evans, M. W., *Physica Scripta* Vol. 39 No. 6, pages 689–693 (1989) References: 17

"Exact Time Evolution of a Quantum Harmonic-Oscillator Chain," Hong, J., *Progress of Theoretical Physics* Vol. 82 No. 2, pages 291–298 (1989) References: 13

"On the Lyapunov Exponent of a Harmonic-Oscillator Driven by a Finite-State Markov Process," Leizarowitz, A., *SIAM Journal on Applied Mathematics* Vol. 49 No. 2, pages 404–419 (1989.) References: 19

"Dynamic Equivalence of a Two-Dimensional Quantum Electron-Gas and a Classical Harmonic-Oscillator Chain with an Impurity Mass," Lee, M. H., Hong, J., and Florencio, J., *Journal of Physics A—Mathematical and General* Vol. 22 No. 8, pages L331–L335 (1989). References: 28

"Squeezing and Frequency Jump of a Harmonic-Oscillator — Comment," Janszky, J. and Yushin, Y., *Physical Review A—General Physics* Vol. 39 No. 10, pages 5445–5446 (1989.) References: 11

"Squeezing and Frequency Jump of a Harmonic Oscillator – Reply," Zaidi, H. R., and Fan, H. Y., *Physical Review A—General Physics* Vol. 39 No. 10, pages 5447–5449 (1989). References: 8

"Limits of Weak Damping of a Quantum Harmonic-Oscillator" Caldeira, A. O., Cerdeira, H. A., and Ramaswamy, R., *Physical Review A—General Physics* Vol. 40 No. 6, pages 3438–3440 (1989). References: 6

"Equivalence of the Propagator of Quasistatical Solutions and the Quantum Harmonic-Oscillator," Bonato, C. A., Thomaz, M. T., and Malbouisson, A. P. C., *Physical Review A—General Physics* Vol. 39 No. 4, pages 2225–2227 (1989). References: 8

"Wigner's Function of the Two-Dimensional Harmonic-Oscillator," Gorshenkov, V. N., (Russian) *Izvestiya Vysshikh Uchebnykh Zavedenii Fizika* Vol. 32 No. 7, pages 105–106 (1989). References: 3

E. F. Nichols and G. F. Hull, *Physical Review* Vol. 17, pages 26, 91 (1903). (Experimental measurement of light pressure using a torsion pendulum.)

Index

A

B

C

D

E

F

M

N

O

P

Q

R

S

T

U

V

W

Y

Z